Applied Mathematical Sciences

EDITORS

Fritz John
Courant Institute of Mathematical Sciences
New York University
New York, NY 10012

J.E. Marsden
Department of Mathematics
University of California
Berkeley, CA 94720

Lawrence Sirovich
Division of Applied Mathematics
Brown University
Providence, RI 02912

ADVISORS

M. Ghil New York University

J.K. Hale Brown University

J. Keller Stanford University

K. Kirchgässner Universität Stuttgart

B. Matkowsky Northwestern University

J.T. Stuart Imperial College

A. Weinstein University of California

EDITORIAL STATEMENT

The mathematization of all sciences, the fading of traditional scientific boundaries, the impact of computer technology, the growing importance of mathematical-computer modelling and the necessity of scientific planning all create the need both in education and research for books that are introductory to and abreast of these developments.

The purpose of this series is to provide such books, suitable for the user of mathematics, the mathematician interested in applications, and the student scientist. In particular, this series will provide an outlet for material less formally presented and more anticipatory of needs than finished texts or monographs, yet of immediate interest because of the novelty of its treatment of an application or of mathematics being applied or lying close to applications.

The aim of the series is, through rapid publication in an attractive but inexpensive format, to make material of current interest widely accessible. This implies the absence of excessive generality and abstraction, and unrealistic idealization, but with quality of exposition as a goal.

Many of the books will originate out of and will stimulate the development of new undergraduate and graduate courses in the applications of mathematics. Some of the books will present introductions to new areas of research, new applications and act as signposts for new directions in the mathematical sciences. This series will often serve as an intermediate stage of the publication of material which, through exposure here, will be further developed and refined. These will appear in conventional format and in hard cover.

MANUSCRIPTS

The Editors welcome all inquiries regarding the submission of manuscripts for the series. Final preparation of all manuscripts will take place in the editorial offices of the series in the Division of Applied Mathematics, Brown University, Providence, Rhode Island.

SPRINGER-VERLAG NEW YORK INC., 175 Fifth Avenue, New York, N.Y. 10010

Applied Mathematical Sciences | Volume 58

Applied Mathematical Sciences

1. John: **Partial Differential Equations**, 4th ed.
2. Sirovich: **Techniques of Asymptotic Analysis.**
3. Hale: **Theory of Functional Differential Equations**, 2nd ed.
4. Percus: **Combinatorial Methods.**
5. von Mises/Friedrichs: **Fluid Dynamics.**
6. Freiberger/Grenander: **A Short Course in Computational Probability and Statistics.**
7. Pipkin: **Lectures on Viscoelasticity Theory.**
9. Friedrichs: **Spectral Theory of Operators in Hilbert Space.**
11. Wolovich: **Linear Multivariable Systems.**
12. Berkovitz: **Optimal Control Theory.**
13. Bluman/Cole: **Similarity Methods for Differential Equations.**
14. Yoshizawa: **Stability Theory and the Existence of Periodic Solutions and Almost Periodic Solutions.**
15. Braun: **Differential Equations and Their Applications**, 3rd ed.
16. Lefschetz: **Applications of Algebraic Topology.**
17. Collatz/Wetterling: **Optimization Problems.**
18. Grenander: **Pattern Synthesis: Lectures in Pattern Theory, Vol I.**
20. Driver: **Ordinary and Delay Differential Equations.**
21. Courant/Friedrichs: **Supersonic Flow and Shock Waves.**
22. Rouche/Habets/Laloy: **Stability Theory by Liapunov's Direct Method.**
23. Lamperti: **Stochastic Processes: A Survey of the Mathematical Theory.**
24. Grenander: **Pattern Analysis: Lectures in Pattern Theory, Vol. II.**
25. Davies: **Integral Transforms and Their Applications**, 2nd ed.
26. Kushner/Clark: **Stochastic Approximation Methods for Constrained and Unconstrained Systems.**
27. de Boor: **A Practical Guide to Splines.**
28. Keilson: **Markov Chain Models—Rarity and Exponentiality.**
29. de Veubeke: **A Course in Elasticity.**
30. Sniatycki: **Geometric Quantization and Quantum Mechanics.**
31. Reid: **Sturmian Theory for Ordinary Differential Equations.**
32. Meis/Markowitz: **Numerical Solution of Partial Differential Equations.**
33. Grenander: **Regular Structures: Lectures in Pattern Theory, Vol. III.**
34. Kevorkian/Cole: **Perturbation Methods in Applied Mathematics.**
35. Carr: **Applications of Centre Manifold Theory.**
36. Bengtsson/Ghil/Källén: **Dynamic Meterology: Data Assimilation Methods.**
37. Saperstone: **Semidynamical Systems in Infinite Dimensional Spaces.**
38. Lichtenberg/Lieberman: **Regular and Stochastic Motion.**

(continued on inside back cover)

Theoretical Approaches to Turbulence

Edited by
D.L. Dwoyer M.Y. Hussaini R.G. Voigt

With 90 Illustrations

Springer-Verlag
New York Berlin Heidelberg Tokyo

D.L. Dwoyer
M.Y. Hussaini
R.G. Voigt
ICASE
NASA Langley Research Center
Hampton, Virginia 23665
U.S.A.

AMS Subject Classification: 76FXX

Library of Congress Cataloging in Publication Data
Main entry under title:
Theoretical approaches to turbulence.
(Applied mathematical sciences; v. 58)
Bibliography: p.
1. Turbulence—Addresses, essays, lectures.
I. Dwoyer, Douglas L. II. Hussaini, M. Yousuff.
III. Voigt, Robert G. IV. Series: Applied mathematical sciences (Springer-Verlag New York Inc.); v. 58.
QA1.A647 vol. 58 510 s [532'.0527] 85-14765
[QA913]

This is to certify that the papers authored by D.M. Bushnell, G.A. Chapman, and P. Moin were prepared in their roles as U.S. Government employees and are thus in the public domain.

Printed and bound by Halliday Lithograph, West Hanover, Massachusetts.
Printed in the United States of America.

9 8 7 6 5 4 3 2 1

ISBN 0-387-96191-7 Springer-Verlag New York Berlin Heidelberg Tokyo
ISBN 3-540-96191-7 Springer-Verlag Berlin Heidelberg New York Tokyo

Preface

Turbulence is the most natural mode of fluid motion, and has been the subject of scientific study for almost a century. During this period, various ideas and techniques have evolved to model turbulence. Following Saffman, these theoretical approaches can be broadly divided into four overlapping categories -- (1) analytical modelling, (2) physical modelling, (3) phenomenological modelling, and (4) numerical modelling. With the purpose of summarizing our current understanding of these theoretical approaches to turbulence, recognized leaders (fluid dynamicists, mathematicians and physicists) in the field were invited to participate in a formal workshop during October 10-12, 1984, sponsored by The Institute for Computer Applications in Science and Engineering and NASA Langley Research Center. Kraichnan, McComb, Pouquet and Spiegel represented the category of analytical modelling, while Landahl and Saffman represented physical modelling. The contributions of Launder and Spalding were in the category of phenomenological modelling, and those of Ferziger and Reynolds in the area of numerical modelling. Aref, Cholet, Lumley, Moin, Pope and Temam served on the panel discussions. With the care and cooperation of the participants, the workshop achieved its purpose, and we believe that its proceedings published in this volume has lasting scientific value.

The tone of the workshop was set by two introductory talks by Bushnell and Chapman. Bushnell presented the engineering viewpoint while Chapman reviewed from a historical perspective developments in the study of turbulence. The remaining talks dealt with specific aspects of the theoretical approaches to fluid turbulence. We now summarize these talks as reported in this volume.

Bushnell focuses attention on the control aspect of the turbulence problem. First, he examines the canonical structure of turbulence in wall-bounded flows gleaned from detailed flow visualization and conditional sampling measurements conducted over the past twenty years. Then he discusses the sensitivities of these flows to various control parameters (such as additives and micro and macro geometric variations, etc.). Finally, he explains how these sensitivities provide the test cases for turbulence theories.

Chapman and Tobak present a rather novel overview of turbulence theories in the context of interactions between observations, theoretical ideas and modelling of turbulent flows. With some precaution and reservation, they assume that turbulence could be studied within the framework of Navier-Stokes equations. After providing a brief historical background for turbulence studies, they make the case for three distinct stages of development in the scientific study of turbulence. They call the earliest period the "statistical movement" when turbulence was looked upon from the non-deterministic point of view. The second stage, called the "structural movement", started in the thirties and is essentially observational. Its principal contribution is the recognition of the presence and importance of structures in turbulence. The third and most recent stage originated in the sixties, and is called the "deterministic movement". This encompasses bifurcation and strange attractor theories, theory of fractals and renormalization group theory. Their article contains comprehensive and pertinent references for anyone who would like to get acquainted with various approaches to turbulence.

Ferziger has been part of the research program in large eddy simulation (LES) of turbulent flows since its inception at Stanford in 1974. His article provides an excellent introduction and overview of LES. After presenting the historical background, he lays down the foundations of the subgrid scale modelling. He then proceeds to a critical review of various models in vogue. The impact of supercomputers on LES is discussed. A number of developments required to advance the field including better models, better derivations of initial conditions, and better treatment of boundary conditions are also discussed.

After a brief introduction to statistical and dynamical methods to treat turbulence, Herring starts with a statement of the moment closures, and then presents a simple calculation illustrating the possible shortcomings of the second order two- and one-point closure. Then some successes of the two-point second order modelling are discussed particularly in the case of turbulent convection at low Reynolds number. The article closes with relevant comments on closure providing the rational framework for the subgrid scale modelling procedure.

Kraichnan's work is one of the most important contributions of the workshop. The first half of this paper presents in a unified manner material not necessarily new, but in his opinion, insufficiently appreciated in the turbulence community. The second half focuses on the technical aspects of what he calls the decimation approach to turbulence. This new nonperturbative approach focuses only on a certain number of modes, the effect of neglected modes being modeled by random forces with specifically imposed dynamical and statistical symmetries.

Landahl's work comes under the category of coherent structure modelling, or what Saffman calls physical modelling of turbulence. His inviscid flat-eddy model for coherent structures in the near wall region of a turbulent boundary layer yields flow structure surprisingly similar to what has been observed in experiments.

Launder's article on phenomenological turbulence models is a superb discussion of the capabilities and limitations of single-point closures. He confines his attention to three types of models in this class -- a two-equation eddy viscosity model (EVM), an algebraic stress model (ASM), and a differential stress model (DSM) -- which are the subject of most current activity in this area. The author makes an honest attempt to give an accurate flavour of what has been achieved and where more needs to be achieved in single-point closure. He also discusses briefly the efforts to develop a split-spectrum model in which the turbulence energy spectrum is divided into two parts with separate equations provided for the energy dissipation rate and the rate at which energy passes from large scales to small scales. He further notes that such efforts will bridge the gap between single-point and sub-grid schemes.

Renormalization group methods (which have proved to be extremely useful in the study of critical phenomena) have been recently applied to the study of transition to turbulence, hydrodynamic turbulence in the similarity spectrum range, and subgrid scale modelling in the numerical simulation of fully developed turbulence. McComb concentrates on his own contribution to the last category. What he calls the "Iterative Averaging" technique appears to be a promising way of applying the renormalization group method to homogeneous isotropic turbulence. It remains to be seen how such techniques could be extended to include wall regions without compromising whatever rigor they lay claim to.

Pouquet's article on "Statistical Methods in Turbulence" starts with the description of the properties of a non-dissipative flow, and goes on to show where the statistical closures have been useful. An interesting part of this paper is its discussion of statistical methods and chaos.

Reynolds and Lee give a flavour of what impact full turbulence simulation (FTS) can have on phenomenological modelling of turbulence. Their recent simulation of homogeneous turbulence subject to irrotational strains and under relaxation from these strains appears to have revealed rather controversial new physics regarding the behaviour of the anisotropy of the Reynolds stress, dissipation and vorticity fields.

Saffman provides cogent reasons for vortex dynamics constituting one of the fundamental theoretical approaches to the understanding of turbulence. Vortex dynamics falls into the category of physical modelling of turbulence, and is based on the surmise that turbulent flows can be thought of as assemblies of vortical states which are exact solutions to the Navier-Stokes equations or the Euler equations. He lists the vortical states (relevant to the physical modelling of turbulence) as (1) two-dimensional array of finite area vortices, (2) three-dimensional stretched vortices, (3) vortex rings, (4) finite amplitude Tollmien-Schlichting waves, and (5) vortex sheets. He confines his attention to the first category of vortical states with a brief discussion of other states.

Spalding notes a number of defects (in the present phenomenological modelling of turbulence) essentially due to the neglect of spottiness of turbulent flows. He provides a theoretical formulation of a two-fluid model of turbulence with preliminary results in the case of the plane wake, the axisymmetric jet, and one-dimensional laminar flame propagation. Although some qualitative agreement with experiment has been obtained with respect to features which other models cannot predict at all, it must be noted that the subject is very much in its infancy and there are a number of open questions. Nevertheless, the two-fluid concept can form a basis for further advancement.

Spiegel dispels any doubts one may have that chaotic solutions of the fluid dynamic equations exist. His article is a clear and cogent exposition of the view point that chaos may not be turbulence but that "the more we learn about chaos, the better we will understand turbulence". He describes the approach which assumes amplitude expansions near to the onset of instability. After providing the background on linear stability, he discusses the amplitude equation for triple instability; i.e., three modes going unstable almost simultaneously. He goes on to show that the onset of instability in a continuous band of wave numbers leads to chaotic coherent structures. His proposal to look at data from experiments or numerical simulations, and calculate Liapunov exponents and dimensions of attractors as functions of relevant parameters such as Reynolds number, is worth serious consideration. This is one way of separating the chaotic aspect of turbulence from its other aspects. This union of two disparate approaches might shed some light on new physics of turbulence.

DLD, MYH, RGV

Contents

		Page
PREFACE		v
CHAPTER I.	Turbulence Sensitivity and Control in Wall Flows Dennis M. Bushnell	1
CHAPTER II.	Observations, Theoretical Ideas, and Modeling of Turbulent Flows -- Past, Present, and Future Gary T. Chapman and Murray Tobak	19
CHAPTER III.	Large Eddy Simulation: Its Role in Turbulence Research Joel H. Ferziger	51
CHAPTER IV.	An Introduction and Overview of Various Theoretical Approaches to Turbulence Jackson R. Herring	73
CHAPTER V.	Decimated Amplitude Equations in Turbulence Dynamics Robert H. Kraichnan	91
CHAPTER VI.	Flat-Eddy Model for Coherent Structures in Boundary Layer Turbulence Marten T. Landahl	137
CHAPTER VII.	Progress and Prospects in Phenomenological Turbulence Models B.E. Launder	155
CHAPTER VIII.	Renormalisation Group Methods Applied to the Numerical Simulation of Fluid Turbulence W.D. McComb	187
CHAPTER IX.	Statistical Methods in Turbulence A. Pouquet	209

Page

CHAPTER X. The Structure of Homogeneous Turbulence
William C. Reynolds and Moon J. Lee 231

CHAPTER XI. Vortex Dynamics
P.G. Saffman 263

CHAPTER XII. Two-Fluid Models of Turbulence
D. Brian Spalding 279

CHAPTER XIII. Chaos and Coherent Structures in Fluid Flows
E.A. Spiegel 303

CHAPTER XIV. Connection Between Two Classical Approaches to Turbulence: The Conventional Theory and the Attractors
R. Temam 337

POSITION PAPERS BY PANEL MEMBERS

CHAPTER XV. Remarks on Prototypes of Turbulence, Structures in Turbulence and the Role of Chaos
Hassan Aref 347

CHAPTER XVI. Subgrid Scale Modeling and Statistical Theories in Three-Dimensional Turbulence
Jean-Pierre Chollet 353

CHAPTER XVII. Strange Attractors, Coherent Structures and Statistical Approaches
John L. Lumley 359

CHAPTER XVIII. A Note on the Structure of Turbulent Shear Flows
Parviz Moin 365

CHAPTER XIX. Lagrangian Modelling for Turbulent Flows
S.B. Pope 369

Contributors

Hassan Aref, Division of Engineering, Brown University, Providence, RI 02912, U.S.A.

Dennis M. Bushnell, NASA Langley Research Center, Hampton, VA 23665, U.S.A.

Gary T. Chapman, NASA Ames Research Center, Moffett Field, CA 94035, U.S.A.

Jean-Pierre Chollet, Institut de Mécanique de Grenoble, 38402 Saint-Martin d'Hères Cedex, France.

Joel H. Ferziger, Department of Mechanical Engineering, Stanford University, Stanford, CA 94305, U.S.A.

Jackson R. Herring, National Center for Atmospheric Research, Boulder, CO 80309, U.S.A.

Robert H. Kraichnan, 303 Potrillo Drive, Los Alamos, NM 87544, U.S.A.

Marten T. Landahl, Department of Aeronautics and Astronautics, Massachusetts Institute of Technology, Cambridge, MA 02139, U.S.A.

B.E. Launder, Department of Mechanical Engineering, University of Manchester, Manchester M60 1QD, United Kingdom.

Moon J. Lee, Department of Mechanical Engineering, Stanford University, Stanford, CA 94305, U.S.A.

John L. Lumley, Sibley School of Mechanical and Aerospace Engineering, Cornell University, Ithaca, NY 14853, U.S.A.

W.D. McComb, Department of Physics, University of Edinburgh, Edinburgh EH9 3JL, United Kingdom.

Parviz Moin, NASA Ames Research Center, Moffett Field, CA 94035, U.S.A.

S.B. Pope, Sibley School of Mechanical and Aerospace Engineering, Cornell University, Ithaca, NY 14853, U.S.A.

A. Pouquet, Centre de la Recherche Scientifique, Observatoire de Nice, 06007 Nice Cedex, France.

William C. Reynolds, Department of Mechanical Engineering, Stanford University, Stanford, CA 94305, U.S.A.

P.G. Saffman, Department of Applied Mathematics, California Institute of Technology, Pasadena, CA 91125, U.S.A.

D. Brian Spalding, Computational Fluid Dynamics Unit, Imperial College of Science and Technology, London SW7 2BX, United Kingdom.

E.A. Spiegel, Department of Astronomy, Columbia University, New York, NY 10027, U.S.A.

R. Temam, Laboratoire d'Analyse Numerique, Université Paris-Sud, 91405 Orsay Cedex, France.

Murray Tobak, NASA Ames Research Center, Moffett Field, CA 94035, U.S.A.

CHAPTER I. Turbulence Sensitivity and Control in Wall Flows

Dennis M. Bushnell

ABSTRACT

Paper briefly reviews the sensitivity of turbulent wall flows to both uniform and non-uniform inputs of a static and dynamic nature. Input parameters considered include pressure gradient, wall mass transfer, compressibility, additives, and both micro and macro geometric variations including wall and outer region devices. These various sensitivities provide potential test cases for the examination of turbulent theories.

INTRODUCTION

The ultimate technological uses of turbulence theory or closure models are design, optimization, and prediction. The variety of technological applications which involve turbulence is immense, and includes scales from atmospheric to near-wall motions of less than a thousandeths of an inch. A recent National Academy of Sciences study of aeronautics in the year 2000 and beyond pinpointed turbulence and turbulence control as a vitally important research area, with possible payoffs which include divergence/departure free flight, drag reduction with resultant energy saving, and improved propulsion integration for optimal control and performance.

Historically, the control aspect of the turbulence problem has not received the same attention as closure, the general attitude being that the turbulence physics must be understood before attempts at control would be worthwhile. Clauser[1] was evidently one of the first to suggest that, if one viewed the unknown turbulence physics as a "black box" problem, the contents of this box (turbulence physics) could be studied by changing various "inputs" (initial/boundary conditions) and observing the resultant flow changes or "output." Data of this type are also directly useful for (a) parameterization of turbulent theories and (b) applications of turbulence control such as drag reduction and flow separation minimization. References 2-4 provide extensive bibliographies for wall flow sensitivity as applied to (a) drag reduction, (b) embedded body-incident turbulence interaction, and (c) spatial variations, respectively. The purpose of the present paper is to briefly summarize qualitatively the known sensitivities of a small subset of the possible turbulent motions, turbulent wall flows, to various control inputs. Also included are selected applications of this information to turbulence (1) theory, (2) understanding, and (3) technology.

TURBULENCE CONTROL PARAMETERS AND CONONICAL TURBULENT STRUCTURE

The equations of motion (continuity, momentum) for the case of low speed flow define the basic control parameters for turbulent wall flows. These parameters include (a) initial and bounding instantaneous velocity field, (b) instantaneous initial and bounding pressure field (constant density flow), and (c) instantaneous bounding geometry.

The most obvious classification of control inputs is into the general categories of steady state and dynamic (time-varying) and the corresponding spatial cases of uniform and one, two, or three space variable inputs. For more general fluid flow cases, controls engendered from additional force fields (such as MHD and EHD body forces), additives and thermodynamic and chemical effects are possible. The more obvious control parameters are listed on Figure 1. These parameters can also be organized (in various combinations) into a listing of classes of flows with roughly similar overall turbulence structure, as has been done in Figure 2 (Ref. 5). It should be noted that steady-state inputs are considerably easier to implement than dynamic ones, especially at high speeds, due to inertia, frequency response, power, fatigue, and scale problems associated with dynamic input systems.

Before discussing the known sensitivities of wall turbulence to various control inputs, it is of interest to briefly examine the structure of the turbulence one is trying to modify. A schematic of the typical wall turbulence flow modules is indicated on Figure 3. This information is a result of both detailed flow visualization and "conditional sampling" measurements conducted over the past 20 years, particularly at Stanford University (e.g., Refs. 5-8).

The turbulence production process in wall flows appears to involve at least three different scales of motion; a large outer scale (which for low Reynolds number is evidently the residue of the Emmons spots (Ref. 9)), intermediate scales, sometimes referred to as typical or Falco eddies with dimension the order of 100 wall units, and a near wall region where the Reynolds stress is produced (in a very intermittent fashion) by a process termed "bursting." Within the wall region quasi-stationary weak longitudinal (counter-rotating) vortices exist with an individual dimension on the order of 30 to 40 wall units, an average transverse spacing of approximately 100 wall units and an average lifetime the order of 1000 wall units. The wall streak structures and the intermittent turbulent production events (or "bursts") in the near wall region are generally referred to as the "coherent structures" in the wall boundary layer. The bursting occurs randomly in space and time, but does have identifiable scale(s) and frequency. The bursts are at least partially induced by the upwelling associated with the counter-rotating wall region streamwise vortices (wall streaks). Turbulence production, which is a violent ejection of fluid from the wall region, is preceded by the somewhat more gradual formation, at approximately 20 to 30 wall units from the surface, of an inflection in the instantaneous longitudinal velocity profile.

What is agreed concerning this turbulence production process is the stages, scales, and frequency of the burst cycle and the presence, scales, and structure of the wall streaks. What is not clear is the origin of the ubiquitous wall streaks, and details of the interrelationship(s) between the three (or more) scales involved (outer, Falco, or "typical" and inner regions).

One approach to the interpretation of the turbulence control experiments is to view them as attempts to interfere either with some component of the turbulence production cycle (e.g., breakup/amplify the large eddies, alter the instantaneous inflectional profile, etc.,) or to change the communication between the various scales. Liepmann in Reference 10 states, "Probably the most important aspect of the existence of deterministic structures in turbulent flow is the possibility of turbulence control by direct interference with these large structures. Such control could lead to very significant technological advances." Narasimha[11] indicates that wall turbulence can indeed be altered fairly easily. "We may conclude by remarking on "How easy it appears to be to surpress turbulence" -- whether you suck or blow, squeeze or bend, heat or cool, or do any of a vast number of other things to it, turbulence can be destroyed, or at least disabled, provided the operation is done properly."

Additional possible variables in the turbulence control problem include (a) combinations of influences and their relative phasing, (b) the rate at which effects are applied or removed (equilibrium/nonequilibrium turbulence structure), and (c) the length scale of the application ($\delta/10$, δ, 10δ, etc.). . Of particular recent interest is the use, as a "control," of non-planar discrete geometries (Ref. 2) (as differentiated from conventional "roughness"). Historically, the usual control parameters have been pressure gradient and wall injection/suction.

SENSITIVITY OF WALL TURBULENCE TO STEADY STATE CONTROLS

Uniform Inputs

Flow/Thermodynamic Parameters - The particular parameters of concern in this section include pressure gradient, wall suction/injection, compressibility, additives, two-phase flow, MHD, EHD, and chemical reactions.

The effects of pressure gradient upon the structure of wall turbulence is one of the few cases studied from the micro (coherent structures) viewpoint.[12] The presence of an adverse pressure gradient increases the wall burst rate (presumably by increasing the strength/biasing the instantaneous near wall inflection points) and, due to an increase in mean shear away from the wall, the turbulence fluctuation amplitude is also greatly increased. The turbulence energy profile broadens appreciably and the position of maximum turbulence energy moves out into the boundary layer away from the immediate near wall region. In spite of this large increase in turbulence level and activity the mean wall shear is reduced, due to the same reason that the turbulence is amplified, direct biasing of the mean velocity profile by the pressure force. This biasing is strongest near the wall (due to the $u \to 0$ boundary condition). Also, the intermittancy region in the outer portion of the boundary layer narrows, probably due to a "filling in" of the "valleys" by the increased wall burst activity. As would be expected, a favorable pressure gradient increases the wall shear but reduces the mean shear over most of the boundary layer, thereby reducing the fluctuation intensity. The wall bursting rate is also reduced (due perhaps to the presence of a mean bias which mitigates against the formation of localized near wall inflections) and the large outer scales are more sharply defined. For large enough favorable pressure gradients the flow will "relaminarize" which, in this case, means Reynolds stress production can fall below the level necessary to maintain a "fully turbulent" profile. However, the presence of large residual fluctuation levels generally causes the turbulence production to increase again quite rapidly once the pressure force field is withdrawn. (See Ref. 11 for an excellent discussion/review of the entire relaminarization issue.)

The gross effects of distributed wall suction/injection are similar to favorable/adverse pressure gradients, respectively. Note that for both the pressure gradient and wall mass transfer boundary conditions changes in turbulence amplitude are the opposite of changes in wall shear stress (increase in turbulence energy occurs concomitently with a reduction in wall shear and visa-versa). This situation is contrary to the conventional expectation, which was probably engendered from the increased shear associated with turbulent vs. laminar flow.

The effects of compressibility in wall flows are notable by their near absence. The results of a decade and a half of conventional closure computations and comparisons with data in wall flows up to Mach numbers the order of 40 indicates that, if the mean density variation and low Reynolds number effects are taken into account, the incompressible closure models work quite well for purposes of engineering prediction.[13] This is quite surprising in view of the presence of a new and potentially large "compressibility" (pressure-dilitation) term in the turbulent kinetic energy equation. Measurements and calculations of pressure fluctuation levels in compressible flows indicate the expected very large increases with Mach number, but computational results indicate very little local correlation between these pressure fluctuations and the dilitation i.e., the individual terms in the correlation are large, but the time averaged spatially localized correlation is not.[14] It should also be noted that the fluctuating Mach number in compressible wall flows is usually less than 1, and therefore, the major new compressibility physics associated with transient "eddy shocklets" (which appears to affect compressible free shear layers to first order) are not a major feature of the wall flows considered herein.

The effects of various additives upon turbulent wall flows are of major importance in terms of engineering application, the wall shear can be reduced up to 80 percent (in water) using such additives as (a) polymers, (b) fibers, and (c) micro bubbles (Ref. 15 and refs. therein). The polymer influence upon the turbulence structure is fairly well documented, with the major effects occurring in the wall region. Smaller scale motions, especially of the tightly wound helical vortex type, are spread out and, to a certain extent, damped. The mean transverse spacing between wall streaks is increased but visualization indicates the <u>same number of wall bursts per streak</u>, just less streaks (due to increased spacing) and therefore less total number of wall bursts (lower burst frequency). The optimal effectiveness for polymer drag reduction (in terms of pounds of drag reduced per pound of polymer) is obtained with distributed injection into the wall region. The polymer is evidently not effective in significantly suppressing the larger outer scale motions. The mechanism for bubble drag reduction is apparently mainly one of reducing the average density in the flow (the air bubbles displace water which has a factor of O(800) larger density[16]). Detailed data on possible turbulence structure alterations due to bubble injection are not yet available.

The use of MHD for turbulence control is particularly interesting because this body force can be applied directly to the turbulence field only (e.g., flow-aligned magnetic field in a developed channel/pipe) and any drag level between fully turbulent and laminar can be obtained simply by altering the strength of the applied field.[17] Obviously a magnetic fluid is required for reasonable field strengths to be effective. A field parallel to the mean flow affects (directly) only the off-axis turbulence components, and therefore tends to "mono-dimensionalize" the turbulence, thereby tending to destroy "vorticity stretching," one of the most important mechanisms responsible for maintaining the turbulence field. To a limited extent a tendency to "mono-dimensionalize" the fluctuation field is also present in the case of additives, where typically the axial turbulence component either remains nearly the same or may actually increase while the other components are reduced.[2]

One application of EHD to turbulence control (which also requires special fluid properties, e.g., almost pure water or dry air) involves the "ion wind." In this case, a corona discharge is established from an array of point electrodes. This discharge induces electrostatic body forces into the momentum equations which effectively converts (through the continuity equation) part of the tangential velocity field into the direction normal to the surface.[18] The resultant influence upon the turbulent field is probably similar to wall blowing (which the surface corona wind resembles, albeit without mass transfer through the surface). The influence of chemical reactions upon detailed turbulence structure is nearly an unknown. To first order the usual closure techniques seem to apply, using the local thermal and concentration fields as influenced by the chemical reactions, in analogy to the case of compressibility.

<u>Geometric Parameters</u> - Surface and streamline geometry has a tremendous influence upon the structure of turbulent wall flows. Obviously such effects can differ appreciably depending upon the

relative scale of any geometric non-uniformity. Small geometry variations (scale < O(δ)) are usually termed "roughness" whereas flow changes due to large scale geometric non-uniformities (scale > δ) are termed "curvature" effects. Considering first these large scale (or curvature) effects, three sub cases can be identified (a) longitudinal, (b) transverse, and (c) in-plane. For turbulent shear flows the sense of the longitudinal curvature is crucial. Convex longitudinal curvature is extremely stabilizing to the outer eddy structures of wall flows and large effects are noted even for quite large curvature values ($\delta/R > O(.05)$). Data indicate that the Reynolds stress can even change sign (become positive) in the outer region due to convex curvature.[19] Concave longitudinal curvature, even in the turbulent flow case, usually induces quasi-stationary longitudinal (Taylor-Gortler) vortices which (a) are located in the outer portion of the boundary layer, and (b) distort greatly the usual turbulence field. For the free field case the streamline curvature induced by such longitudinal vortex structures is stabilizing (for scales smaller than the vortex). Whether any such stabilization occurs in wall flows with imbedded longitudinal vortices is the subject of current study.

Transverse curvature has a much weaker influence upon turbulence structure. Values of δ/R of O(1) are required to produce the types of changes in macroscopic boundary layer parameters which occur in the longitudinal curvature case for δ/R of O(.05). Skin friction is increased in the external flow case (e.g., boundary layer on a flow aligned needle). What is perhaps not generally appreciated is that pipe flows (one of the cononical wall flows) possesses an (internal flow) transverse curvature effect ($\delta/R = 1$). The information concerning the influence of in-plane curvature is both fairly recent and incomplete. Typical realizations include (a) the out-flow (shoulder) portion of vertical strut-boundary layer intersection regions and (b) the initial transverse rotation region on rotating hubs with a stationary upstream portion. The limited information available indicates that in-plane curvature decreases the scale of the outer region motions.[20]

The influence of the roughness, or small scale, variable geometry case upon the wall turbulence structure is confined primarily to the near wall/near roughness region. Both experiment and theory indicate that rough surfaces under laminar boundary layers do not generally cause an increase in net drag, the pressure drag associated with the localized flows about and between the roughness elements is almost exactly equal to the skin friction drag on the corresponding flat surface. When the flow is turbulent the passage of eddy motion over these separated regions causes localized erruptions and increased vertical mass/momentum interchange, which alters the magnitude but not the nature of the flow dynamics very near the roughness elements.[21] Data indicate that the outer eddies over a rough surface are more clearly defined. Classically, roughness is characterized in relation to an "equivalent sandgrain" surface. More recent studies indicate that each type/class of rough surface produces its own set of dynamics. Examples of "non-sandgrain" behavior include net heat transfer reduction for some spire geometries and different drag behavior for transverse and longitudinal grooved surfaces. The transverse grooves, termed a D-type roughness (as opposed to the sandgrain or K-type roughness) can, in some cases (depending upon groove aspect ratio) reduce the non-dimensional Reynolds stress near the surface while the longitudinal grooves (or riblets) produce local shear reductions of up to 50 percent and net drag reductions of O(10%).[22] The physics associated with the riblets involves imposition of a strong transverse viscous force which displaces the various turbulence production processes away from the physical wall and reduces their intensity. The burst frequency and process does not seem to be affected, only the burst intensity. The longitudinal flow within the riblet grooves is observed to be very quiescent (due presumably to the strong damping effect of the transverse viscous force).

Passive porous walls constitute another micro-geometry varient for turbulent wall flows. The basic concept is to remove the wall impenetrability condition without the imposition of net mass transfer (suction/injection) through the surface. Conditionally sampled data taken over several types of such surfaces with various sub-surface plenums indicate that the basic near wall turbulence production processes are unchanged.[23] Total surface drag is observed to increase due to a net pressure drag on the sides of the holes through the surface. Recent studies indicate such surfaces can be used as a passive device to delay separation and reduce compressibility drag on transonic airfoils.

<u>Summary Remarks - Uniform Inputs</u> - The various boundary conditions, whose effects are briefly mentioned herein are, even for the "uniform" case, far from an exhaustive listing of the various parameters which can be used to either control or examine wall turbulence. As a further limitation on the available information, the

bulk of the coherent structure/conditional sampling/turbulent structure studies thus far have been carried out for the basic flat plate case. Such detailed information is extremely scanty for other (control) conditions. Of considerable interest, and particularly rich in possible new physics and applications, are combinations of various boundary conditions. Possibilities include not only pairs, but up to N simultaneous effects (where N is the number of possible independent boundary conditions). For example, passive porous walls can be used to control flow separation in adverse pressure gradients and riblets become more effective in favorable pressure gradients (and far less so in adverse). Some information on several combinations is available from the extensive research program carried out at Stanford University in their boundary layer channel.

Spatially Varying Inputs

Flow/Thermodynamic Parameters - The boundary conditions considered in this section are applied over a limited spatial extent (rather than uniformly in space) and then altered or removed. Of major interest is the relaxation processes which occur between one set of boundary conditions and another. In general, experimental studies indicate that such relaxation occurs in the order of 100 length scales of the affected flow. Applying this to a wall flow (for an input condition which has altered the outer region eddy structure) the flow would require the order of 100 δ for the turbulence structure to again resemble the standard flat plate case once the condition is removed. For a boundary condition which effects primarily the near wall flow (out to $O(.2\ \delta)$), the corresponding relaxation region would be the order of 20 δ. Correspondingly, conditions which affect only the immediate sublayer region relax essentially instantaneously, on a scale compared to δ. Also, in general the mean flow relaxes first, then the second-order correlations, then third-order correlations, etc., i.e., the fine details and dynamic balance of the turbulence require much longer than 100 length scales to re-establish. A distinct possibility and additional concern in relaxation studies is that the boundary condition one is relaxing from may not have been applied long enough for the structure to be in equilibrium at the start of the process i.e., there is an initial (set-up) relaxation that must be either satisfied or recognized and studied as a parameter in the problem before the main part of the study (the back relaxation) can be well defined. The prediction of the relaxation distance in such flows is a particularly stringent test of turbulence theories.

One of the most rapid (e.g., step) changes in boundary conditions which can occur is peculiar to high-speed flows-shock/turbulence interaction. Recent research has indicated, from both experiment and theory, that turbulence levels are amplified the order of a factor of 2 by passage through a shock wave.[24] This increase comes about from several mechanisms including the partial conversion, across the shock, of entropy and acoustic fluctuations into vorticity perturbations, as well as direct amplification of vorticity and, probably the most important, direct conversion of mean flow energy into fluctuation energy through shock unsteadiness/oscillations (which is actually a dynamic input).

A spatial relaxation process which occurs in all turbulent wall flows is the recovery from the transition process and the attainment of an asymptotic turbulent structure. This recovery process involves primarily the large scale outer flow motions and requires the usual O(100 δ) downstream of the nominal end of transition. The physics of the transition region involves the growth and subsequent agglamoration of Emmons spots. The large scale turbulent structures which grow within these "islands of turbulence" (Emmons spots) have been traced well into the fully turbulent (merged) region.[9] Studies indicate that the scale of these outer motions is reduced the order of 30 percent by this relaxation process. From the summaries in Reference 4, the relaxation from sudden changes in pressure gradient and surface injection typically occurs fairly quickly, on the order of 30 δ , as these boundary conditions affect primarily the inner turbulence structure, i.e., the burst rate.

<u>Geometric Parameters</u> - As would be expected from the discussion thus far, data indicate that a step change in roughness will relax quickly, on the order of 20 δ .[4] Changes in convex body curvature, due to their influence upon the outer turbulence structures, relax much more slowly, i.e., O(100 δ).[25] For the case of concave curvature changing to a flat surface the longitudinal vortices produced in the concave region will also persist downstream. Their presence has still been detected the order of 100 δ from their point of introduction/production. Recent research at NASA Langley indicates that such embedded longitudinal vortices can be removed through an "unwinding" process using a fixed vortex generator (small aspect ratio plate at yaw) with the opposite circulation and its tip located near the center of the embedded vortex.

Wall waviness is essentially a case of oscillatory longitudinal curvature (convex to concave and back to convex). These curvature variations are obviously accompanied by corresponding variations in pressure gradient. In fact, very innocuous looking short wavelength waves can, due to the inverse quadratic relationship between pressure gradient and wavelength, produce very large nondimensional pressure gradients. However, the oscillatory nature of these gradients ensures, for wavelength the order of the boundary layer thickness, that the turbulence field is always out of equilibrium (to varying degrees) with the local boundary conditions of both pressure and curvature.[26]

There also exists a sizable, and growing, body of data concerning the influence of various embedded bodies upon the structure of turbulent wall flows (Ref. 3 and Refs. therein). Typical geometric configurations include (a) vertical screens/honeycombs, (b) transverse cylinders, and (c) plates and airfoils orientated parallel to the air flow. These various bodies produce quite different effects upon the incident turbulence field. The screens and honeycombs typically break up the larger scale motions across the entire layer with a concomitant very long relaxation distance. Limited experiments at NASA Langley indicate that the honeycomb can also be responsible for the introduction of quasi-stationary longitudinal vortical structures, i.e., for certain size ranges the spanwise wakes from the honeycomb elements evidently produce an instability in the boundary layer flow downstream of the honeycomb out of which arises large scale longitudinal vortices which can increase the mean entrainment rate of the downstream flow. The transverse cylinders produce varients of the usual transverse Karman vortices which, in the outer region, are remarkably persistent and can essentially replace the usual large scale structures. If the cylinder is in the near field of the wall then one sign of the alternating Karman vorticity is suppressed, giving rise, in the limit, to a train of one-sided "control vortices." Current studies at NASA Langley are focusing on the use of a transverse cylinder with an additional control plate mounted above it to bias the "one-sided" shed vorticity toward the sense opposite that of the inate boundary layer turbulence.

The transverse and flow-aligned embedded plates and airfoils have the effect of suppressing the largest scale motions. Studies indicate (a) downstream skin friction reductions the order of 30 percent which can last the order of 100 δ . (b) less boundary layer growth/entrainment, and (c) a lower wall burst rate (along with smaller outer scale motions). Net drag reductions as large as 20 percent or more have been measured for these embedded plate devices, the actual level attainable in a given application is a function of the element drag/robustness/loading. These results provide a strong indication of an important outer-to-inner layer communication link in the wall turbulence/burst production, as these geometric control devices have a direct effect upon the outer flow and yet produce large changes in the near wall region. In this regard, it is of interest to compare the boundary layer and pipe flow cases. The large outer eddy structures in these two cases are quite different and yet the near wall region dynamics are known to be extremely similar. For a given average viscous layer thickness the outer region motions appear therefore to be a modulator, rather than a determiner, of the near wall motions.

SENSITIVITY OF WALL TURBULENCE TO DYNAMIC INPUTS

Flow Parameter Dynamics

The existing studies of dynamic inputs to turbulent wall flows constitute, up to this point, only a weak beginning to the problem. To first order the effect of most experimental attempts at flow control through narrow-band dynamic inputs is to produce a series of quasi-steady results, i.e., the flow is usually nearly in equilibrium with the instantaneous flow values.[8] (See Reference 27 for an excellent summary of the effects of time-dependent stream velocity.) One of the major reasons for this is the difficulty of producing a dynamic input at the scale, frequency and amplitude of the turbulence field. Generally (due in the air case to inertia problems and high velocities) the frequency of the input is so low that only quasi-steady effects occurs. This is particularly true of the case where the free stream is altered by a narrow-band input. On the other hand, with free stream turbulence it is quite easy to produce a small scale and high frequency input (using the flow over fixed bodies, i.e., screens/honeycombs, etc.) but this motion is accompanied by other scales/frequencies and is therefore quite wide band. The influence of such wide-band inputs is to increase the scale of the outer eddies and amplify the turbulence intensity in the outer region.[3] The quantitative effects on the wall flow (and the quantitative increase in wall shear) are a function of both scale and intensity (and structure) of the stream turbulence.

Studies of dynamic blowing and suction have indicated only marginal effects upon turbulence structure, even for cases where relatively high frequency inputs were possible.[28] The possible reasons for this will be discussed under the section on "Summary

Remarks for Dynamic Inputs."

For turbulent bounded wall flows, far downstream of transition, experimental results indicate an apparent insensitivity to free stream sound (stream pressure fluctuations). An acoustic input does have, of course, a profound influence upon the transition process/location but not, apparently, upon a "fully turbulent" wall flow.

Geometrical Dynamics

The effect of a wave convection velocity on the influence of wall waviness is to reduce the effective pressure gradient and therefore the influence upon the turbulence. An attempt at turbulence control using a combined monopole-dipole wall motion with wavelength the order of the boundary layer thickness was a failure.[29] The Reynolds stress field produced by the wall motion input (a) was of large amplitude, and (b) essentially added linearly to the pre-existing turbulence motions, i.e., the two dynamical fields evidently co-existed nearly independently, in spite of the presence of large (non-linear) input amplitudes.

Summary Remarks - Dynamic Inputs

The probable key to increasing the effectiveness of dynamic inputs for control of wall turbulence structure is correct phasing. The turbulence dynamics, due to phase jitter and variability in size/strength of the various motions, occur randomly in both space and time. Therefore, even if an input is applied at the wall burst frequency, there is little likelihood of an interaction because of the random phasing. In fact, from Reference 27, stream velocity forcing does not appear to couple with the bursting process. Thus any dynamical input (as opposed to a static input which, due to its continuous presence, satisfies the phase concurrence problem -- but also provides out-of-phase contributions) should probably include a sensor to trigger the input. This greatly complicates the control problem but is actually well within the emerging state of the art in both sensors and microprocessors. The major difficulty lies in developing suitable (size, dynamical range) actuators or "effectors" which can alter the flow, once a trigger signal is received from the sensors and processed. The microelectronic advances are not really addressing the fundamental actuator difficulties/requirements of ever smaller size and more rapid response (within reasonable power requirements) as flow speed is increased toward practical applications. An additional problem in the feedback-control area is the control trigger (due to the variability and three-dimensional nature of the various processes). Research is currently underway on such active (feedback) wall control concepts at both NASA Langley (in air, using driven transverse wall waves) and Princeton (in water, using longitudinal heating strips). Mention should also be made of the apparent lack of effectiveness of passive devices such as passive porous walls and compliant (moving/flow actuated) walls in connection with flow actuated micro control of wall bursting. To a first approximation, such devices have a destabilizing influence upon burst production. The pre-burst flow involves moving adverse pressure gradients and low pressure regions.. Such low pressure regions tend to induce positive vertical velocities over passive porous walls and upwelling over compliant surfaces. Both of these passive reactions are destabilizing to the pre-burst profile as even deeper profile inflections are engendered. Therefore, the indications thus far are that energy must be added to the system for control (albeit perhaps less than would be saved through wall burst control).

It should be noted that, from the discussion in the section on relaxation behavior contained herein, any change in the immediate wall region relaxes fairly rapidly and therefore an appreciable fraction of the surface would probably have to be covered with sensors and actuators. On the other hand, dynamic devices which interacted with the outer region would have the advantages of (a) larger scale, (b) lower frequency, and (c) much longer relaxation regions and therefore large scale intermittent (spatial) operation.

CONCLUDING REMARKS

The bulk of the detailed turbulence structure data currently available is primarily for the simplist of cases only -- flat plate boundary layers and free shear layers. These data indicate that each disparate geometry has its own set of dominant (non-linear) instabilities. The effects of various boundary/input conditions is to modify these instabilities for low input levels but, for stronger inputs, the basic instability modes/structures which sustain the turbulence field could be altered. Steady state inputs are extremely effective in altering turbulence structure (for both amplification and dimunition) and provide excellent test cases for turbulence theories. The research in the area of dynamical inputs is in a very early stage, with sensor-feedback control loops evidently required for treating the random-phase problem in the turbulence dynamics. (Random as to spatial and temporal occurrence, there is a definable average frequency and the various stages of the wall burst process are only quasi-random.)

References

1. Clauser, F. H.: The Turbulent Boundary Layer. Advances in Applied Mechanics, V. 14, 1956, pp. 1-51.

2. Bushnell, D. M.: Turbulent Drag Reduction for External Flows. AIAA Paper 83-0227, presented at AIAA 21st Aerospace Sciences Meeting, January 10-13, 1983.

3. Bushnell, D. M.: Body-Turbulence Interaction. AIAA Paper 84-1527, presented at the AIAA 17th Fluid Dynamics, Plasmadynamics and Lasers Conference, Snowmass, CO, June 25-27, 1984.

4. Hefner, J. N.; Anders, J. B.; and Bushnell, D. M.: Alteration of Outer Flow Structure for Turbulent Drag Reduction. AIAA Paper 83-0293, presented at AIAA 21st Aerospace Sciences Meeting, January 10-13, 1983.

5. Lilly, G. M.: Vortices and Turbulence. Aeronautical Journal, December 1983, pp. 371-393.

6. Coherent Structure of Turbulent Boundary Layers. Editors, C. R. Smith and D. E. Abbott. AFOSR/Lehigh Univ. Workshop, November 1978.

7. Cantwell, B. J.: Organized Motion in Turbulent Flow. Annual Reviews of Fluid Mechanics, 1981, V. 13, pp. 457-515.

8. Hussain, A. K. M. F.: Coherent Structures - Reality and Myth. Physics of Fluids, V. 26, No. 10, October 1983, pp. 2816-2850.

9. Zilberman, M.; Wygnanski, I.; and Kaplan, R. E.: Transitional Boundary Layer Spot In a Fully Turbulent Invironment. IUTAM Symposium 1976, Physics of Fluids, V. 20, No. 10, 1977, pp. S258-S271.

10. Liepmann, H. W.: The Rise and Fall of Ideas in Turbulence. American Scientist, V. 67, 1979, pp. 221-228.

11. Narasimha, R.; and Sreenivasan, K. R.: Relaminarization of Fluid Flows. Adv. in Applied Mechanics, V. 19, Ed. by Chia-Shun Yih, Academic Press, 1979.

12. Schraub, F. A.; and Kline, S. J.: A Study of the Structure of the Turbulent Boundary Layer With and Without Longituinal Pressure Gradients. Report MD-12 Thermo-Sciences Div., Dept. of Mech. Eng., Stanford Univ., Stanford, CA, March 1965.

13. Bushnell, D. M.; Cary, A. M., Jr.; and Harris, J. E.: Calculation Methods for Compressible Turbulent Boundary Layers. NASA SP-422, 1977.

14. Feiereisen, W. J.; Reynolds, W. C.; and Ferziger, J. H.: Numerical Simulation of a Compressible Homogenous, Turbulent Shear Flow. Report No. TF-13, March 1981, Thermo-Sciences Div., Dept. of Mech. Eng., Stanford University.

15. White, A.; and Hemmings, J. A. G.: Drag Reduction by Additives, Review and Bibliography. BHRA Fluid Engineering, 1976.

16. Madavan, N. K.; Merkle, C. L.; and Deutsch, S.: Numerical Investigations Into the Mechanisms of Microbubble Drag Reduction. ASME Symp. on Laminar-Turbulent Boundary Layers, Energy Resources Technology Conference, New Orleans, LA, Feb. 12-16, 1984, FED-Vol. 11, pp. 31-41.

17. Fraim, F. W.; and Heiser, W. H.: The Effect of a Strong Longitudinal Magnetic Field on the Flow of Mercury in a Circular Tube. J. Fluid Mechanics, V. 33, Pt. 2, 1968, pp. 397-413.

18. Malik, M. R.; Weinstein, L. M.; and Hussaini, M. Y.: Ion Wind Drag Reduction. AIAA Paper 83-0231, presented at AIAA 21st Aerospace Sciences Meeting, Reno, NV, January 10-13, 1983.

19. So, R. M. C.; and Mellor, G. L.: An Experimental Investigation of Turbulent Boundary Layers Along Curved Surfaces. NASA CR-1940, April 1972.

20. Lohmana, R. P.: The Response of a Developed Turbulent Boundary Layer to Local Transverse Surface Motion. ASME J. Fluids Engineering, September 1976, pp. 354-363.

21. Grass, A. J.: Structural Features of Turbulent Flow Over Smooth and Rough Boundaries. J. Fluid Mechanics, 1971, V. 50, Pt. 2, pp. 233-255.

22. Walsh, Michael J.: Turbulent Boundary Layer Reduction Using Riblets. AIAA Paper 82-0169, presented at the 20th AIAA Aerospace Sciences Meeting, Orlando, FL, January 1982.

23. Wilkinson, S. P.: The Influence of Wall Permeability on Turbulent Boundary Layer Properties. AIAA Paper 83-0294, presented at 21st AIAA Aerospace Sciences Meeting, Reno, NV, Jan. 10-13, 1983.

24. Anyiwo, J. C.; and Bushnell, D. M.: Turbulence Amplification in Shock Wave-Boundary Layer Interactions. AIAA J. V. 20, No. 7, July 1982.

25. Simon, T. W.; Moffat, R. J.; Johnston, J. P.; and Kays, W. M.: Turbulent Boundary Layer Heat Transfer Experiments: Convex Curvature Effects Including Introduction and Recovery. Report No. HMT-32, Stanford University, November 1980.

26. Lin, John C.; Weinstein, L. M.; Watson, R. D.; and Balasubramanian, R.: Turbulent Drag Characteristics of Small Amplitude Rigid Surface Wakes. AIAA paper 83-0228, presented at 21st AIAA Aerospace Sciences Meeting, Reno, NV, January 10-13, 1983.

27. Cousteix, J.; and Houdeville, R.: Effects of Unsteadiness on Turbulent Boundary Layers. Lecture Series No. 1983-03, "Turbulent Shear Flows" von Karman Institute for Fluid Dynamics.

28. Narasimha, R.: The Turbulence Problem; A Survey. J. Indian Institute of Science, 64 (A), January 1983, pp. 1-59.

29. Weinstein, L. M.: Effect of Driven Wall Motion on a Turbulent Boundary Layer. IUTAM Symposium on Unsteady Turbulent Shear Flows, Toulouse, France, May 5-8, 1981.

FIGURE 1

KNOWN PARAMETERS
HAVING A FIRST ORDER INFLUENCE UPON
TWO-DIMENSIONAL TURBULENT BOUNDARY-LAYER STRUCTURE

- O PRESSURE GRADIENT
- O CORIOLIS FORCES
- O WALL CURVATURE
- O WALL ROUGHNESS
- O COMPLIANT WALLS (WALL MOTION)
- O ENERGY RELEASE/ CHEMICAL R_X
- O PROXIMITY TO TRANSITION/REYNOLDS NO.
- O SHOCK INTERACTION
- O DENSITY STRATIFICATION (E.G., BUOYANCY PROBLEM)
- O ADDITIVES (POLYMERS, FIBERS)
- O COMPRESSIBILITY (DENSITY VARIATION)
- O TWO-PHASE FLOW
- O EHD AND MHD FORCES
- O STREAM OSCILLATING
- O WALL PERMEABILITY/MICROGEOMETRY
- O WALL MASS TRANSFER

FIGURE 2

KNOWN STRUCTURAL FLOW ZONE OMITTING
HYPERSONICS AND WAVE PHENOMENA

"INVISCID FLOW"
LAMINAR BOUNDARY LAYER
TRANSITION
HOMOGENEOUS TURBULENT FLOWS
ATTACHED TURBULENT BOUNDARY LAYERS, 2-D
ATTACHED TURBULENT BOUNDARY LAYERS, 3-D
REATTACHING/DETACHING ZONES
MIXING LAYER
AXISYMMETRIC WAKE
PLANE JET
AXISYMMETRIC WAKE
PLANE WAKE
RECIRCULATION ZONE (FULLY STALLED ZONE)
SECONDARY FLOW, TYPE 1
SECONDARY FLOW, TYPE 2
MACH NO. EFFECTS
SHOCK/BOUNDARY-LAYER INTER, 2-D
SHOCK/BOUNDARY-LAYER INTER, 3-D
TRAILING EDGE INTERACTIONS
LARGE-SCALE VORTICAL MOTION

FIGURE 3

WALL TURBULENCE PRODUCTION FLOW MODULES AND EVENTS

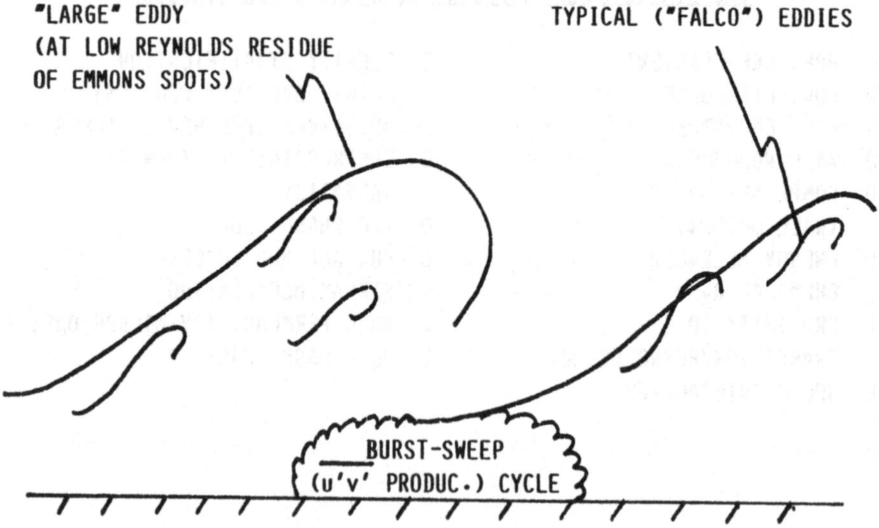

DOMINANT 3-D FLOW MODULES ARE HORSESHOE OR LONGITUDINAL VORTICES AND RESULTING THIN SHEAR LAYERS.

CHAPTER II. Observations, Theoretical Ideas, and Modeling of Turbulent Flows—Past, Present, and Future

Gary T. Chapman and Murray Tobak

INTRODUCTION

Turbulence (or chaos) is one of the oldest and most difficult open problems in physics. Although the subject of this review is turbulence in the field of fluid dynamics, the problem of turbulence pervades many other fields; e.g., cosmology, the structure of the universe. At one time or another, it has occupied the minds of many of the great physicists, particularly in the early part of this century. The problem is so difficult that it has even defied the formulation of a consistent and rigorous definition. In this paper, we shall review the history of the subject and point to recent developments, as well as postulate future directions. To avoid merely enumerating a succession of isolated research events and accomplishments, the review will be presented in a context of the interactions between observations, theoretical ideas, and the modeling of turbulent flows. The context needs further elaboration, but first a few comments are in order regarding a basic premise of the review.

The basic premise is that turbulence can be understood within the framework of the continuum assumption of fluid dynamics. Accepting the premise implies accepting that the Navier-Stokes equations are a complete mathematical description of fluid flows, and hence, capable of describing turbulent flows. There are some experimental facts that might shed doubt on the validity of the assumption. For example, small amounts of long-chain polymers in water have a significant effect on turbulent properties, even though the polymers are dispersed and have dimensions significantly smaller than the dissipation scales of turbulence. Additional recent developments also may raise questions regarding the continuum assumption. Although it needs constant re-examination, the continuum assumption has formed the basis for the study of turbulence over its entire formal history. With precautions and reservations duly noted, the assumption will be accepted as the basis for this review as well. Acceptance of the Navier-Stokes equations also may raise criticism on mathematical grounds, inasmuch as existence has not been proven for solutions of the three-dimensional initial-value problem. Although no examples are known, the possibility cannot be ruled out that solutions become singular, especially at Reynolds numbers representative of fully-developed turbulence. This would imply that additional principles need to be introduced to ensure a complete theory. We shall proceed as if this is not the case. Lanford [1] has compiled an excellent list of the presuppositions entailed by adoption of the Navier-Stokes equations as the framework for understanding turbulence.

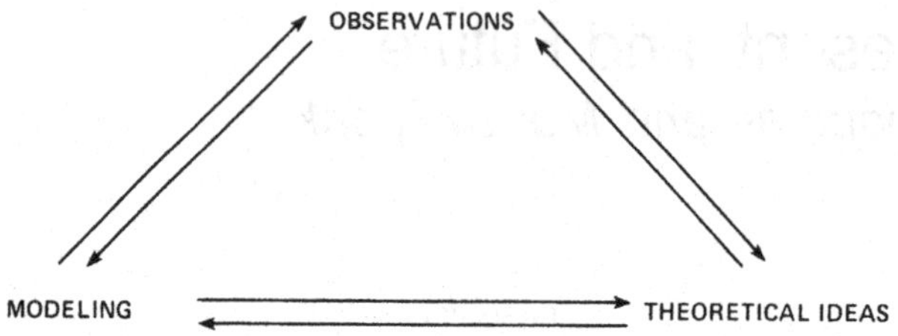

Fig. 1. Observations, theoretical ideas, modeling, and their interactions.

To aid in the discussion, the context of interactions between observations, theoretical ideas, and modeling is illustrated in figure 1. First, let us define the terms: The term "observations" here includes not only empirical data from observations of the physical (real) world, but also empirical data from observations of computer simulations representing solutions of the full Navier-Stokes equations (hence the above-mentioned need for the continuum assumption) or other suitable simulations to be noted later. The term "theoretical ideas" is used here to denote the realm in which observations are transformed into (normally nonmathematical) idealizations or conceptualizations: e.g., the concept of a continuum or of an incompressible fluid. Conceptualizing or theorizing is vital to the study of turbulence, as it is to the study of any scientific discipline, but it represents both a positive and a negative aspect. On the positive side, it is essential that theoretical ideas be postulated, both to further the mathematical steps which follow, as well as to provide hypotheses against which to cast the observations. Theoretical ideas literally provide "a way of seeing." It is this aspect that also may be negative, for a way of seeing may color or bias our observations. These two aspects of the realm of theoretical ideas will surface as major points in the discussions to follow. Finally, the term "modeling" is used to denote the realm in which theoretical ideas are placed within a formal system by means of mathematics.

Information flows back and forth between each of the three realms. For example, observations typically lead to a theoretical idea which provides a basis for both new observations and a mathematical model. The mathematical model can be tested against observations and also provides implications against which to test the theoretical idea. Observations used to test the model can also lead to changes in the model. Once the study of a discipline has begun, it is difficult to tell in which of the three realms it originated. It is clear, however, that the role of theoretical ideas is a very powerful one in that it shapes our approach to the other realms in the sense of dictating "a way of seeing." (On this important point, see also Liepmann's essay [2], to which our own views are much indebted.) In this review, we shall try to show how the "way of seeing" has influenced the study of turbulence, first, by examining the subject in an historical context, and second, by examining recent developments. Finally, some future developments will be postulated.

HISTORICAL PERSPECTIVE

Observations of turbulence are as old as recorded history. The Bible, for example, contains several references to turbulence or chaos. Leonardo da Vinci was

Fig. 2. Very early observations of turbulence--sketch by Leonardo da Vinci, circa 1500.

intrigued by turbulence, as his sketch reproduced in figure 2 (circa 1500) [3] indicates. But, the modern scientific study of turbulence dates from the late 1800s with the work of Osborne Reynolds. In reviewing the subject from that date to the present, one is struck by the appearance of three distinct movements, each of which (despite some overlap), can be characterized by a definite point of view with a reasonably well-defined beginning. The earliest of these, which has a strong nondeterministic flavor, will be referred to as the statistical movement; the next, which is predominantly observational, will be referred to as the structural movement; and the most recent will be called the deterministic movement. The three movements, with a few key events noted, are sketched in figure 3.

Statistical Movement

As noted above, Osborne Reynolds' observations of transition in pipe flow in 1883 [4] mark the beginning of the scientific study of turbulence. Illustrative results of a repetition of Reynolds' experiments [5] are shown in figure 4.

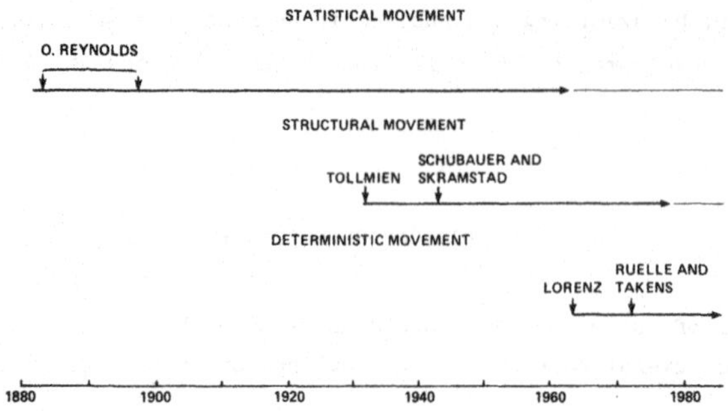

Fig. 3. Three movements in the history of turbulence research: statistical, structural, and deterministic.

Reynolds' observations led him to decompose the velocity field into a mean flow plus a perturbation. Considering the perturbation flow either too complicated or incomprehensible, he time-averaged the Navier-Stokes equations on the basis of the decomposition and arrived at what are today called the Reynolds-averaged Navier-Stokes equations [6]. The averages of products of the perturbation terms appear in the mean-flow equations as (what are now called) Reynolds stresses. Reynolds' view that

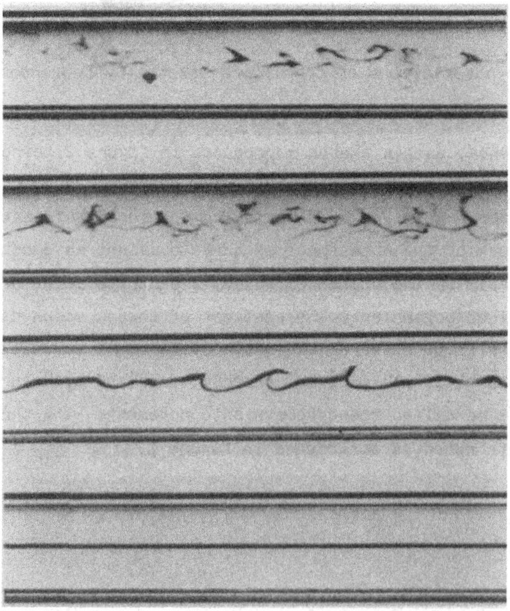

Fig. 4. First systematic study of turbulence--a repetition of O. Reynolds' 1883 dye experiments (from [5]).

the perturbations (and hence, the observations of the flow's chaotic structure), were unpredictable or incomprehensible in detail was not merely the result of observations, but was a view that would find support in the generally accepted world-view of the time in which turbulence or chaos was considered synonymous with disorder (the term used in Webster's dictionary) or unpredictability (nondeterministic) or incomprehensibility. This viewpoint, which is formalized in the statistical theory of random perturbations, was bolstered by the success of the statistical mechanics approach to the kinetic theory of gases (e.g., Jean's book on the topic [7]). The statistical point of view gained even further support by the great success achieved in theoretical physics on the introduction of quantum mechanics.

The statistical view of turbulence (a theoretical idea) had two principal effects: First, it encouraged focusing further observations of the flow on means and various averages. Second, because the Reynolds-stress terms consisted of averages of products of perturbation quantities, it suggested that the extension required to complete the modeling had to involve the next higher moment, requiring in turn an even higher moment, and so forth, leading, of course, to the celebrated problem of closure. The closure problem has proven so formidable in modeling that, in the face of practical exigencies, it has been more or less set aside in favor of phenomenological modeling. The latter has had considerable success in the practical realm, but has had essentially no impact in the realm of theoretical ideas.

The emphasis on observations of means and various averages led to significant development in instrumentation, particularly with regard to hot-wire anemometry. Development of the laser in the 1960s made possible the introduction of laser-doppler velocimetry (LDV), heralded as a great new tool in the study of turbulence. But nothing fundamentally new in the way of theoretical ideas has yet resulted from its introduction. Our contention is that the principal impediment is the statistical viewpoint itself; the use of the tool was prescribed too narrowly in accordance with its perceptions.

There is an extensive body of literature on the mathematical study of turbulence within the statistical theoretical framework. Included is not only the work of Reynolds, but that of Taylor, von Karman, Heisenberg, Kolmogoroff, Loitsianskii, Kraichnan, to mention only a few. The book by Hinze [8] contains an excellent summary of this work. Summaries of more recent work can be found in Monin and Yaglom [9] and in the contribution to this conference by Pouquet [10]. Kolmogoroff's "five-thirds law" [11], formulated for the statistical regime consisting of "locally isotropic turbulence," is an example of a particularly successful result of theoretical analysis within the statistical framework. It is successful in part because it ignores the regime of scales where most of the energy resides in most turbulent flows, and where the nature of the source of the turbulence is still evident. Even in the regime to which it ostensibly applies, the "five-thirds law" has been criticized on the basis that the turbulence is not truly isotropic, but intermittent. Hence, the law needs to be corrected for the nonspace-filling nature of turbulence [12]. Its blindness to these structural facts is precisely the disability of the statistical theoretical idea.

In our view, the principal shortcoming of the statistical approach is that the introduction of the statistical idea (predicated on a nondeterministic theoretical basis) at such an early stage of the study inhibits the interactions which otherwise would occur between observations, theoretical ideas, and modeling. The consequence is a paucity of imagery or structure about which to conceptualize. Our argument is not that statistical or averaging methods should have no role in the study of turbulence, but only that their introduction at the beginning of the study tends to stifle the flow of information and prevent conceptualization. Historically, this situation, coupled with the difficulties arising from the closure problem, encouraged the introduction of a line of research having an even more detrimental effect on theoretical ideas: phenomenological modeling.

Boussinesq's replacement of the molecular viscosity coefficient in the Navier-Stokes equations by a turbulent viscosity coefficient (eddy viscosity) in 1877 [13] and Prandtl's subsequent modeling of that term by means of the mixing-length idea in 1925 [14] had the effect of discouraging further theoretical ideas. First, Boussinesq's introduction of the eddy-viscosity concept entailed an inextricable confounding of flow properties and material properties. Second, introduction of the mixing-length idea through analogy with ideas from the kinetic theory of gases inadvertently gave the impression that a powerful and consistent statistical approach had been established. The widespread adoption of these ideas in modeling the equations governing turbulent mean flows broke the link with the framework of the original Navier-Stokes equations, so that ideas from the latter framework concerning, e.g., new structures arising out of instabilities, became more or less irrelevant to turbulence studies. It was in this sense that adoption of the eddy-viscosity and mixing-length ideas effectively discouraged further theoretical ideas. Despite their great success in practice, the eddy-viscosity and mixing-length ideas, in fact, have serious shortcomings. The eddy-viscosity idea, for example, raises difficult conceptual problems which are, in effect, artificial, being the consequence of the nonphysical confounding of flow and material properties. Additionally, the conditions justifying the mixing-length idea rest on the assumptions of kinetic theory (small units traveling relatively long distances between interactions), and these conditions are not completely fulfilled by the properties of turbulent eddies. A more thorough review of phenomenological models and their application can be found in [15], and in papers at this conference by Launder [16] and W. C. Reynolds [17].

In the 1940s and 1950s, observations increasingly pointed to the structural content of turbulence. There was increasing evidence of intermittency, for example, and the realization that vorticity tended to concentrate in localized intervals of space and time (e.g., Batchelor and Townsend [18]). But the power of the statistical viewpoint--the ruling theoretical idea of the period--was sufficient to shape even the perception of the evidence of structure. An example is Landau's influential description of the origins of turbulence [19]. The study of laminar instability of flow past a flat plate had led to the discovery of the Tollmien-Schlichting waves and their measurement by Schubauer and Skramstad [20]. Although their occurrence was a deterministic, perfectly predictable event, in 1944 Landau postulated that turbulence was the result of an indefinitely large sequence of such events, and hence, in effect, unpredictable in detail.

The example illustrates how a ruling theoretical idea may shape our approach to the realms of observations and modeling in the sense of dictating "a way of seeing." Our severest criticism of the statistical movement is that it has resulted in a structureless theory having little power of conceptualization.

Structural Movement

This movement is dominated by observations. Early experimentalists studying turbulence noticed that their observations were not entirely in keeping with the purely nondeterministic statistical viewpoint. Observations of Tollmien-Schlichting waves as evidence of initial instability in a transitional flow [20] have already been mentioned, as have been the observations of Batchelor and Townsend [18] regarding nonuniformities of vorticity in homogeneous isotropic turbulent flows. From the late 1950s to the present, a virtual flood of observations has been published concerning the structures that occur in turbulent flows. The following are outstanding

Fig. 5. Top and side views of Emmons turbulent spot (from [22]).

examples: (1) The turbulent spot, first reported by Emmons [21]. Figure 5, taken from the work of Cantwell, Coles, and Dimotakis [22], is illustrative of this structure in turbulence. More recent detailed measurements have brought out additional structural features [22]. (2) Structures in wall-bounded shear flows, first

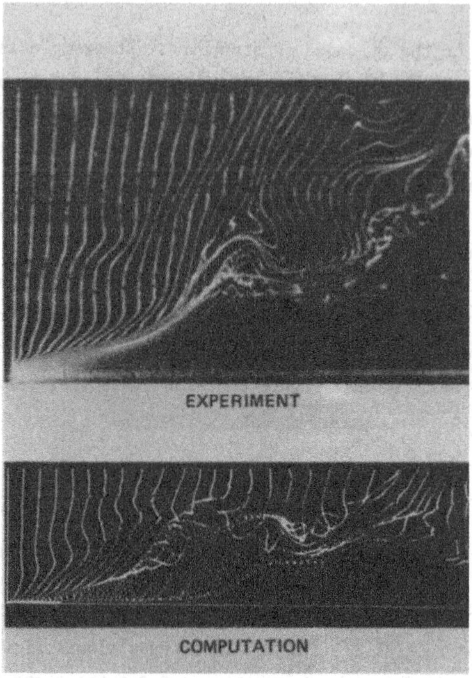

Fig. 6. Structures in turbulent wall shear flows (from [30]).

observed by Kline and Runstadler [23]. Figure 6 illustrates their appearance [24]. (3) Structures in turbulent free shear layers, investigated extensively by Roshko [25] and others (cf. fig. 7 from [5] and [26]). Cantwell [27] has published an excellent review of these experimentally observed turbulent structures.

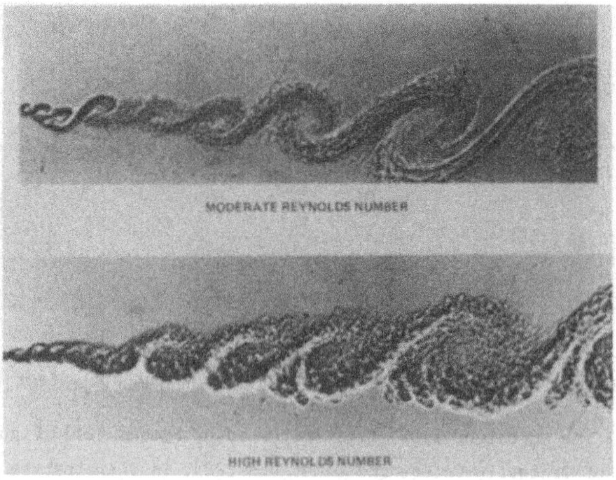

Fig. 7. Coherent structures in turbulent free shear layers (from [5]).

The structural movement has also included observations of computer simulations of turbulent flows. Examples are the results of Rogallo [28] for homogeneous isotropic turbulence, and those of Moin and Kim [29] for turbulent channel flow. A simulated hydrogen-bubble observation of the latter flow [30] is shown in figure 6, where it is compared with the already noted experimental observation of Kim, et al., [24]. The use of computer simulations has a deterministic character inasmuch as, strictly speaking, the computations are based on deterministic equations. However, many of the early analysts of these simulations adopted the same classical statistical methodology as that employed by the experimentalists. That is, for the most part, they measured only means and various averages. Some experimentalists were actively trying to measure and characterize the structures they were observing by developing a measurement methodology that reflected both the presence of coherent structures and their apparently random occurrence. These included conditional sampling techniques such as the method developed by Blackwelder and Kaplan [31], and the proper orthogonal decomposition method developed by Lumley [32,33]. The methods differ in the degree of subjective bias imposed by the experimenter, with the latter method having essentially none. Computer simulations of turbulent flows also have been analyzed by the same methods (e.g., [33] and [34]). Although they admit the presence of coherent structures, all of these methods contain the implicit assumption that the occurrence of structures is governed by random (and, thus probably incomprehensible) events. Full realization of the possibility of a deterministic and comprehensible chaotic flow behavior does not yet seem to have occurred within the structural movement.

The principal contribution of the structural movement has been the recognition of the presence and importance of structures in turbulence. Dryden [35] had recognized the possibilities in an early review (cf. Roshko's attribution [25]) as the following quotation makes clear: "It is necessary to separate the random process from nonrandom processes. It is not yet fully clear what the random elements are in turbulent flow." Despite its promise, the structural movement to date has had two major shortcomings. The first is that no new theoretical ideas have emerged that could be translated into formal mathematical models. The prevailing viewpoint remains one of unpredictability or indeterminism, in which the coherent structures are conceived as having been sprinkled about randomly in time and space. In addition, the structures that have been observed cannot be fitted easily into the statistical models currently in use. Landahl [36] (cf. also his contribution to this collection [37]) is one of the few theoreticians who have attempted to incorporate structures within a compatible statistical model. Bushnell (cf. [38], this collection) has been instrumental in finding ways to use structural observations practically to improve aircraft performance. The second shortcoming is the Achilles' heel of the structural movement: the jungle of observational detail, lacking the ordering hand of theory. The difficulty stems from the unassimilated mixture of random and deterministic elements in the generally accepted "way of seeing" the coherent structures. While the structures are assumed to be randomly distributed in time and space, each occurrence is assumed governed by a locally deterministic cause (e.g., a local instability). A consistent, overall theory has been lacking which has the possibility of assimilating the random and deterministic elements into a single viewpoint allowing, e.g., deterministic chaos.

In summary, the structural movement has demonstrated the presence and importance of structures in turbulence, but so far, has not resulted in new theoretical ideas having the power to abet modeling. The principal criticism is precisely the inverse of that of the statistical movement: in place of theory without structure, the result to date has been structure without theory.

Deterministic Movement

This is the most recent movement, although its origins date from the pioneering essays of Poincaré, the principal one bearing the title "On the Curves Defined by Differential Equations" [39]. The particular way of seeing inspired by Poincaré, strongly geometric as expressed in the language of topology, became known as the qualitative theory of differential equations. The field was advanced by the research of Andronov and his colleagues [40] who introduced the useful notions of "topological structure" and "structural stability." From the same line stems bifurcation theory, showing how structures may change with changes in conditions. The comprehensive review of Sattinger [41] demonstrates the extraordinary range of scientific disciplines in which bifurcation theory now plays a role. Applications to hydrodynamics are exemplified by the works of Joseph [42] and Benjamin [43]. Taken together, "nonlinear dynamical systems" is perhaps the best descriptive title for this body of theory. Its origins and some of its philosophical implications are traced in a recent interesting essay by M. W. Hirsch [44].

For the purposes of this review, dedicated to turbulence studies, the deterministic movement will be dated by the work of Lorenz in 1963 [45] and that of Ruelle and Takens in 1971 [46]. Lorenz discovered, via numerical computation, that a simple dynamical model of a fluid system yielded flow properties having bounded aperiodic behavior in time of a form apparently so chaotic, yet deterministic, that (in current terminology) it is said to indicate presence of a "strange attractor" (see fig. 8). Sparrow's book [47] contains a thorough study of the Lorenz equations (see also a thoughtful review of the book by Guckenheimer [48]). Unaware of Lorenz's results (but expert in the theory of dynamical systems), Ruelle and Takens independently proposed strange-attractor behavior as a model for turbulence, and argued via mathematical analysis that the turbulent state would be reached after the fluid system had undergone a finite and small number of bifurcations. A more accessible and updated account of the Ruelle-Takens thesis, fittingly first presented at

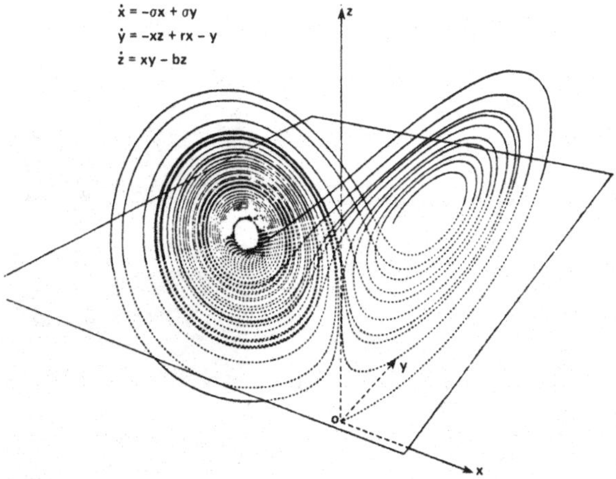

Fig. 8. The Lorenz attractor.

a symposium honoring the mathematical heritage of Poincaré, recently has been published by Ruelle [49]. An example of a fluid system typifying the sequence proposed by Ruelle and Takens is the Taylor-Couette flow shown in figure 9 [50]. Within the same period (1967) Mandelbrot [51] posed the idea of non-space-filling curves (which he had named fractals) as a model for explaining intermittency in turbulence. Also during this period ideas were proposed for handling problems involving rapid transitions in the important scales of structures, directed particularly towards application to phase transitions in solids (cf., e.g., Wilson and Fisher [52]). This methodology goes by the name of "renormalization group theory"; its relevance to turbulence studies has been noted by several authors (e.g., Siggia [53]).

The above account notes the principal elements of the deterministic movement; they will be discussed in greater detail in the next section. Here, their historical significance will be touched on briefly to complete our historical perspective. First, the work on strange attractors has shown that chaotic behavior can occur in even simple deterministic systems (as few as three nonlinear ordinary differential equations). Second, deterministic chaotic behavior can occur after just a few bifurcations of a dynamical system. (Other routes to deterministic chaotic behavior involving, e.g., period-doubling, are possible as well, cf. [54].) Third, the ideas underlying bifurcation theory, strange attractors, fractals, and renormalization group theory provide a rich body of imagery, containing a considerable potential for conceptualization of turbulence structures. Taken together, the first and second points offer the possibility that chaos or turbulence in fluid dynamics can be understood as a state of a simple deterministic system. The third point suggests a basis for the construction of models.

Viewed from the framework of observations, theoretical ideas, and modeling

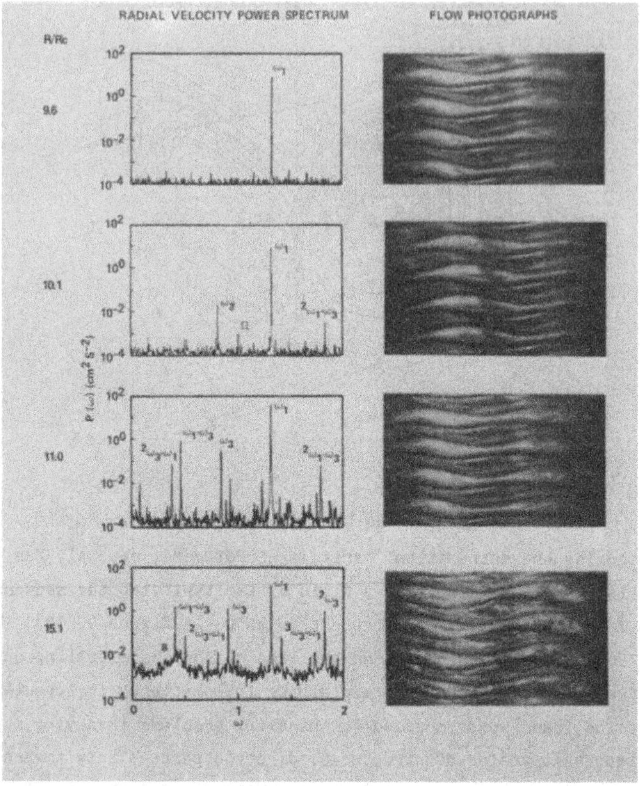

Fig. 9. Stages in Taylor-Couette flow with increasing Reynolds number R; R_c = critical R for onset of Taylor instability (from [50]).

already described, the impact of the deterministic movement can be summarized as follows: It has provided a basis for new observations, as exemplified by the work of Swinney and his colleagues on Taylor-Couette flow (fig. 9 and [50]). It has provided an impetus for mathematicians to return to the Navier-Stokes equations themselves as a basis for modeling turbulent flows (e.g., Temam, et al., [55]) as well as simpler systems that exhibit chaotic behavior. Finally, the extensive studies of the Lorenz equations have shown the synergistic power of the computer when computational studies are carried out in conjunction with general mathematical theory. This is the eloquent point of Guckenheimer's review [48] of Sparrow's book [47]. The shortcoming of the movement (if what follows can be called a shortcoming for so new a movement) is that the effort to date has focused on simple systems; and, in fluid dynamics, principally on the mechanisms of transition. The application of current ideas to fully-developed turbulent flows has not yet been seriously undertaken, nor is it clearly evident yet that the understanding which has been gained can be converted into successful models for turbulent flows of practical interest. On the latter point, and to offset potential criticism from researchers focusing perhaps too exclusively on the ultimate goal of predictive power, we can do no better than cite the following from Hirsch's essay [44]: "The end result of a successful mathematical model may be an accurate method of prediction. Or it may be something quite different but not necessarily less valuable: a new insight . . ."

RECENT DEVELOPMENTS

Our account of the deterministic movement took note of four recent developments which need further elaboration. These are bifurcation theory, strange attractors, fractals, and renormalization group theory. As indicated earlier, bifurcation theory and strange attractors emerged as part of the study of dynamical systems that originated with Poincaré. The study of fractals has roots in a number of areas, e.g., the study of Brownian motion and the meteorological studies of Richardson. Mandelbrot's book "The Fractal Geometry of Nature" [56] contains an excellent discussion of these origins. Finally, the study of phase transitions in condensed-phase matter was a principal source of inspiration for renormalization group theory. Each of these developments can be considered only briefly here; there are entire papers devoted to them elsewhere in this collection (cf. the contributions of Spiegel [57], Mandelbrot [58], and McComb [59]). Although they are often treated separately, and the perspective each brings to the subject is important, it is becoming clear that they cohere, and together offer the possibility of reflecting the greater part of many of the complex and chaotic processes in nature.

Bifurcation Theory

Generally speaking, bifurcation theory is the study of equilibrium solutions of nonlinear evolution equations and how they change with changes in the parameters of the problem. In fluid-dynamic applications, we are interested in equilibrium solutions of evolution equations of the form

$$\vec{U}_t = H(\vec{U},\lambda), \qquad (1)$$

where $\vec{U}$ is the velocity vector and λ is a parameter (e.g., Reynolds number, angle of attack, Mach number). An equilibrium solution is taken to mean the solution to which $\vec{U}(t)$ evolves after the transient effects associated with the initial values have died away. Equilibrium solutions may be time-invariant, time-periodic, quasi-periodic, or chaotic depending on conditions.

Changes in equilibrium solutions can occur at two levels. The first occurs as a result of instability in equation (1). As the parameter λ is varied, a critical value λ_c can be reached beyond which the original solution becomes unstable. New solutions, called bifurcating solutions, appear, some of which may be stable, and some unstable to small perturbations. By stable and unstable we mean the following: If a small perturbation of the solution decays to zero as $t \to \infty$, the solution is said to be asymptotically stable; if the perturbation grows, the solution is said to be asymptotically unstable. Stable branches of bifurcating solutions can be either local or global. A bifurcation solution is said to be local if it can be mapped onto the original solution without cutting the solution space; if it cannot, the bifurcation solution is said to be global. In addition, the bifurcation can be

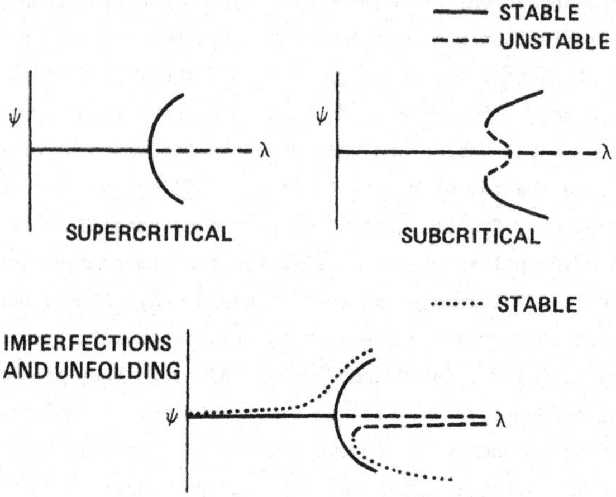

Fig. 10. Supercritical and subcritical bifurcations and the unfolding of a bifurcation curve due to imperfections.

supercritical or subcritical, as illustrated in figure 10. In a supercritical bifurcation (shown by the pitchfork bifurcation), there is at least one branch of stable bifurcating solutions that is continuous with the original solution at the bifurcation point λ_c. Thus, for a small change in λ across λ_c, there is a stable bifurcating solution that is $O(\Delta)$ close to the original solution such that as $\lambda - \lambda_c \to 0$, $\Delta \to 0$. This is not the case for a subcritical bifurcation shown on the right of figure 10. Here, for a small change in λ across λ_c, there is no branch of stable bifurcating solutions that is continuous with the original branch. This type of bifurcation normally leads to hysteresis behavior because the critical point for the upper branch in the case shown does not occur at the same value of λ_c as it does for the lower branch. The symmetrical bifurcation curves shown in figure 10 often result from idealized problems. In practice, there is less enforced symmetry, or there is a boundary condition, or a scale that was suppressed in the idealized problem. When these are brought into consideration, the idealized bifurcation diagram may undergo an unfolding. This is illustrated in figure 10 with the pitchfork. The idealized pitchfork has the following form (to leading order)

$$\psi^3 - \lambda\psi = \psi(\psi^2 - \lambda) = 0 , \tag{2}$$

whereas the general (unfolded) bifurcation to this order has the form

$$\psi^3 + a(\lambda)\psi^2 + b(\lambda)\psi + c(\lambda) = 0 . \tag{3}$$

For the case shown in figure 10, $a = 0$, and b represents the effect of a small imperfection. Bifurcation in the case of Taylor-Couette flow between rotating cylinders has this form, in which the c term is the result of including ends in the concentric cylinders, rather than treating the idealized problem in which the ends are at plus-and-minus infinity. The first stage of the (idealized) Taylor-Couette flow problem typifies a common type of bifurcation in which an original time-invariant equilibrium solution is replaced at the bifurcation point by another time-invariant equilibrium solution, in this case one describing the Taylor vortices. A second type of bifurcation is the "Hopf" bifurcation in which the original time-invariant equilibrium solution is replaced by a branch of stable equilibrium solutions which are time-periodic solutions. The Hopf-type of bifurcation is common in aerodynamics, for example, the Karman vortex street in the wake behind a circular cylinder for $Re > Re_c \approx 50$. A third type of bifurcation of great interest occurs when a quasi-periodic equilibrium solution is replaced by a bounded aperiodic solution having chaotic properties. Taylor-Couette flow for $R/R_c = 23.5$ [50] illustrates this type of behavior. It is suggested that this behavior indicates the presence of a strange attractor--the subject to be treated in the next section.

The second level at which changes in equilibrium solutions occur focuses on the class of equilibrium solutions that is time-invariant. Here, we concentrate attention on the singular points in the equilibrium flows where $\vec{U} = 0$. With $\vec{U}_t = 0$ in equation (1), we can recast equation (1) to directly describe particle trajectories or streamlines:

$$\vec{U} = \vec{X}_t = G(\vec{X},\lambda) , \tag{4}$$

where $\vec{X}$ is the spatial coordinate of the fluid element. Here, as λ crosses λ_c, a singular point may bifurcate into multiple singular points, or a new pair of singular points may appear, or a pair disappear. However, bifurcation at this level need not imply nonuniqueness in the governing flow equations. Equilibrium solutions may remain stable and unique on either side of λ_c. The bifurcation of singular points in the flow will be referred to as structural bifurcation. All structural bifurcations are global in the mapping sense described earlier. Structural bifurcations in fluid flows are described with examples in [60].

The general topic of bifurcation theory has received considerable attention in the past few years with development of an extensive body of literature. Examples of this genre of work are given in [41,61,62].

Strange attractors

The recognition that bifurcation to a bounded aperiodic solution can occur, indicating presence of a strange attractor, represents a significant step in the study of turbulence. Strange attractors appear in forced dissipative systems and can occur with relatively small nonlinearities (cf. the Lorenz system in fig. 8). The following account is an attempt to give a more geometric sense to the term.

First, we need to introduce some additional terminology. By the state of a fluid-dynamic system, we shall mean a complete specification of the velocity field at an instant in time. The space of all states is the state space. The term orbit will refer to a solution of the differential equations determining points in state space (i.e., the Navier-Stokes equations), regarded as a curve in state space, and the solution flow will refer to the motion on the state space that advances each point along its respective orbit. The term equilibrium solution that was introduced earlier in connection with bifurcation theory, represents the long-term behavior of a solution flow after transients have died away. Time-invariant equilibrium solutions can be represented as fixed points in state space. Time-periodic equilibrium solutions can be represented as closed paths (i.e., circles) in state space. Equilibrium solutions having two incommensurable periods are representable on tori (called 2-tori) in state space. These orbits are called attractors if orbits start-

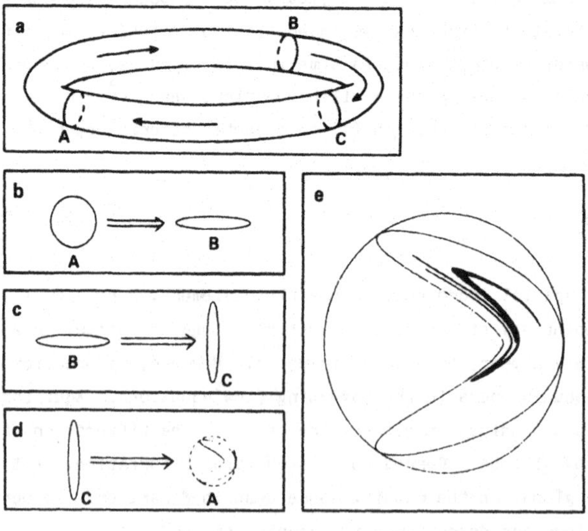

Fig. 11. Qualitative behavior of a model strange attractor.

ing sufficiently close to them converge to them. Convergence in this sense is equivalent to the notion of asymptotic stability introduced earlier.

Whereas the Landau theory of turbulence [19] supposed that tori of increasing dimension (n-tori) would succeed each other in an indefinite sequence of supercritical bifurcations, Ruelle and Takens [46] argued that beyond a 2-torus, a "strange attractor" would appear on the next bifurcation. Lanford [1] has presented a useful qualitative description of how a model strange attractor might succeed a 2-torus. His description is reproduced in figure 11. The solution flow shown in part (a) is primarily in the direction of the arrows around the torus with relatively small transverse motion. Plotted against time, a solution would appear as noisily periodic. Parts (b), (c), and (d) show successive intersections of a set of solution curves that undergo squeezing and stretching (b), rotation (c), and folding (d) in one circuit of the torus. Four iterations of this "return mapping" produce the complex layered structure shown in (e). As Lanford has noted, the separate phases (a)-(d) would occur simultaneously in real examples. In fact, the outcome depends only on the nature of the mapping which takes a point in the cross-section A into the next place where the solution curve through that point recrosses A. This mapping is called the return mapping or the Poincaré mapping. One assumes that the sequence of return mappings has a limit, and the limit is representative of the strange attractor. It must have the following characteristics: First, solutions "phase-mix" (cf. Joseph [42]). Unlike periodic or quasi-periodic functions, the strange attractor has an autocorrelation function which decays rapidly in time. This is a property shared by all observations of turbulent flows. Second, the strange attractor has a sensitive dependence on initial conditions. Solutions which start out even infinitesimally close together must eventually depart from each other. In Lanford's description of a model strange attractor (fig. 11), it is the stretching property that ensures the eventual departure of adjacent solutions. Finally, each member of the set of solutions comprising the strange attractor occupies zero volume in state space. It is this characteristic that forces the strange attractor to have noninteger dimensionality, or, in Mandelbrot's terms [63], fractal measure. In Lanford's model attractor, the folding property (d) gives rise to a multilayered structure that does not occupy any volume, and it is this property that accounts for the attractor's noninteger dimensionality. It is also this property that manifests itself in terms of intermittency.

For a more comprehensive study of the various connections between strange attractors and turbulence, the reader is referred to Lanford's review articles [1,64]. Ott's review [65] of strange attractors in a dynamical systems framework has an excellent section on the connection with fractal dimensionality. The connection as a possible property of solutions of the incompressible Navier-Stokes equations was apparently first aired by Foias and Temam [66], and has been recently sharpened [67,68]. The formulation of appropriate measures and dimensional descriptions of strange attractors is a subject of intense current interest [69-71].

Fractals

The fractal idea as a description of turbulence precedes that of the dimensionality of the strange attractor and, in the geometric form put forward by Mandelbrot [56], has considerable conceptual power. Hence, it will be in that context that fractals and fractal dimensionality will be briefly described. A fractal curve is a curve that is everywhere continuous but nowhere differentiable. An example of a fractal curve is the Brownian motion of a particle. Lest one choose to disregard the fractal idea too quickly, one should recall that the Navier-Stokes equations can be considered an ensemble average over a set of Brownian-motion curves. To illustrate these curves and their properties, two examples have been selected. These are shown in figure 12. The simplest example is the Koch curve. This curve is constructed by the following recursive procedure: take a line one unit long and divide it into three equal segments. Remove the center segment and replace it with two

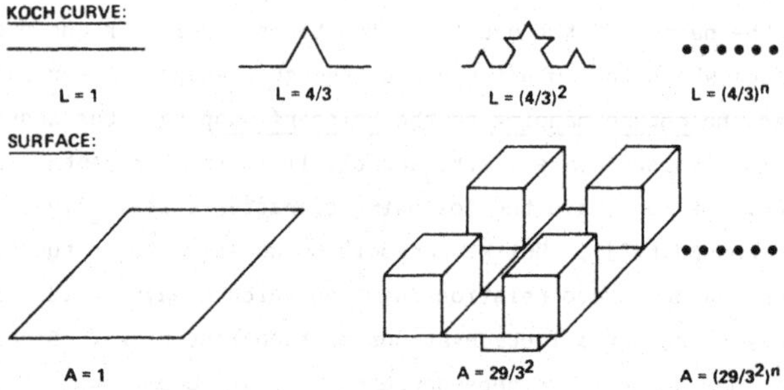

Fig. 12. Examples of fractals.

equal segments to form a hat (see fig. 12). This process is repeated recursively on each of the new segments that are formed at successive steps. Now the length of the resulting curve increases without limit as the number of repetitions (n) increases without limit, but the curve does not fill up any space. Only a line with apparent texture results. Another way to think of this is to take three of these Koch curves and form an equilateral triangle. Now note that we have a finite area enclosed (an island) by a perimeter (coastline) that is infinitely large. This point is well described in Mandelbrot's book [56] in the chapter entitled "How long is the Coast of Britain?" The following question arises: Is there a way to form a relationship between the line length and the unit of size at any point in the iteration? There is, and it is as follows: Let

$$L = s^{\mu} \tag{5}$$

where L is the length of the line and s is the length of the element used to construct L. Hence, for the Koch curve we have

$$\left(\frac{4}{3}\right)^n = \left[\left(\frac{1}{3}\right)^n\right]^{\mu}$$

or

$$\mu = 1 - \frac{\ln 4}{\ln 3}$$

Now μ can be interpreted as the difference between the Euler dimension (D_E) of the element (s), which in this case is 1, and the dimension of the line L, which is called the fractal dimension (D_f). Hence,

$$\mu = D_E - D_f$$

or, for the Koch curve, $D_f = \ln 4/\ln 3$ which is about 1.28. Note that if $D_f = D_E$, the length of the line does not depend on the size of the unit of construction, which is what one expects for smooth curves.

A better example, and one more closely related to turbulence, is the second example in figure 12, namely, that of a surface. One may think of this surface as a surface of vorticity. It is distorted in a recursive manner as follows: Divide the unit square into nine small squares; now, remove the four corner squares and the center square and replace them by building a small square box over the open squares (center box down for convenience). With each of the 29 sides of the new figure, repeat the process. The surface becomes more and more distorted with each step, and the actual surface area increases without limit as the number of iterations (n) increases. Hence, the surface in two dimensions becomes more and more distorted, but never fills up space in three dimensions. In a manner similar to that used for the Koch curve, the fractal dimension is found to be $\ln 29/\ln 9 + 1$ or about 2.54. Now, this is a rather simplistic model for a sheet of vorticity that has been distorted into a parcel of turbulence because of instabilities. Even though the model is simplistic, it is true that a hot wire passing the distorted sheet would exhibit intermittency. The observations that high-Reynolds-number turbulent flows exhibit intermittency go back to a paper by Batchelor and Townsend [18]. These authors noted that in high-Reynolds-number homogeneous turbulent flows, the vorticity was not distributed uniformly but was concentrated on sheets or other localized regions of space. Attempts have been made to derive a fractal dimension for this turbulence based on higher-order statistical information. Values between 2.0 and 3.0 have been derived. However, the reduction of information on the higher-order statistics to a fractal dimension requires specification of a topological form of the turbulence, and this step has led to considerable disagreement. Mandelbrot [56] suggests that reasonable topologies bound the value of the fractal dimension between 2.5 and 2.7. In a recent attempt to establish a basis for these values, Chorin [72] calculated the distortion of a vortex tube using the Euler equations. That calculation showed that the vorticity contracted (in an L_2 norm sense) to a fractal dimension of about 2.5, in reasonable agreement with the lower bound postulated by Mandelbrot.

Renormalization group theory

As noted earlier, the concept of the renormalization group has its roots in the study of condensed phase matter, in particular, certain crystal problems. The concept behind it is easiest to understand in that context. An example is shown at

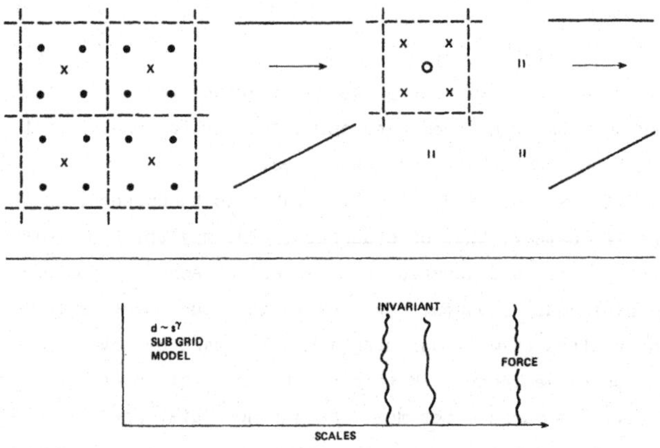

Fig. 13. Renormalization group ideas.

the top of figure 13. Here, a portion of a square lattice of molecules (dots) is shown on the top left. The interactions of four of the molecules are computed, so that they can be replaced by a supermolecule (X) which has the combined property of the four. Then, four of the supermolecules treated together form a still larger supermolecule and so on. One seeks scaling properties of the system that become invariant with repetitions of the process, so that only a small number of reclusterings may be required. A more detailed description of this process can be found in the paper by Wilson and Kogut [73]. The idea has been used by Feigenbaum et al. [74] in a novel way that calls attention to the universal scaling numbers for period-doubling. The idea has been used by Siggia [53] as follows: A fluid-dynamics problem with turbulence is being solved computationally by the use of finite-difference methods. In order to conserve computational time, a coarse grid is preferred. In that case, a subgrid turbulence model is required to handle the dissipation that occurs below the resolution of the grid. At the bottom of figure 13, a way to develop this subgrid turbulence model is sketched by means of a scale description. A problem is set up where the large-scale structures are forced. A fine grid is used so that dissipation by the subgrid scales is not important. A slightly coarser grid is then used with a dissipation term chosen so as to keep the midscales unchanged. The problem is repeated with a still coarser grid and dissipation term that, again, keeps the midscales unchanged. Several repetitions of the process should suffice to reveal the existence of a dissipation-scaling relation such as $d \sim s^{\gamma}$, where d is the dissipation, s is the grid size, and γ an exponent to be determined. This is a simple numerical application of a renormalization group idea. Another example is presented by McComb ([59], in this collection). Additional attempts should be forthcoming.

FUTURE DIRECTIONS

In our view, future directions for the study of turbulence will reflect the recent developments of the deterministic movement, together with statistical elements and structural observations that are consistent with the deterministic approach. In a previous paper [60], the authors suggested a framework for studying nonlinear problems in flight dynamics. That same framework is proposed here as being suited to the deterministic approach to the study of turbulence. The framework has four premises involving the elements structure, change, chaos, and scale:

(1) All flows have structure.

(2) Structures change in systematic ways with changes in parameters.

(3) Some changes lead to a special class of structures, chaos.

(4) Structures have various scales.

The premises allow the following interpretation of observations: Flow structures are interdependent. Changes in structures occur in discrete ways at definite and repeatable values of parameters. Chaos in fluid systems can occur after a finite number of bifurcations (discrete changes). Chaos is deterministic and can be represented by a strange attractor of finite dimensionality. Finally, the various scales of turbulence interact (a restatement of the observation that the structures are interdependent).

Taken together with a corresponding set of mathematical ideas, the four premises form a strong theoretical framework (a way of seeing) for the understanding, and (potentially) the modeling of turbulent flows. The premises of the theoretical framework are:

(1) Structures are describable in topological terms.

(2) Changes in structure are describable by bifurcation theory.

(3) Chaos is describable by the theory of strange attractors and fractals.

(4) Scales are describable by group theory ideas.

Where this body of theory will lead in the modeling of turbulent flows is not yet completely obvious. However, at least the following seems likely: (1) Some form of averaging of the Navier-Stokes equations probably still will be required. Whatever its form, the averaging will be carried out such that: (a) it allows the representation of at least the major structural and subcritical bifurcations that occur both in the outer flow (away from boundary layers) and within the turbulence itself, and (b) it incorporates chaotic information. (2) The chaotic portion of the problem probably will be modeled by a finite-dimensional strange attractor along the lines of current developments in dynamical systems theory. Progress here hinges on the formulation of appropriate measures of the strange attractor which will allow a rational finite-dimensional representation of its essential nature. The representation will be driven by the mean flow and, in turn, supply information to the mean-flow equations to be used in forming the Reynolds stresses. This is where renormalization-group theory ideas will be required to resolve only the essential scales of the problem. Topological ideas will be used to replace the formerly ubiquitous "turbulent eddies" (and the currently, perhaps equally nonspecific, "small-scale chaotic structures") by a concise and descriptive grammar of structural forms.

The ideas which have been discussed perhaps can be illustrated more clearly by examples. Two flows have been selected to illustrate some of the ways in which flow properties change with changes in the governing parameters. Our purpose here is twofold: First, to demonstrate the extent to which descriptions based on the language of the proposed theoretical framework are compatible with observations; and second, to demonstrate some of the principal structural features of flows which we believe should be capturable by a mathematical model constructed according to the prescription above.

The first example is that of the two-dimensional cylinder immersed in an incompressible crossflow that is uniform and steady far upstream (fig. 14). The main parameter is Reynolds number (Re). We examine a sequence of flows as Re is increased from very low values (Re < 7) to values of the order of 10^6. Sketch (a) in figure 14 depicts the regime resulting from a Hopf-type bifurcation occurring at Re ≈ 50, in which the previously steady flow is replaced by a time-periodic flow (cf. [75]). As Re is increased by increments, at least two additional bifurcations occur, leaving the wake with a quasi-periodic large-scale structure and with chaotic small-scale structures superposed on the free-shear layers (sketch (b)). The small-scale structures work their way forward in the wake as Re increases, until they are in the vicinity of the separation points on the cylinder. At a definite and repeatable value of Re (Re ≈ 3.5×10^5 [76,77]) an antisymmetric interaction occurs having opposite effects on the separation points, which leaves the mean flow asymmetric and disrupts the previously quasi-periodic form of the wake flow (sketch (c)). As has been documented in [77], this event has all of the characteristics of a subcritical bifurcation, including hysteresis, inasmuch as the return to a symmetric mean flow with reduction in Re from higher values occurs at a lower critical value of Re than had the onset of asymmetry. With a small further increase in Re, a second interaction restores the symmetric character of the mean flow (sketch (d)). This event also involves the hysteresis with reduction in

Re from above, characteristic of a subcritical bifurcation. Finally, at sufficiently high values of Re (sketch (e)) the chaotic structures have moved forward of the separation points, rendering the boundary-layer flow on the cylinder turbulent. A periodic structure reasserts itself in the wake, now with a new fundamental period reflecting the scale of the chaotic structures characteristic of both the boundary layer and the free-shear layers. The example confirms first that descriptions based on the language of the theoretical framework are in fact compatible with observations. Starting with the onset of flow separation at Re ≈ 7 (describable as a structural bifurcation, not shown in fig. 14), the onset of each of the flows sketched in figure 14 can be described as either a subcritical bifurcation or a

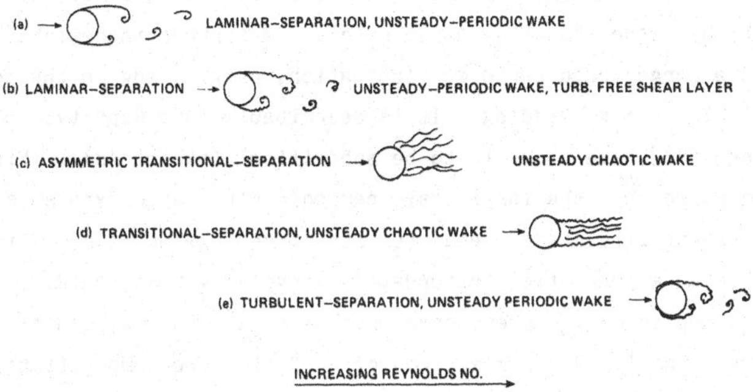

Fig. 14. Stages of incompressible flow past a circular cylinder with increasing Reynolds number.

structural bifurcation of the mean flow, occurring at a definite and repeatable critical value of Re. Second, with regard to computations, the example suggests that a mathematical model capable of capturing the sequence of important bifurcations will have to acknowledge the existence of small-scale chaotic structures and how they may interact with large-scale periodic structures.

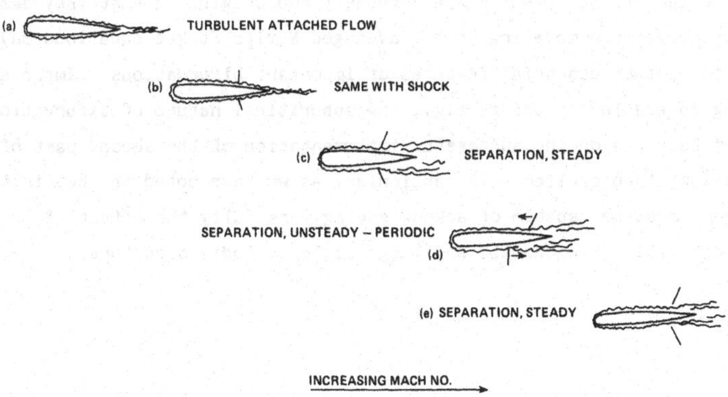

Fig. 15. Stages of transonic flow past a symmetric airfoil at zero angle of attack with increasing Mach number.

The second example is that of a two-dimensional airfoil immersed at zero angle of attack in a compressible flow that is uniform and steady far upstream (fig. 15). The main parameter here is free-stream Mach number (M_∞); we examine a sequence of flows as M_∞ is increased in the transonic speed range, with Reynolds number (Re) held fixed. The value of Re is supposed sufficiently high to maintain a turbulent attached boundary layer over the major part of the airfoil at the lowest value of M_∞. In sketch (a) of figure 15, M_∞ is low enough so that no shock waves develop. As M_∞ increases, shock waves develop (sketch (b)) but the boundary-layer flow remains attached. An additional increment in M_∞ strengthens the shock waves and the boundary-layer flow separates behind the shocks but remains essentially steady except for the presence of small-scale chaotic structures. This is a structural bifurcation of the mean flow (sketch (c)). A further incremental change in M_∞ results in a large-scale periodic fluctuation, originating in the separated-flow regime and akin to vortex-shedding. It is describable as a Hopf-type bifurcation, and observations [78] have shown it to be subcritical (sketch (d)). Finally, with another increment in M_∞, the large-scale periodic fluctuation vanishes and the flow returns to an essentially steady separated structure (sketch (e)). This event is also describable as a subcritical bifurcation, involving hysteresis.

Extensive studies of the phenomenon have been carried out both experimentally [78] and computationally [79] for a biconvex airfoil. The computations, initiated by Levy [79], were based on the Reynolds-averaged Navier-Stokes equations with closure achieved by means of a phenomenological (mixing-length) turbulence model. Computational results were able to capture essential features of the bifurcations, including the critical values of M_∞ signalling both the onset and the termination of periodic fluctuations as well as the form and frequency of the fluctuations themselves. The computational results were not sufficiently accurate, however, to demonstrate the subcritical nature (i.e., the presence of hysteresis) of the bifurcations.

Levy's computational results are extremely encouraging in that they demonstrate that even rather simple modeling of the averaged Navier-Stokes equations may be sufficient to capture essential features of important bifurcations. Improvement of the modeling to enable capturing, e.g., the subcritical nature of bifurcations, would appear to hinge on the successful implementation of the second part of the prescription outlined earlier. In particular, as we have noted in the first example, the model must be capable of acknowledging more fully the effect of interactions between small-scale chaotic and large-scale periodic structures.

REFERENCES

1. O. E. Lanford: "Strange Attractors and Turbulence." Hydrodynamic Instabilities and the Transition to Turbulence, Topics in Applied Physics, 45, H. L. Swinney and J. P. Gollub, eds. Springer, Berlin, Heidelberg, New York, 7-26 (1981).

2. H. A. Liepmann: The Rise and Fall of Ideas in Turbulence. American Scientist, 67, No. 2, March-April, 221-228 (1979).

3. J. J. Cornish, III: Vortex Flows, the Eighth Quick-Goethert Lecture Series, University of Tennessee Space Institute, Tullahoma, Tenn. (Oct. 1982).

4. O. Reynolds: An experimental investigation of the circumstances which determine whether the motion of water shall be direct or sinuous, and of the law of resistance in parallel channels. Phil. Trans. R. Soc. London, Ser. A, 174, 935-982 (1883).

5. M. Van Dyke: An Album of Fluid Motion, Parabolic Press (1982).

6. O. Reynolds: On the dynamical theory of incompressible viscous fluid and the determination of the criterion. Phil. Trans. R. Soc. London, Ser. A., 186, 123-161 (1895).

7. J. Jeans: An Introduction to the Kinetic Theory of Gases, Cambridge Univ. Press (1940).

8. J. O. Hinze: Turbulence. An Introduction to Its Mechanism and Theory, McGraw-Hill Book Co. (1959).

9. A. S. Monin and A. M. Yaglom: Statistical Fluid Mechanics: Mechanics of Turbulence, 1, MIT Press, Cambridge, Mass. (1971).

10. A. Pouquet: Statistical Methods in Turbulence. Proceedings of the ICASE/NASA Workshop on Theoretical Approaches to Turbulence (Oct. 1984).

11. A. N. Kolmogoroff: The Local Structure of Turbulence in Incompressible Viscous Fluid for Very Large Reynolds Numbers. C. R. (Doklady), Acad. Sci. U.R.S.S., 30, No. 4, 301-305 (1941).

12. B. B. Mandelbrot: "Intermittent Turbulence and Fractal Dimension: Kurtosis and the Spectral Exponent 5/3 + B," Turbulence and Navier-Stokes Equations, Lecture Notes in Mathematics, Vol. 565, R. Temam, ed., Springer, Berlin, Heidelberg, New York, 121-145 (1976).

13. J. Boussinesq: Essai sur la théorie des eaux courantes, l'Acad. Sci., Paris, t. 23, No. 1, pp. 1-680, 1877; Théorie de l'écoulement tourbillonnant et tumultueux des liquides dans les lits rectilignes á grande section, I-II, Gauthier-Villars, Paris (1987).

14. L. Prandtl: Bericht über Untersuchungen zur ausgebildeten Turbulenz, ZAMM, **5**, 2, 136-139 (1925).

15. M. W. Rubesin: Numerical Turbulence Modeling. Computational Fluid Dynamics, AGARD LS-86, Paper No. 3, 1977.

16. B. E. Launder: Progress and Prospects in Phenomenological Turbulence Models. Proceedings of the ICASE/NASA Conference on Theoretical Approaches to Turbulence (1984).

17. W. C. Reynolds: Impact of Large-Eddy Simulation on Phenomenological Modeling of Turbulence. ICASE/NASA Conference on Theoretical Approaches to Turbulence (1984).

18. G. K. Batchelor and A. A. Townsend: The Nature of Turbulent Motion at High Wave Number. Proc. R. Soc. London, Ser. A, **199**, 238-255 (1949).

19. L. Landau: On the Problem of Turbulence. C. R. Acad. Sci., U.R.S.S., **44**, 311 (1944).

20. G. B. Schubauer and H. K. Skramstad: Laminar Boundary-Layer Oscillations and Transition on a Flat Plate. NACA Rep. 909 (1948).

21. H. W. Emmons: The Laminar-Turbulent Transition in a Boundary Layer. Part I. J. Aero Sci., **18**, No. 7, 490-498 (1951).

22. B. J. Cantwell, D. E. Coles, and P. E. Dimotakis: Structure and Entrainment in the Plane of Symmetry of a Turbulent Spot. J. Fluid Mech., **87**, Part 4, 641-672 (1978).

23. S. J. Kline and P. W. Runstadler: Some Preliminary Results of Visual Studies of the Flow Model of the Wall Layers of the Turbulent Boundary Layer. J. Appl. Mech., Trans. ASME, Ser. E, 26, No. 2, 166-170 (1959).

24. H. T. Kim, S. J. Kline, and W. C. Reynolds: The Production of Turbulence Near a Smooth Wall in a Turbulent Boundary Layer. J. Fluid Mech., 50, Part 1, 130-160 (1971).

25. A. Roshko: Dryden Research Lecture. Structure of Turbulent Shear Flows: A New Look. AIAA J., 14, No. 10, Oct., 1349-1357 (1976).

26. G. L. Brown and A. Roshko: On Density Effects and Large Structure in Turbulent Mixing Layers. J. Fluid Mech., 64, Part 4, 775-816 (1974).

27. B. J. Cantwell: Organized Motion in Turbulent Flow. Ann. Rev. Fluid Mech., 13, 457-515 (1981).

28. R. S. Rogallo: Numerical Experiments on Homogeneous Turbulence. NASA TM-81315 (1981).

29. P. Moin and J. Kim: Numerical Investigation of Turbulent Channel Flow. J. Fluid Mech., 118, 341-377 (1982).

30. P. Moin: Probing Turbulence Via Large Eddy Simulation. AIAA Paper 84-0174, (1984).

31. R. F. Blackwelder and R. E. Kaplan: On the Wall Structure of the Turbulent Boundary Layer. J. Fluid Mech., 76, Part 1, 89-112 (1976).

32. J. L. Lumley: "The Structure of Inhomogeneous Turbulent Flows," in Atmospheric Turbulence and Radio Wave Propagation, A. M. Yaglom and V. I. Tatarsky, eds., NAUKA, Moscow, 166-178 (1967).

33. J. L. Lumley: "Coherent Structures in Turbulence," in Transition and Turbulence, R. E. Meyer, ed., Academic Press, New York, 215-242 (1981).

34. J. Kim: On the Structure of Wall-Bounded Turbulent Flows. Phys. Fluids, 26, No. 8, 2088-2097 (1983).

35. H. L. Dryden: "Recent Advances in the Mechanics of Boundary Layer Flow," in Advances in Applied Mechanics, 1, Academic Press, New York, 1-40 (1948).

36. M. T. Landahl: Dynamics of Boundary Layer Turbulence and the Mechanism of Drag Reduction. Phys. Fluids, 20, No. 10, Part 11, Oct. S55-S63 (1977).

37. M. T. Landahl: Flat-Eddy Model for Coherent Structures in Boundary Layer Turbulence. Proceedings of ICASE/NASA Conference on Theoretical Approaches to Turbulence (1984).

38. D. Bushnell: Turbulence Sensitivity and Control in Wall Flows. Proceedings of ICASE/NASA Conference on Theoretical Approaches to Turbulence, Oct. 1984.

39. H. Poincaré: Sur les courbes définiés par les équations differentielles, I-VI, Oeuvres, Vol. 1, Gauthier-Villars, Paris, 1880-1890.

40. A. A. Andronov, E. A. Leontovich, I. I. Gordon, and A. G. Maier: Qualitative Theory of Second-Order Dynamical Systems, Wiley, New York, 1973.

41. D. H. Sattinger: Bifurcation and Symmetry Breaking in Applied Mathematics. Bull. (New Series) Amer. Math. Soc., 3, No. 2, 779-819 (1980).

42. D. D. Joseph: Stability of Fluid Motions. I. Springer-Verlag, Berlin, Heidelberg, New York (1976).

43. T. B. Benjamin: Bifurcation Phenomena in Steady Flows of a Viscous Fluid. I, Theory. II, Experiments. Proc. R. Soc. London, Ser. A, Vol. 359, 1-43 (1978).

44. M. W. Hirsch: The Dynamical Systems Approach to Differential Equations. Bull. (New Series) Amer. Math. Soc., Vol. 11, No. 1, July 1984, 1-64.

45. E. N. Lorenz: Deterministic Non-periodic Flow. J. Atmos. Sci., 20, No. 2, Mar., 130-141 (1963).

46. D. Ruelle and F. Takens: On the Nature of Turbulence. Commun. Math. Physics, 20, No. 3, 167-192 (1971).

47. C. Sparrow: The Lorenz Equations: Bifurcations, Chaos, and Strange Attractors. Springer-Verlag, New York, Heidelberg, Berlin (1982).

48. J. Guckenheimer: (Rev.) The Lorenz Equations: Bifurcations, Chaos, and Strange Attractors. By Colin Sparrow. Amer. Math. Monthly, 91, No. 5, 325-326, May (1984).

49. D. Ruelle: Differentiable Dynamical Systems and the Problem of Turbulence. Bull. (New Series) Amer. Math. Soc., 5, No. 1, July 1981.

50. P. R. Fenstermacher, H. L. Swinney, and J. P. Gollub: Dynamical Instabilities and the Transition to Chaotic Taylor Vortex Flow. J. Fluid Mech., 94, Part 1, 103-128 (1979).

51. B. B. Mandelbrot: "Sporadic Turbulence," Boundary Layers and Turbulence (Kyoto International Symposium, 1966), Supplement to Phys. Fluids, 10, S302-S303 (1967).

52. K. G. Wilson and M. E. Fisher: Critical Exponents in 3.99 Dimensions. Phys. Rev. Lett., 28, 240-243 (1972).

53. E. D. Siggia: Numerical Studies of Small-Scale Intermittency in Three-Dimensional Turbulence. J. Fluid Mech., 107, 375-406 (1981).

54. J. P. Eckmann: Roads to Turbulence in Dissipative Dynamical Systems. Rev. Mod. Phys., 53, No. 4, Pt. 1, 643-654 (Oct. 1981).

55. R. Temam, ed.: Turbulence and the Navier-Stokes Equations. Lecture Notes in Mathematics, 565, Springer-Verlag, New York (1976).

56. B. B. Mandelbrot: The Fractal Geometry of Nature. W. H. Freeman, San Francisco (1982).

57. E. A. Spiegel: Bifurcation and Chaos-Relevance to Turbulence. Proceedings of ICASE/NASA Workshop on Theoretical Approaches to Turbulence, Oct. 1984.

58. B. B. Mandelbrot, Fractal Geometry of Turbulence. Proceedings of ICASE/NASA Workshop on Theoretical Approaches to Turbulence, Oct. 1984.

59. W. D. McComb: Renormalisation Group Methods and Turbulence. Proceedings of ICASE/NASA Workshop on Theoretical Approaches to Turbulence, Oct. 1984.

60. G. T. Chapman and M. Tobak: Nonlinear Problems in Flight Dynamics. Proc. Berkeley-Ames Conference on Nonlinear Problems in Control and Fluid Dynamics, to appear in Math. Sci. Press, 1984. Also NASA TM-85940, May 1984.

61. M. Golubitsky and D. Schaeffer: Imperfect Bifurcation in the Presence of Symmetry. Commun. Math. Phys., 67, No. 3, 205-232 (1979).

62. J. Guckenheimer and P. Holmes: Nonlinear Oscillations, Dynamical Systems, and Bifurcations of Vector Fields. Springer-Verlag, New York (1983).

63. B. B. Mandelbrot: Fractals in Physics: Squig Clusters, Diffusions, Fractal Measures, and the Unicity of Fractal Dimensionality. J. Stat. Phys., 34, Nos. 5/6, 895-930 (1984).

64. O. E. Lanford, III: The Strange Attractor Theory of Turbulence. Ann. Rev. Fluid Mech., 14, 347-364 (1982).

65. E. Ott: Strange Attractors and Chaotic Motions of Dynamical Systems. Rev. Modern Phys., 53, No. 4, Pt. 1, 655-671 (1981).

66. C. Foias and R. Temam: Some Analytic and Geometric Properties of the Solutions of the Evolution Navier-Stokes Equations. J. Math. Pures Appl., 58, 339-368 (1979).

67. P. Constantin, C. Foias, O. Manley, and R. Temam: Connexion entre la théorie mathématique des équations de Navier-Stokes et la théorie conventionelle de la turbulence, C. R. Acad. Sci., Paris, t. 297, Serie I, 599-602 (1983).

68. R. Temam: Attractors and Determining Modes in Fluid Mechanics. Physica 124A, 577-586 (1984).

69. D. Ruelle: Five Turbulent Problems. Physica 7D, 40-44 (1983).

70. D. Farmer, E. Ott, and J. A. Yorke: The Dimension of Chaotic Attractors. Physica 7D, 153-180 (1983).

71. L. S. Young: Entropy, Lyapunov Exponents, and Hausdorff Dimension in Differentiable Dynamical Systems. IEEE Transactions on Circuits and Systems, 30, No. 8, (1983).

72. A. J. Chorin: The Evolution of a Turbulent Vortex. Commun. Math. Phys., 83, No. 4, 517-535 (1982).

73. K. G. Wilson and J. Kogut: The Renormalization Group and the ε Expansion. Phys. Rep. C., 12, No. 2, 75-200 (1974).

74. M. J. Feigenbaum, L. P. Kadanoff, and S. J. Shenker: Quasi-Periodicity in Dissipative Systems: A Renormalized Group Analysis. Physica 5D, 370-386 (1982).

75. M. Nishioka and H. Sato: Mechanism of Determination of the Shedding Frequency of Vortices Behind a Cylinder at Low Reynolds Numbers. J. Fluid Mech., 89, Part 1, 49-60 (1978).

76. N. Kamiya, S. S. Suzuki, and R. Nishi: On the Aerodynamic Force Acting on Circular Cylinder in the Critical Range of the Reynolds Number. AIAA Paper 79-1475, Williamsburg, Va. (1979).

77. G. Schewe: On the Force Fluctuations Acting on a Circular Cylinder in Crossflow from Subcritical up to Transcritical Reynolds Numbers. J. Fluid Mech., 133, 265-285 (1983).

78. J. B. McDevitt, L. L. Levy, Jr., and G. S. Deiwert: Transonic Flow About a Thick Circular-Arc Airfoil. AIAA J., 14, No. 5, 606-613 (1976).

79. L. L. Levy, Jr.: Experimental and Computational Steady and Unsteady Transonic Flows About a Thick Airfoil. AIAA J., 16, No. 6, 564-572 (1978).

CHAPTER III. Large Eddy Simulation: Its Role in Turbulence Research

Joel H. Ferziger

I. Historical Introduction

The origins of large eddy simulation (LES) lie in the early global weather prediction models. The grids used in those codes could hardly resolve the largest atmospheric structures. Modeling was, and still is, required for the unresolved scales.

The first engineering application of LES was Deardorff's (1970) paper on channel flow which provided many of the foundations of the subject and influenced much of the more recent work. Deardorff's work was followed by Schumann's (1973) thesis; Schumann later led a group at Karlsruhe that specialized in the simulation of convective heat transfer. Reynolds and the present author began in 1972 and have concentrated on developing the fundamentals of the subject, systematic extension to more complex flows, and application to turbulence modeling. The NASA-Ames group, beginning in 1975, specialized in accurate simulation of simple flows and on full simulation (see below). Leslie and his group in London have, since 1976, followed similar lines, and have used turbulence theories to develop subgrid scale models. In the past few years, several French groups, including Electricite de France, ONERA-Chatillon, University of Lyon, and CERT, have used LES.

Full turbulence simulation (FTS) simulates turbulent flows with no modeling at all. The number of flows which can be simulated by this technique is limited, and the Reynolds numbers must be small. The pioneering work in this field was done by Orszag and his group at MIT in 1972. Full simulation has since been used by other groups, principally those mentioned above.

The number of LES and FTS cases run is restricted by scarcity of resources. Consequently, runs have to be selected with care. Engineering applications of LES are almost entirely indirect, i.e., they are aimed at improving models used in time averaged simulations rather than the simulation of technologically important flows. However, recent advances in VSLI technology are reducing the cost of a given computation, and large computers will be more widely available in the future. This will allow more groups to use LES and will widen the range of applications.

In this paper we shall explore some of the accomplishments of these methods and the problems that need to be solved to extend their engineering usefulness.

II. Foundations of Large Eddy Simulation

A. Homogenous Flows

Reviews of LES and FTS are given in the papers of the present author (Ferziger, 1983), and Rogallo and Moin (1984), and will not be repeated here. In this chapter, we shall concentrate on introducing the issues to be discussed later in this paper. The emphasis is on LES. FTS is the preferred approach when its application is possible, but in this paper we shall regard it as a tool to guide and assist in the development of models used in LES. More on FTS and its applications to time average modeling can be found in the companion paper by Reynolds (1984).

Central to LES is the definition of the large-scale part of the velocity and pressure fields which the method will attempt to simulate. The unresolved parts of the fields need to treated by an approximation we call a model (meteorology calls it a parameterization). For homogeneous flows, the large scale fields are most conveniently defined by filtering, a concept introduced by Leonard (1974). The large-scale field is defined as:

$$\overline{u}(\underline{r}) = \int G(\underline{r} - \underline{r}', \Delta)\, u(\underline{r}')\, d\underline{r}' \tag{1}$$

where u is the instantaneous velocity and Δ is a width that separates the large and small eddies. Various filters, G, have been used including a Gaussian, a box in physical space, and a sharp cutoff in Fourier space. Although arguments favoring one or the other can be made, for homogeneous flows the choice is not of great importance.

The definition (1) allows one to divide the velocity (and pressure) field into the resolved and unresolved components:

$$u_i = \overline{u}_i + u_i' \tag{2}$$

Filtering a linear equation with constant coefficients using Eq. (1) produces an exact equation for the filtered field. For homogenous flows, the principal difficulty arises from filtering the nonlinear term in the Navier-Stokes equations. This yields:

$$\overline{u_i u_j} = \overline{\overline{u}_i \overline{u}_j} + \overline{\overline{u}_i u_j'} + \overline{u_i' \overline{u}_j} + \overline{u_i' u_j'} \tag{3}$$

The first term on the right-hand side, the only one that depends entirely on the resolved field, can be computed directly. In recent years, most simulations have done so. Prior to this, an alternative approach in which one writes:

$$\overline{\bar{u}_i\bar{u}_j} = \bar{u}_i\bar{u}_j + (\overline{\bar{u}_i\bar{u}_j} - \bar{u}_i\bar{u}_j) \qquad (4)$$

was used. The first term on the right side of Eq. (4) causes no problem while the second, which has been called the Leonard stress, can be approximated using Taylor series, as suggested in Leonard's original paper.

The last three terms in Eq. (3) require modeling. One writes:

$$R_{ij} = \overline{\bar{u}_i u_j'} + \overline{u_i' \bar{u}_j} + \overline{u_i' u_j'} \qquad (5)$$

and further divides R_{ij} into traceless and diagonal components:

$$R_{ij} = (R_{ij} - \frac{1}{3} R_{kk}\delta_{ij}) + \frac{1}{3} R_{kk}\delta_{ij} = \tau_{ij} + \frac{1}{3} R_{kk}\delta_{ij} \qquad (6)$$

The diagonal component is normally combined with the pressure:

$$P = p + \frac{1}{3} R_{kk} \qquad (7)$$

so that the pressure computed in LES is P, not the true pressure; corrections for the subgrid scale energy need to be applied to obtain the true pressure.

The subgrid scale Reynolds stress tensor, τ_{ij} is the principal quantity requiring modeling. If, somehow, an exact representation of this tensor were available and there were no numerical errors, LES would produce the exact large-scale field and we would also have the unresolved contributions. Such a simulation would yield all the data one could hope for. Thus an accurate subgrid scale model is the key to quality large eddy simulations. As the Reynolds number increases, the fraction of the total field that is unresolved also increases, the model is required to represent a larger range of turbulence scales, and the accuracy of a simulation becomes more sensitive to the quality of the subgrid scale model. For these reasons, a large portion of this paper is devoted to the issue of subgrid scale modeling.

Homogeneous flows, by definition, are ones in which every location in the flow is equivalent (statistically) to every other. In these flows, it is sufficient to study a representative piece of fluid. This piece must be large enough to contain all of the significant scales of motion of the turbulence; it may also deform with the applied mean field (if any). The surroundings may be assumed identical to the considered piece of fluid (this is correct in a statistical sense), and periodic boundary conditions may be applied. These have two important advantages: firstly, the difficult issue of how to specify the conditions at inflow boundaries can be avoided, and, secondly, spectral methods which are inherently more accurate than finite difference approximations become the natural method of computing spatial derivatives.

B. Inhomogeneous Flows

Inhomogeneous flows, i.e., ones in which the statistical properties of the turbulence vary with location, are more difficult to treat by LES. There are a number of reasons why this is so.

It is more difficult to define the resolved portion of the field in these flows. Deardorff (1970) and Schumann (1973) defined the filtered field as an average over a finite volume over which the equation is discretized. This definition simplifies the derivation of the equations to be solved numerically, but it also makes the significance of the subgrid Reynolds stress less transparent and the study of its modeling more difficult. The issue is the subject of research at the present time, but a few results can be cited.

The flows simulated to date are inhomogeneous in only one direction; work on flows which are inhomogeneous in two directions is in progress. The homogeneous directions in these flows can be treated as inhomogeneous flows, so it is the inhomogeneous directions on which we must concentrate.

For channel flow, most of the simulations to date have used finite difference methods in the inhomogeneous direction. Kim and Moin (1981) showed that, in this method, the filter in the inhomogeneous direction could be regarded as an asymmetric box filter. More recent simulations used spectral methods based on Chebychev polynomials. Moser (1984) did a full simulation of flow in a curved channel for which filtering is not an issue. To define a filter, one can note that, in its Fourier representation, Eq. (1) becomes:

$$\hat{\bar{u}}(k) = \hat{G}(k)\,\hat{u}(k) \tag{8}$$

which can be generalized to any spectral representation. Thus, for Chebychev polynomials, the filtered variable can be defined by multiplying its Chebychev representation by an appropriate filtering function. This approach is not without difficulties, however, as the filtering operation no longer commutes with differentiation, thus introducing terms into the filtered equations that are not present in homogenous flows.

Another difficulty is that in inhomogeneous flows one cannot, in general, simulate a small region of the flow in isolation from its surroundings. As a result, the ratio of the computational domain size to the size of the small eddies is larger. For a given computational resource, this requires the model to represent a larger fraction of the turbulence, and the results are more dependent on the quality of the subgrid model. The Reynolds numbers of the flows that can be simulated may also be more limited.

By definition, inhomogenous flows have boundaries that cannot be treated with periodic boundary conditions, which introduces several new difficulties:

1. It may be necessary to extend the computational domain a long distance in some direction.

2. The sizes of the important eddies may vary greatly from one part of the flow to another. Near solid boundaries, turbulence structures become very small. In free shear flows, the size of the largest eddies grows rapidly with time or downstream distance.

3. It may be necessary to specify inflow conditions that represent the large three-dimensional, time-dependent structures of a single realization of a turbulent flow.

4. It may be necessary to deal with complex geometry.

All of these problems increase the difficulty of doing LES. Some progress has been made on the first two problems, but little has been done on the last two. We shall study these issues further in this paper.

III. Subgrid Modeling

The need for subgrid modeling was demonstrated in the preceding chapter. Here, we shall look at what has been done in this area and at some of the remaining problems. A number of approaches to constructing models have been taken; these vary from intuitively writing down a model to application of the most sophisticated turbulence theories. In some cases, several approaches lead to the same place.

A. Eddy Viscosity Models

As is the case for time average modeling, the eddy viscosity concept is the most commonly basis for models. The intuitive derivation of eddy viscosity models is based on the observation that the subgrid scale Reynolds stress tensor is symmetric and traceless. The simplest tensor functional of the velocity field that has these properties is the rate of strain tensor, so it is natural to assume proportionality between the two:

$$\tau_{ij} = -2\nu_T S_{ij} = \nu_T \; \frac{\partial U_i}{\partial x_j} + \frac{\partial U_j}{\partial x_i} \tag{9}$$

The connection between the two tensors has been assumed a scalar (the eddy viscosity) but it could be a second or fourth rank tensor. The eddy viscosity has dimensions L^2T^{-1} and must be constructed from appropriate dimensional quantities. For the subgrid scale field, the natural length scale is the one associated with the filter, Δ. The velocity scale can be approximated by $S\Delta$, where $S = (S_{ij}S_{ij})^{1/2}$ is the strain rate of the large scale field; this is the questionable assumption. These estimates lead to Smagorinsky's (1963) model:

$$\tau_{ij} = (C_S\Delta)^2 \; S \; S_{ij} \tag{10}$$

which has been widely used.

More sophisticated derivations of this model are based on dynamic equations for the subgrid scale turbulence. The further assumptions required to reduce these equations to a model as simple as Eq. (10) display the limitations of the model. The first such derivation was made by Lilly (1966) who argued that, if the cutoff between the large and small scales lies in the inertial subrange, where the energy spectrum has a -5/3 dependence, one could derive Eq. (10) in a way that connects the model constant with Kolmogoroff's constant. In this way, he predicted the constant to be 0.21.

A more accurate analysis of the time scale of the subgrid scale leads to a variation on the Smagorinsky model:

$$\nu_T = C \; \Delta^{4/3} \; L^{2/3} \; S \tag{11}$$

where L is the integral scale of the turbulence. This has been derived by a number of authors (Ferziger and Leslie, 1979, Yoshizawa, 1982, among others) but has not been used in simulations. For the parameter range used in most simulations to date, the difference would probably not be significant.

Smagorinsky's model works well in predicting the decay of isotropic turbulence. Despite the simulations having Reynolds numbers too low to permit the existence of an inertial subrange in the spectrum, the values of the constant required to match the experimental decay are close to Lilly's prediction.

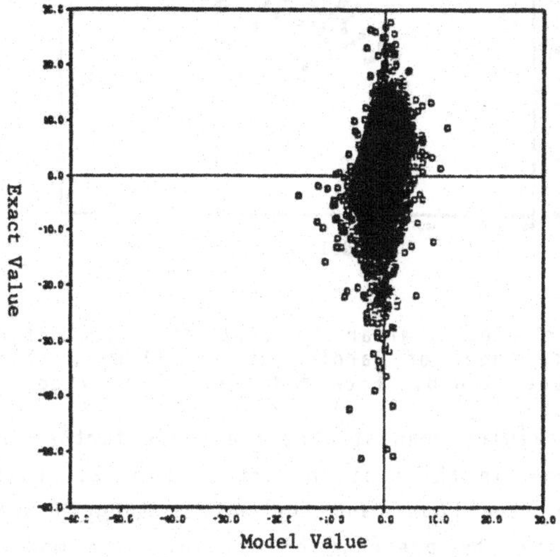

Fig. 1. A scatter plot of the exact subgrid-scale Reynolds stress and the Smagorinsky model representation of it. Each point represents a component of the stress at one point in the flow at a particular time. A total of 4096 points have been plotted; most of them lie within the cross-hatched area in the center. The correlation coefficient between the exact and model variables is approximately 0.4. From McMillan and Ferziger (1980).

The model can be tested by accepting the results of a full simulation as data. One can then compute all of the quantities defined above and thereby simultaneously determine both τ_{ij} and the model's prediction of it. A scatter plot, such as the one shown in Fig. 1, then provides a test of the model. As the figure shows, for isotropic turbulence, the Smagorinsky model does correlate some of the subgrid-scale Reynolds stress, but its performance is far from ideal. In fact, the correlation coefficient is approximately 0.4, indicating that the model represents only about 16 percent of the subgrid scale Reynolds stress.

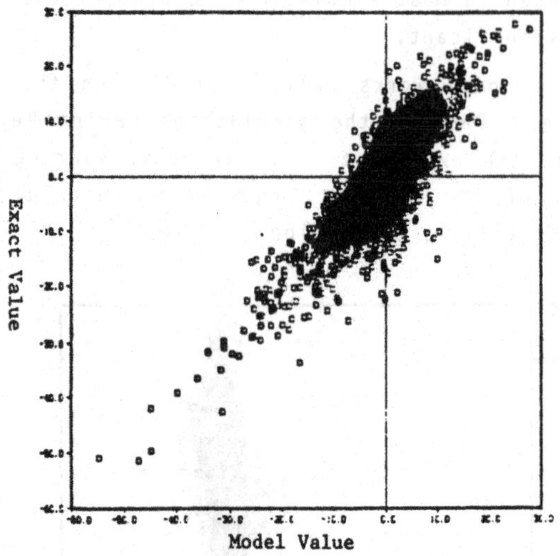

Fig. 2. A scatter plot similar to that of Fig. 1 for the scale-similarity model of Bardina et al. (1981). The correlation coefficient is 0.8. From McMillan and Ferziger (1980).

For flows containing mean strain, there are further difficulties. A major one is shown in the scatter plot of Fig. 2. Essentially all correlation between the model and the actual subgrid scale Reynolds stress disappears in the presence of strain. The model is clearly inaccurate here, and as most technological flows are shear or strain flows, this is a serious problem. Another issue is whether the mean strain ought to be included in the model. Experiments and simulations clearly show that the mean strain should not be included in Eq. (9). McMillan and Ferziger (1980) further showed that the model constant does not change if the mean strain is not included in the expression for the eddy viscosity; they recommended that the model be entirely independent of the mean strain. Results for inhomogeneous flows (see below) also suggest this solution.

In anisotropic flows, isotropic filters are not appropriate. The filter ought to reflect the structure of the turbulence. This requires a different length scale in each direction; the ratio of the length scales is very large in the vicinity of a wall. If a scalar eddy viscosity model is to be used, the length to be used in the model needs to be defined. Most authors have used the cube root of the filter volume:

$$\Delta = (\Delta_1 \Delta_2 \Delta_3)^{1/3} \tag{12}$$

but McMillan and Ferziger (1980) showed that the diagonal cell length:

$$\Delta = (\Delta_1^2 + \Delta_2^2 + \Delta_3^2)^{1/2} \tag{13}$$

might be a better choice. Equation (13) has not been tested in simulations to date.

The only inhomogeneous flow to which this model has been applied is the flow between two parallel walls, i.e., channel or turbulent Poiseuille flow. Such simulations have been made by Deardorff (1970), Schumann (1973), Moin et al. (1978) and Kim and Moin (1981). All of these authors found that the Smagorinsky model could not be applied without modification to this flow. In particular, it is necessary to decrease the model constant to about half of what is needed in homogeneous flows. Furthermore, Schumann and Kim and Moin used a two-component model which has different dependence on the mean flow and the turbulence:

$$\tau_{ij} = -\nu_{T1} S_{ij} - \nu_{T2} \frac{\partial \langle u_i \rangle}{\partial x_2} \tag{14}$$

where $\langle u_1 \rangle$ is the mean value of the streamwise velocity over a plane parallel to the wall and is essentially the time mean velocity. The constant in the first term, which is Smagorinsky's model, is approximately 0.09 or half of the value found in homogenous flows. In the second term, called the inhomogeneous term by these authors, the eddy viscosity used is:

$$\nu_{T2} = \left[\min\,(\Delta_3, X_2)\right]^2 \frac{\partial \langle u_1 \rangle}{\partial X_2} \tag{15}$$

which bears a close resemblence to the mixing length model used in time average calculations.

B. One-Equation Models

We saw above that, although it is commonly used, Smagorinsky's model has severe deficiencies. There is a need for an improved model. Constructing such a model is no simple matter. A number of approaches to constructing models have been suggested. These will be reviewed in this and the following sections, but none has yet proved its merit in extensive tests.

The deficiencies of the Smagorinsky model fall into two categories. The first and more serious one is that the assumed proportionality between the subgrid Reynolds stress and the resolved scale strain may be incorrect. The second is that the assumption used for the velocity scale is of questionable validity.

For time average modeling, two-equation models are currently popular. In these, equations for the turbulence energy and its rate of dissipation are solved along with the equations for the mean field. The energy provides a velocity scale; in combination with the dissipation, it yields a length scale; from these the eddy viscosity can be constructed. For subgrid models, there is a natural length scale, the filter width, Δ. This removes a major difficulty of two-equation modeling, and a model with a single equation for the subgrid turbulence energy appears attractive.

A model of this type has been used by Schumann (1973), who showed that the results it produces are not a significant improvement on those produced by the model presented above.

Clark et al. (1975), in tests of subgrid models using full simulation, investigated a model in which an optimum eddy visosity was chosen at each point of the flow. While this model cannot be used in actual simulations, it provides a standard against which other eddy viscosity models can be measured. It was found that this optimum model provides only a relatively modest improvement on the standard Smagorinsky model, suggesting that the deficiency of the model lies more in the assumption of stress-strain proportionality than in the formulation of the eddy viscosity.

C. Reynolds Stress and Algebraic Models

To remove the assumption of proportionality between subgrid Reynolds stress and resolved scale strain that is the principal deficiency of eddy viscosity models, time average modelers use partial differential equations for the Reynolds stresses themselves. Application of this idea to modeling subgrid stresses seems a natural.

The derivation of the equations for the subgrid Reynolds stresses is straightforward; we shall not write these equations here. This model adds six Reynolds stress equations to the system that must be solved. For time average models, an additional equation for the dissipation is required to provide length scale; the existence of natural length scale makes this unnecessary in LES. The addition of six equations to the set of four conservation equations means that the already high cost of LES is multiplied by 2.5. It is not surprising that this approach is not often used.

The only applications of this method have been made by Deardorff (1972, 1973) in meteorological applications. Unfortunately, he had no means of determining the model constants and adopted the values used in time average modeling. Since the relative importance of various terms in the Reynolds stress equations is different in the two cases, the validity of this approach is open to question. Indeed, Deardorff found that the new model produced a smaller improvement in the quality of the results than he expected.

Time average modelers have used algebraic modeling (Rodi, 1976) as an alternative to full Reynolds stress modeling. This approach reduces the differential Reynolds stress equations to algebraic equations. A similar approach for LES has been suggested by Findikakis and Street (1983). This approach is attractive but, as sufficient data to fix the model constants do not now exist, it cannot be recommended at present.

D. Scale Similarity Model

The filtering procedure that defines the resolved component of the velocity field provides a clear definition of what is to be computed and what needs to be modeled. In particular, it shows that the subgrid scale Reynolds stress represents the effect of the unresolved scales of motion on the resolved scales. It is generally believed that turbulence interactions occur principally between neighboring scales. This suggests that the most important interactions between the resolved and unresolved scales must involve the smallest resolved scales and the largest unresolved scales. Furthermore, it is reasonable to assume that the motions in these scales are similar. This led Bardina et al. (1980) to suggest a scale similarity subgrid scale model:

$$R_{ij} = \bar{u}_i \bar{u}_j - \bar{\bar{u}}_i \bar{\bar{u}}_j \qquad (16)$$

They found that this model correlates the subgrid scale Reynolds stress much more accurately than the Smagorinsky model of Fig. 2. However, it is only minimally dissipative. Thus, they suggested that a model which combines the Smagorinsky and scale similarity models would have the best features of both. This turns out to be the case. The combined model has both the excellent correlation of the scale similarity model and the dissipative property required of any subgrid model. No constant is used in Eq. (16), because the addition of this form to the filtered Navier-Stokes equations makes the latter

Galilean invariant. Their studies also showed that, unlike the Smagorinsky model, it does not lose its ability to correlate the data in strained and sheared flows.

The combined model was tested in homogeneous turbulence by Bardina et al. (1983). For these flows, the Smagorinsky model performs reasonably well so the differences are not great. However, the combined model gives a more rapid transition from the artificial initial conditions to realistic turbulence, more accurate higher-order statistics, and slightly more accurate energy and Reynolds stress histories. Thus the model is promising, but it needs to be tested in inhomogeneous flows before it can be accepted as a solid improvement on the Smagorinsky model.

E. Group Renormalization Methods

A number of turbulence theories have been developed over the last twenty-five years. In the main, these are based on ideas and techniques from quantum field theory and statistical mechanics. Considerable progress has been made but it has been slowed, as has progress in every other approach to turbulence, by the essential nonlinearity of the phenomenon.

These theories are characterized by an emphasis on spectra and wave interactions. The essential approximations relate to the wave interactions; in a typical example, the relative phases of interacting waves are assumed random. These theories have had considerable success in predicting the properties of isotropic turbulence and some success in other homogeneous flows. The reader interested in the details of these theories is referred to the book by Leslie (1973).

It has proven difficult to extend these theories to inhomogeneous flows. However, because the small scales of turbulence are more universal and more nearly isotropic than the large scales, the theories may be more applicable to the small scales than to the large ones. Hence, there has been a considerable effort to apply turbulence theories to subgrid modeling in the last ten years. In the last few years, the renormalization group (a method borrowed from statistical mechanics), has been the subject of a number of papers. This subject is reviewed by McComb (1984) and Herring (1984).

These theories usually produce subgrid models which can be expressed as a kind of spectral eddy viscosity; i.e., in Fourier space they take the form:

$$\tau_{ij}(k) = -\nu(k)\, k^2 E_{ij}(k) \tag{17}$$

where E_{ij} is the turbulence spectrum function. Simulations which have been made with these models have been quite successful. The major issue facing them is how well they will extend to inhomogeneous flows for which spectral representations are more difficult to use than for homogeneous flows.

IV. Initial and Boundary Conditions

The selection of initial and boundary conditions for flow simulations often receives less than its fair share of attention. This is true of all flow computation, not just LES, but we shall consider only the latter.

A. Initial Conditions

LES is by definition a time-accurate method. The flows which have been simulated with it can be divided into time-developing flows and statistically steady flows. In the former, the ensemble average state is a function of time, while in the latter it is not. Time-developing flows include all of the homogeneous flows and most of the free shear flows that have been simulated. The statistically steady flows that have been simulated include channel and annular flows and some free shear flows and boundary layers.

Each turbulent flow contains evolving three-dimensional structures peculiar to it. A principal aim of LES and FTS is to simulate these structures. It is not easy to determine whether this has been accomplished, but statistical quantities of order three and higher (e.g., skewness and flatness) are often used as indicators. The optimum initial condition for an LES is a snapshot of a realization of the flow to be simulated which, by definition, contains the appropriate structures. Such initial conditions can be derived only from other simulations and are usually unavailable; they cannot be constructed by statistical methods. Unfortunately, in many cases, statistical methods are the only ones available. In some cases, the eigenmodes of a stability calculation are used in formulating initial conditions. Whatever method is used, the initial conditions do not accurately represent the actual state of a turbulent flow, thus causing the early parts of large eddy simulations to be unphysical.

In homogeneous flows, the turbulence length scales increase with time. When these flows are simulated, these scales eventually become larger than the considered region. Periodic boundary conditions (see below) are then inappropriate, and the simulation becomes unphysical and must be stopped. Thus, simulations of homogeneous flows are rendered unphysical in the early stages by the initial conditions and in later stages by the boundary conditions. This may leave only a short time span in which the results represent realistic physics; this span increases with the number of grid points used and, therefore, with the size of the computer used.

For time-developing free shear flows, the simulations fall into two categories. In some, the aim is to simulate the transitional phase of the flow. Others try to simulate fully developed flows. The latter suffer from the same problems as the homogeneous flows; there are unphysical time spans at the beginning and end of the simulations which limit the accurately simulated time span. In the transitional flows, the choice of initial conditions depends on the interests of the computor. For transitional flows, the initial conditions consist of a laminar free shear flow and perturbations; the differences among various authors lie in their choices of perturbations. Some authors choose perturbations which contain just a few modes derived from linear stability theory. Others choose random perturbations. Unfortunately, in most laboratory experiments, the perturbations are neither accurately controlled nor measured. Consequently, comparisons with experiments suffer from the inability of the simulations to match the experimental initial conditions accurately. There are also upstream feedback effects in the experiments that are both hard to measure and hard to simulate.

For statistically steady flows, and the channel and annular flows in particular, the principal issue impacting on the initial conditions is the relatively slow development of the flow. This means that, unless initial conditions that accurately represent a snapshot of the flow can be formulated, a long initial development time is required before the results can be accepted as data. The initial conditions used in these flows are typically superpositions of a mean profile (usually a turbulent profile), eigenmodes of the Orr-Sommerfeld equation (usually with finite amplitude) and noise. Omission of the Orr-Sommerfeld modes increases the time required for the simulation to come to equilibrium and may even cause the turbulence to decay. An alternative is to take the initial conditions from the results of another simulation. Furthermore, long periods are required to obtain stable time averages after the flow has developed. For all of these reasons, these flows require longer run times than any of the others simulated to date.

As the complexity of flows simulated increases, the construction of appropriate initial conditions becomes more difficult. Considerable attention needs to be devoted to this issue.

B. Boundary Conditions

In any flow computation, it is necessary to simulate only a portion of the flow. Artificial boundaries that separate the considered region from the surroundings must be introduced; their location may have an important effect on the quality of the results. This is as true for LES as for any other flow simulation method.

In directions in which the flow is statistically homogeneous, periodic boundary conditions are the natural choice. They also eliminate the need for user-provided information and the subjectivity that it may contain. One must ensure that the selected region is large compared to any eddy in the flow; otherwise, improper self-influence is introduced and the simulation is invalid. Homogeneous turbulence was selected as the first object of study in part because periodic boundary conditions could be applied.

At solid walls, no-slip boundary conditions on the velocity are well established; they have been applied by the NASA-Ames and Stanford groups. However, the boundary conditions to be applied to the pressure (if any) are not established. Indeed, Moser and Moin (1984) have suggested that the pressure may have non-analytic behavior near the wall. Inclusion of no-slip wall boundary conditions in LES requires that a large fraction of the grid points be placed in close proximity to the wall.

To avoid the waste of resource associated with putting so many points near the wall, Deardorff (1970), Schumann (1973), and other members of the Karlsruhe group have used artificial boundary conditions. They place the first grid point outside the region in which viscous effects are important. Schumann's boundary condition has the form:

$$u(\underline{r}, t) = C\Delta \frac{\partial u}{\partial y} (r, t) \quad \text{at} \quad y = y_1 \tag{18}$$

i.e., a linear relationship between the fluctuating components of the stress and the velocity. Experimental evidence indicates that such a relationship is more correct if it contains a time lag.

Inflow boundary conditions are the most difficult to generate and have not been used in LES to date. Inflow conditions influence the simulated flow in a manner similar to initial conditions. In particular, inappropriate inflow conditions may contaminate the flow for a considerable distance downstream of the boundary on which they are applied. However, accurate conditions must contain appropriate structures which can be obtained only from another simulation. (Use of periodic boundary conditions is essentially a way of doing this.) This is possible for transitional flows for which the inlet condition consists of a laminar profile and perturbations determined either form an Orr-Sommerfeld equation, random noise, or both. Finding a method of generating inflow conditions for fully developed flows is an important issue that must be faced in the near future.

IV. Future Directions of Large Eddy Simulation

A. Advances in Computer Technology

Large eddy simulation has come a long way but is still hardly beyond its infancy. The future directions will be determined largely by the rate of development of large computers. The supercomputer projects sponsored by the governments of the U.S. and Japan have been well publicized, and the machines they produce will play an important role in determining the future of LES. Fluid dynamics computations will consume a large fraction of the resources of these machines. We therefore anticipate major changes in the kinds of problems that are attacked with LES in the next five years.

Another important and related trend is the development of VSLI technology. This technology dominates fast computer memory applications and is beginning to become an important factor in secondary memories. Microprocessors, in just over ten years, have progressed from four-bit chips through eight- and 16-bit processors to 32-bit microcomputers. Chips will soon make it feasible to build a desk-top computer with the power of a mainframe of ten years ago. When these chips are used in clever architectures, significant reductions in the cost of a computation can be expected.

B. LES on Supercomputers

The largest present computers--the Cray-1 and the CYBER 205--operate at more than 100 million floating-point operations per second and have 2-4 million words of fast memory. In the next few years, machines with ten times the speed and memory will become available. Machines with yet another order of magnitude in speed (with as yet unspecified memory sizes) may become available not long after that, say in five years. These machines will cost as much as the present machines, so access to them will be restricted. They will therefore be research machines. The tasks assigned to them will be natural extensions of current work. We shall look at some of the possibilities.

For homogeneous turbulence, the new machines will make it possible to use grids as large as 128 × 128 × 128 and, in a few cases, 256 × 256 × 256. One will then be able to simulate Reynolds numbers which match the experimental ones. More importantly, it will be possible to simulate flows with an inertial subrange, thus permitting the study of subgrid scale modeling under conditions which resemble those in the flows that one really wishes to simulate. It will also become possible to study sheared and strained turbulence over long enough time spans to answer questions that have long perplexed turbulence theorists. The new machines should thus permit exciting advances in turbulence theory.

It should become possible to simulate free shear flows from the inception of instability to the fully developed turbulent state. This will permit the study of the effect of various perturbations on the early stages of free shear flows, a possibility with many applications. We should also be able to simulate spatially developing free shear flows and the pressure feedback effects that are currently difficult to deal with. Sound generation by turbulent flows, which can be done crudely at present, will be open for investigation. Finally, it may be possible to study the impingement of free shear flows on solid walls.

Application of LES to combusting flows has been slow in coming, for several reasons. Chemical reaction requires mixing at the molecular level, so the small scales must be treated accurately in reacting flows. It is also necessary to carry additional equations representing the conservation of chemical species and equations for the chemical kinetics. The task is formidable. Nevertheless, the new supercomputers should make it possible to begin to simulate turbulent combusting flows.

For wall-bounded flows, the new generation of computers will bring with it the ability to simulate fully turbulent channel flows at moderate Reynolds numbers, temporally developing boundary layers, and possibly, spatially developing boundary layers. It will also be possible to simulate transition without need of a model. The addition of other phenomena (for example, unsteadiness and heat transfer) will play an important role in many applications.

Another area with great potential for LES is meteorology. Although many LES ideas were borrowed from this field, meteorology could probably benefit from the systematic approaches developed by engineers. The much higher Reynolds numbers force meteorology to model far more of the physics than engineering, and experimental data are harder to come by. An approach which systematically uses data from LES at one scale to model at the next larger scale seems especially attractive. Applications in oceanography should also be possible.

C. Wider Use of LES

VSLI technology will reduce the cost of machines equivalent to the present supercomputers and make them more widely available. LES may become a tool used occasionally for final checking and testing of designs. There must first be exploratory investigations, followed by a slowly increasing use of LES as an applications tool. For this to happen, a number of developments will have to take place first.

Flows which are inherently three-dimensional and time-dependent should be the first targets. For these, three-dimensional, time-dependent calculations are required and could as easily be large eddy simulations as ensemble average calculations. The choice depends on the questions one is trying to answer. Consider, for example, the in-cylinder engine flow. Some questions, such as the average pressure history in a cycle and, perhaps, optimum spark timing, can be answered with an ensemble average model. Others, such as the prediction of misfire, probably require analysis of individual cycles via LES. Operationally, the major difference between the two approaches lies in the length scale appearing in the turbulence model; the same code can be used in both approaches!

There are many free shear flows of technological interest, including combusting flows. For these, a number of research issues need to be resolved, including the development of accurate boundary conditions for both the computational inflow and outflow surfaces and a method of dealing with the rapid growth of length scale in free shear flows.

One has to face many of the same issues in wall-bounded flows. Simulation of spatially developing wall bounded flows requires methods of dealing with the in- and out-flow boundaries. These flows are less intensely turbulent than free shear flows, so the problem is one of simultaneously considering a sufficiently large part of the flow and using small enough grids to capture the important eddies in the flow. This will probably require elimination of the layers closest to the wall from consideration by means of artificial boundary conditions of the type used by Deardorff and Schumann. It should be possible to develop these conditions by using LES at relatively low Reynolds numbers. LES applications in wall bounded flows will also be expanded to include many "extra effects", including heat transfer, blowing, suction, curvature, and rotation.

These are some of the principal directions in which LES can make important contributions in the near future. There is much to do. If this program is to come to fruition, the number of people in the field will need to increase.

V. Conclusions

Large eddy simulation has made important strides in the past ten years. A number of simulations have been made which display the capability of the method to complement laboratory experiments both in providing physical explanation of the phenomena involved in turbulence and in providing the correlations used in turbulence modeling.

To further extend the capability of LES, a number of developments need to take place. These include the development of better models for the unresolved scales, better methods of deriving initial conditions, and better treatment of the boundary conditions. Several directions in which these developments may take place have been laid out in this paper.

The coming explosion in computer capacity driven by advances in VSLI technology will provide both much larger supercomputers and cheaper large computers. This will lead to growth of LES as a tool for turbulence research and as a top-line tool for applications.

Acknowledgments

This work has been supported in part by NASA-Ames Research Center and the Office of Naval Research. The contributions of many colleagues and students to this work deserve mention, but, in order to avoid omitting someone, I shall mention only my research partner of the past twelve years, W. C. Reynolds, whose influence has been enormous.

References

Bardina, J., Ferziger, J. H., and Reynolds, W. C., "Improved Subgrid Models for Large Eddy Simulation," AIAA paper 80-1357, 1980.

Bardina, J., Ferziger, J. H., and Reynolds, W. C., "Contributions to Large-Eddy Simulation of Turbulent Flows," Report TF-19, Dept. of Mech. Engrg., Stanford Univ., 1983.

Bardina, J., Ferziger, J. H., and Reynolds, W. C., "Contributions to Large-Eddy Simulation of Turbulent Flows," Report TF-19, Dept. of Mech. Engrg., Stanford Univ., 1983.

Cain, A. B., Reynolds, W. C., and Ferziger, J. H., "Simulation of the Transition and Early Turbulent Regions of a Free Shear Flow," Report TF-14, Dept. of Mech. Engrg., Stanford Univ., 1981.

Clark, R. A., Ferziger, J. H., and Reynolds, W.C., "Direct Simulation of Homogeneous Turbulence and its Application to Testing Subgrid Scale Models," Report TF-9, Dept. of Mech. Engrg., Stanford Univ., 1975.

Clark, R. A., Ferziger, J. H., and Reynolds, W. C., "Direct Simulation of Homogeneous Turbulence and its Application to Testing Subgrid-Scale Models," Report TF-9, Dept. of Mech. Engrg., Stanford Univ., 1975.

Clark, R. A., Ferziger, J. H., and Reynolds, W. C., "Evaluation of Subgrid Scale Turbulence Models Using a Fully Simulated Turbulent Flow," J. Fluid Mech., C91D, 92, 1979.

Deardorff, J. W., "Numerical Simulation of Turbulent Channel Flow," J. Fluid Mech. 1970.

Deardorff, J. W., "A Numerical Study of Three-Dimensional Turbulent Channel Flow at Large Reynolds Number," J. Fluid Mech., C41D, 452, 1970.

Deardorff, J. W., Three-Dimensional Numerical Study of Turbulence in an Entrained Mixing Layer," AMS Workshop in Meteorology, 271, 1973.

Deardorff, J. W., "Three-Dimensionsl Numerical Modeling of the Planetary Boundary Layer," Boundary-Layer Meteorology, C1D, 191, 1974.

Feiereisen, W. J., Reynolds, W. C., and Ferziger, J. H., "Computation of Compressible Homogeneous Turbulent Flows," Report TF-13, Dept. of Mech. Engrg., Stanford Univ., 1981.

Ferziger, J. H., "Higher Level Simulations of Turbulent Flow, in Computation of Transonic and Turbulent Flows" (J. A. Essers, ed.) Hemisphere, 1983.

Ferziger, J. H. and Leslie, D. C., "Large-Eddy Simulation: An Accurate Approach to Turbulence Simulation," Proc. AIAA Conf. Comp. Fluid Mech., 1979.

Findikakis, A., and Street, R. L., "An Algebraic Model for Subgrid-Scale Turbulence in Stratified Flows," J. Atm. Sci., 36, 1934, 1979.

Herring, J. R., "Theoretical Approaches to Turbulence Modeling," this volume.

Kim, J. J., and Moin, P., "Large Eddy Simulation of Turbulent Channel Flow," J. Fluid Mech., 1981.

Leonard, A., "Energy Cascade in Large-Eddy Simulations of Turbulent Fluid Flows," Adv. in Geophys., C18AD, 237, 1974.

Leslie, D. C., Theories of Turbulence, Oxford U. Press, 1973.

Lilly, D. K., "On the Computational Stability of Numerical Solutions of Time-Dependent, Nonlinear, Geophysical Fluid-Dynamic Problems," Mon. Wea. Rev., C93D, 11, 1965.

McComb, W., "Applications of Group Renormalization Methods to Turbulence Modeling," this volume.

McMillan, O. J. and Ferziger, J. H., "Tests of New Subgrid-Scale Models in Strained Turbulence," AIAA paper 80-1339, 1980.

McMillan, O. J. and Ferziger, J. H., "Tests of New Subgrid-Scale Models in Strained Turbulence," AIAA paper 80-1339, 1980.

Moin, P., Reynolds, W. C., and Ferziger, J. H., "Large-Eddy Simulation of Turbulent Channel Flow," Report TF-12, Dept. of Mech. Engrg., Stanford Univ., 1978.

Moin, P. and Kim, J. J., "Large-Eddy Simulationof Turbulent Channel Flow," J. Fluid Mech., 1981.

Moser, R. D. and Moin, P., "Direct Simulation of Turbulent Flow in a Curved Channel," Rept. TF-20, Dept. of Mech. Engrg., Stanford Univ., 1984.

Orszag, S. A., "Numerical Simulation of Incompressible Flow within Simple Boundaries. I. Galerkin (Spectral) Representations, Stud. Appl. Math., C50D, 293, 1971.

Orszag, S. A., and Patterson, S., "Direct Simulation of Isotropic Turbulence," Phys. Rev. Lett., 1972.

Reynolds, W. C., "Application of Turbulence Simulation to Phenomenological Models," this volume.

Riley, J. J., and Metcalfe, R. W., "Simulation of Free Shear Flows," Proc. Third Symp. Turb. Shear Flows, Davis, 1981.

Rodi, W., "A New Algebraic Relation for Calculating the Stress," ZAMM, 56, 1976.

Rogallo, R.J., "Experiments in Homogeneous Trubulence," Report, NASA-Ames Research Center, 1981.

Rogallo, R. G. and Moin, P., Numerical Simulation of Turbulent Flows, Ann. Revs. Fluid Mech., 1984.

Schumann, U., "Ein Untersuchung über der Berechnung der Turbulent Strömungen im Platten- und Rinspalt-Kanelen," dissertation, Karlsruhe, 1973.

Shirani, E., Ferziger, J. H., and Reynolds, W. C., "Simulation of Homogeneous Turbulent Flows including a Passive Scalar," Report TF-15, Dept. of Mech. Engrg., Stanford Univ., 1981. F66.

Smagorinsky, J., "General Circulation Experiments with the Primitive Equations. I. The Basic Experiment," Mon. Wea. Rev., C91D, 99, 1963.

Yoshizawa, A. "A Statistical Investigation upon the Smagorinsky Model in the Larfe-Eddy Simulatio of Turbulence, J. Phys. Soc. Japan, to be published.

CHAPTER IV. An Introduction and Overview of Various Theoretical Approaches to Turbulence

Jackson R. Herring

1. INTRODUCTORY COMMENTS

Theoretical methods to treat turbulence may be organized along two principal lines: (1) statistical, in which the distribution function of the flow (or equivalently, various moments of the velocity or vorticity field) serve as the basic ingredient; and (2) the dynamical, in which the mechanisms of instability and turbulent coherent structures serve as the primary focus. We include in the former second-order modeling, the two-point moment closures, and methods based on the distribution function. Central to this approach is the assumption that complete knowledge of the flow is not necessary for an approximate knowledge of low-order moments, or other simple features of the flow's distribution function. The dynamical approach focuses upon some key aspect of turbulence such as coherent structure (vortices) and follows their evolution in detail by employing suitable but perhaps heavily approximate dynamical equations. Practitioners argue that averaging (ensemble or time) may be a final needed step, but may legitimately be done only after developing some understanding of how to represent the relevant structures.

The statistical methods work well if the ensemble of flows is close--in some sense--to Gaussian; the second is essential if the flow is far from Gaussian. From a practical standpoint, Gaussian means here that the structures of the flow are space-filling, and vice versa. The first approach is frankly computational; in the second, it is assumed that--given the nature of turbulence--insight must precede sensible computations. Finally, there are hybrid methods, such as the large-eddy simulation and the renormalization group approach which compute the details of the large-scale flow, but assume the effects of omitted small scales may with some legitimacy be replaced by suitable averages over their rapid fluctuations. Had the statistical approach succeeded dramatically, interest in the second would likely now be much less.

We examine certain of these proposals, discussing their successes and indicating their respective significance in the face of our growing capacity for numerical simulations without any approximations. We begin with a statement of the moment closures, and then present a simple calculation illustrating possible failing of second-order (two- and one-point) closure. The example is drawn from two-dimensional turbulence, where ideas of second-order modeling are simple to posit without complicating tensor aspects. We then indicate some successes of the two-point second-order modeling, drawn from the example of turbulent thermal convection.

2. TWO-POINT CLOSURE

We begin with some considerations of two-dimensional flows. Interest in this problem exists not only because of its simplicity, but also because it serves as a prototype of the large scales of the earth's atmosphere, in which the two-dimensional dynamics stems from the rapid rotation of the earth. We are interested in predicting the spectrum of kinetic energy;

$$E(k,t) = 2\pi k U(k,t) = \langle \underline{u}'(\underline{k},t)\ \underline{u}'(-\underline{k},t)\rangle / 2 \tag{2.1}$$

$$= (1/2)\int d(\underline{x}-\underline{x}')\ \langle \underline{u}'(\underline{x},t) \cdot \underline{u}'(x',t)\rangle \exp(i\underline{k} \cdot (\underline{x}-\underline{x}')) \quad ,$$

where $\underline{u}'(\underline{k},t)$ is the Fourier transform of the velocity fluctuation field $\underline{u}'(\underline{x},t)$. We suppose the domain on which the Navier Stokes equation,

$$(\partial/\partial t + \nu(\nabla))\ \underline{u}'(\underline{x},t) = -\nabla P - \underline{u}' \cdot \nabla\underline{u}' + \underline{f}(\underline{x},t) \quad , \tag{2.2}$$

is specified as a large square on whose boundaries u' is periodic. Here, $\underline{f}(\underline{x},t)$ is a possible (random) driving force to maintain the flow. The dissipation operator, ν, is as yet unspecified; in some numerical studies it selectively damps the small scales, as for example would

$$\nu(\nabla) = \nu_4 \nabla^4 \quad . \tag{2.3}$$

For the time being, we take the (ensemble) mean flow (= $\underline{V}(\underline{x},t)$) to be zero.

The simplest closure for U(k,t) has the form,

$$[d/d(2t) + \nu(k)]\, U(k) =$$

$$\int_\Delta dpdq\; B(k,p,q)\;\theta\,(k,p,q)\; U(q)\{U(p)-U(k)\} + \langle \underline{f}(\underline{k}) \cdot f(-\underline{k})\rangle \quad , \tag{2.4}$$

where, for two-dimensional isotropic flows,

$$B(k,p,q) = 2\{(k^2-q^2)(p^2-q^2)/k^2\}\;(1 - x^2)^{1/2} \tag{2.5}$$

In (2.4), the dpdq integration ranges over all (p,q) such that (k,p,q) can form a triangle and x is the cosin of the interior angle opposite leg k. The triad-relaxation factor $\theta(k,p,q)$ represents the effective life-time of the straining which transfers energy among the available scale sizes. For practical purposes it is the (R.M.S.) strain by large scales,

$$\theta(k,p,q) = \eta(k) + \eta(p) + \eta(q) \quad ,$$

$$\eta(k) \sim \{\int_0^k E(p)p^2dp\}^{1/2} + \nu(k) \quad . \tag{2.6}$$

The equations (2.3)-2.6) are known as the eddy damped quasi-normal Markovian approximation (EDQNM), and are structurally similar to the quasi-normal approximation first proposed for two-dimensional turbulence by Ogura (1963). The EDQNM is closely related to the Test Field Model (TFM) proposed by Kraichnan (1971). The latter is that used here. The equations may be easily modified to incorporate anisotropic and inhomogeneous flows (Herring, 1978). If this is done, we obtain a formalism which contains much of "second-order modeling" as the limit $(\underline{x}-\underline{x}') \to 0$.

How well does (2.3)-(2.6) represent the second-order moments of actual fluid flow satisfying (2.2)? Initial indications (Herring, et al., 1974) were promising. However, there is a recent growing body of evidence that its forecast may--in certain instances--be in serious error (Fornberg, 1976; Herring and McWilliams, 1984). Briefly, if the flow is forced randomly at a particular scale, k_0, the steady state E(k) is given fairly well by the closure. However, if the flow is allowed to decay, and if the Reynolds number is sufficiently high, the closure can seriously overestimate the energy transfer to large scales, and the transfer of squared vorticity (enstrophy) to small. Experimentally, a key to assessing the applicability of the closure is the kurtosis of the vorticity field:

$$K = \langle[\nabla x u']^4\rangle \,/\, [\langle(\nabla x u')^2\rangle]^2 \quad .$$

As long as this quantity remains close to its Gaussian value (of 3.0) the closure is satisfactory; however, for decay experiments at large Reynolds numbers (10^{+4} for an integral scale) the kurtosis grows from its initial value of 3 to about 30. Figure 1 summarizes the results of a numerical study by Herring and McWilliams (1984) illustrating this point.

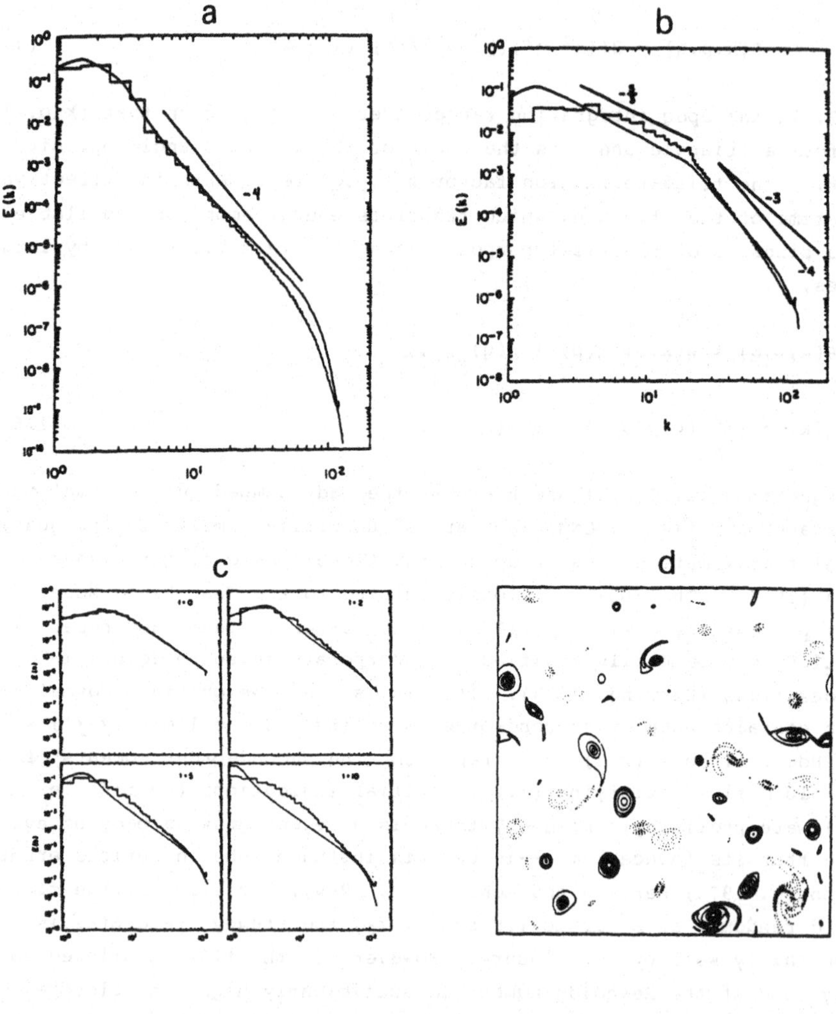

Fig. 1. Results of numerical comparison of two-point closure with direct numerical simulation for two-dimensional turbulence: (a) and (b) randomly driven, stationary turbulence with spike-forcing at $k = 3$ and at 20. Resolution is (128x128), Reynolds no. = 50; (c) decay experiment for same resolution but much higher Reynolds number ($10^4 \to 10^5$); (d) contour plots of vorticity in the direct simulation of the decay experiment at $t = 10$. After Herring and McWilliams (1984).

Notice that at late stages the real flow consists of isolated vortex kernels, separated by wide regions in which the flow is quiescent. This is patently a strongly non-Gaussian field; for example, $K(t=10) = 30$.

To indicate possible sources of error in the closure, let us try to examine--via the inhomogeneous form of the closure procedure--how vorticity may be amplified into intense kernels in two-dimensional flow. To do this we must record the turbulence production for the small-scale Reynolds stress (for example, the sub-grid scale stress) that arises from their interactions with larger-scale strain. We then split the flow into ensemble mean and fluctuating part, and make the usual (W.K.B.) arguments about the scale of $\langle \underline{u}'(\underline{x})\underline{u}'(\underline{x}')\rangle$ being smaller than those of $\langle \underline{V}(\underline{x},t)\rangle$. This is the spectral form for the rapid distortion terms. We then may regard regions of space in which the turbulent Reynolds stress grows as regions in which there is transfer to small scales and consequent dissipation. Conversely, if there is no growth of the small scales of the Reynolds stress, we have a region of a stable structure (i.e., the intense vortex regions). We recall that since two-dimensional flow admits no mean-gradient induced instability, whatever intensification of small-scale structures must come at the expense of large scales. The equation for $\langle \underline{u}'_i(\underline{k})\, u'_j(-\underline{k})\rangle(\underline{k})$ may be conveniently written in terms of complex scalars $\langle |u'_1(k) + iu'_2(k)|^2, (u'_1(k) + iu'_2(k))^2\rangle/2 = \{U_0(k), U_2(k)\}$, where $U_0(k)$ is the energy spectral density, and $U_2(k)$ represents the turbulent Reynolds stress (actually the second angular harmonic of $U(\underline{k})$). These quantities satisfy Herring (1975),

$$\{D/Dt + 2\nu(k)\}\, U_0(k) = S^*(1 + k\partial/\partial\{4k\})\, U_2 + cc + \ldots \tag{2.7}$$

$$\{D/Dt + 2\nu(k)\}\, U_2(k) =$$

$$i\zeta U_2(k) + S(1/2 + k\partial/\partial\{4k\})U_0(k) + \ldots, \tag{2.8}$$

$$S = -\partial V_1/\partial X_1 + \partial V_2/\partial X_2 + i\{\partial V_1/\partial X_2 + \partial V_2/\partial X_1\} \quad ,$$

$$\zeta = -\partial V_1/\partial X_2 + \partial V_2/\partial X_1 \quad . \tag{2.9}$$

Here S represents the strain, and ζ is the vorticity, as computed in the coordinates $\{(\underline{x}_1 + \underline{x}_2)/2 = \underline{X}\}$, and D/Dt is the substantial derivative in that coordinate frame. We here recorded only the mean fluctuating interactions; the "..." represent the omitted turbulence terms, for which the inhomogeneous form of the closure (2.3)-(2.6) is required (they give the usual diffusion and return to isotropy terms).

Parenthetically, we note that (2.7)-(2.8) may be used to derive the enstrophy inertial range ($E(k) = 2\Pi k U_0{}^{-}k^{-3}$), if we assume that a constant and large Rotta-like term $-\mu U_2$ is the main omitted term in (2.8). The resulting equation for U_0 is, approximately,

$$(D/Dt + 2\nu(k)\ U_0 = \frac{|S|^2\mu}{\mu^2+\zeta^2}\ (1 + k\ \partial/\partial\{4k\})\ (1 + k\ \partial/\partial\{2k\})\ U_0 + \ldots.$$

This expression clearly shows the role of the vorticity in suppressing energy transfer.

Using (2.7)-(2.9), we may estimate the growth of the small scale Reynolds stress, which we define as,

$$E_i(k) = 2\Pi\int_k^{\infty} p\,dp\,U_i(p) \quad . \tag{2.10}$$

For orientation, we may take $E_0(k) \sim k^{-\gamma_0{}^{-3}}$ exp (Λt) and $E_2(k) \sim k^{\gamma_2{}^{-3}}$ exp (Λt). Then, neglecting the closure terms, the growth rate for (2.7)-(2.8) is:

$$\Lambda^2 = [|S|^2(\gamma_2(1+\gamma_0) - \xi^2] \tag{2.11}$$

In other words, small scales grow in regions where the strain exceeds vorticity, and conversely there is no growth of small-scale turbulence if the large-scale vorticity exceeds--in absolute value--the large-scale strain. Of course, the omitted terms (diffusion and return to isotropy) will act to dampen the excesses of vorticity in different spacial regions. Similar arguments concerning the amplification of enstrophy based on Lagrangian dynamics instead of the more approximate treatment given here have been proposed by Weiss (1981) and by McWilliams (1982). Their analysis suggests that the local value of $[|S|^2 - \zeta^2]$ determines the growth of small-scale turbulence.

Let us return now to the homogeneous problem. At any given time in the flows development, we should observe fluctuations in the local values of strain and vorticity. If we now assign the "large scales" the role of mean field in the above inhomogeneous calculation, then the analysis suggests a tendency for high vorticity regions not to erode, and those of large strain to dissipate the small-scale energy. This would further accentuate the intermittency (active vortex regions separated by quiet regions). Note that at the level of second-order moments, there is no way--in the context of homogeneous flow--to know of this development: squared vorticity and squared strain have the same spectrum.

The problem discussed here--the development of strongly non-Gaussian flow--is not necessarily a universal attribute of two- and quasi-two-dimensional flows. We have seen examples of randomly driven flows which are well forecast by closure, and for which measures of non-Gaussianity are small. However, the example given here of non-Gaussian behavior is sobering, in that the closure--at least in its homogeneous context--has no way of diagnosing such a development. The qualitative aspect of this problem seems to be that the initial vorticity concentrations are not eroded, but that other regions of large strain are; there is no explosive growth of vorticity as may perhaps develop in three dimensions.

The numerical study described here may be compared to the recent calculation of Bracket and Sulem (1984) who report a k^{-3} spectrum in a very high resolution initial value problem. Their calculation begins with an initially Gaussian state concentrated at small wave numbers. They report that rolled up convoluted secondary structures or isolated vortices begin to form at long times, but that vorticity gradient sheets (which are more space filling) are still present. Whether the isolated vortices will eventually dominate their calculation is not clear. With respect to the k^{-3} range difference, there may well be an initial value dependence that is not yet well understood. Bracket and Sulem also note that lower resolutions ($< 512^2$) are unable to resolve a viscous instability essential for the formation of the k^{-3} inertial range.

As the above discussion suggests, the nature of the two-dimensional enstrophy inertial range is not completely understood. However, we should stress that our numerical simulation--an equivalent resolution comparison--is a fair test of closure and the emergence of the isolated vortices in this case implies serious error in the closures. Conversely, in the randomly forced cases--for which the role of the vortices is much less significant--the agreement is much better.

The above example is an extremely idealized prototype of large-scale atmospheric (planetary-scale) motion. A more realistic model is that of quasi-geostrophic equations, which allows for vertical (stable) density stratification. These equations ((2.12) below) admit the same inviscid constraints of energy (kinetic plus potential) and enstrophy (squared total [including planetary] vorticity). This similarity led Charney (1972) to propose an analogous spectrum for the enstrophy. He was at that time able to marshall some observational support for the theory. There is also another set of observations of atmospheric motion which invites comparison to ideas of quasi-two-dimensional turbulence. In Figure 2 we show observations of atmospheric velocity fields gathered by aircraft (747) during (mainly) transoceanic flights.

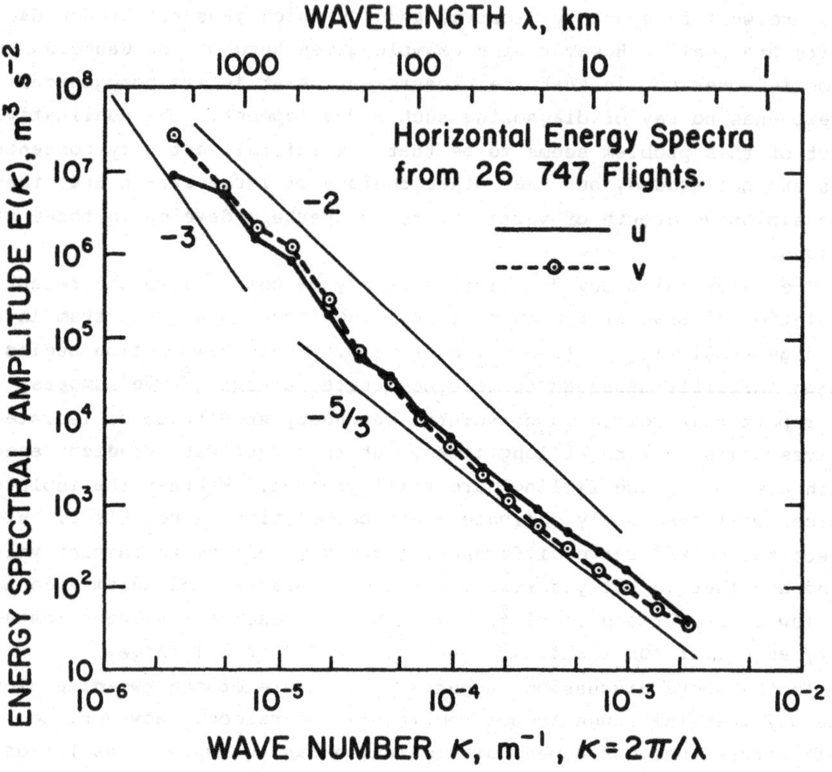

Fig. 2. Observed atmospheric spectrum over scales ($10^{-6} < k < 10^{-2}$) m-1, as inferred from aircraft (747) observations. After Lilly and Petersen (1984).

The data is consistent with a -5/3 law, and observations like this earlier led Gage (1979) to hypothesize this spectral to be an "inverse cascade range" of (quasi-) two-dimensional turbulence, in which energy is input at scales just larger than that of large convective storms. Energy is then cascaded back to larger scales, where it eventually merges with the planetary scale k-3 range. The inverse cascade can be seen to emerge in the large k randomly forced run (shown here in Figure 1) of Herring and McWilliams (1984).

The interpretation of this range of "mesoscale" motion as two-dimensional turbulence is not secure; a counter proposal is that it is an internal wave spectrum possibly generated from the same source (Van Zandt, 1982) similar to the internal wave spectrum in the ocean meso-scale, hypothesized by Garrett and Munk (1979). To decide this issue, we need some way to discriminate between the two dynamics. Gage and Nastrom (1984) have suggested that if the Charney hypothesis is extended into the -5/3 range--including the proposal that kinetic and potential energy are equipartitioned in this range--then the observed spectrum is more consistent with the inverse cascade hypothesis. The equations to be considered here simply state the conservation of potential vorticity,

$$(\partial/\partial t + u \cdot \nabla)\zeta = 0 \ , \ \underline{u} = (-\partial\Psi/\partial x_2 \ , \ \partial\Psi/\partial x_1) \quad ,$$

$$\zeta = - \nabla^2 \Psi \quad . \tag{2.12}$$

The closure may in fact be shown to yield an equipartitioning in this range (equivalent here to a three-dimensionally isotropic spectrum for $\langle\zeta(\underline{x})\zeta(x')\rangle$) (Herring, 1980). This result is not obvious, since the dynamics of (2.12) are not isotropic. Thus, questions of simple two-dimensional flow are of some importance in attempting to sort out the dynamics of the mesoscale of the earth's atmosphere.

3. APPLICATION OF CLOSURE TO THERMAL CONVECTION

We now turn to a very different topic: thermal convection at low Reynolds numbers. For this problem, there is some evidence that turbulent flows need not be strongly non-Gaussian. Two-point closures may then be a more useful tool. This evidence comes from direct numerical simulations. Figure 3 shows some results for slip-boundary Boussinesq convection at Ra = 7Rc (Rc = 657.5), and a Prandtl number Pr. = 10. The aspect ratio (lateral dimensions of the horizontal periodic square box to vertical depth of fluid) is $4\sqrt{2}$.

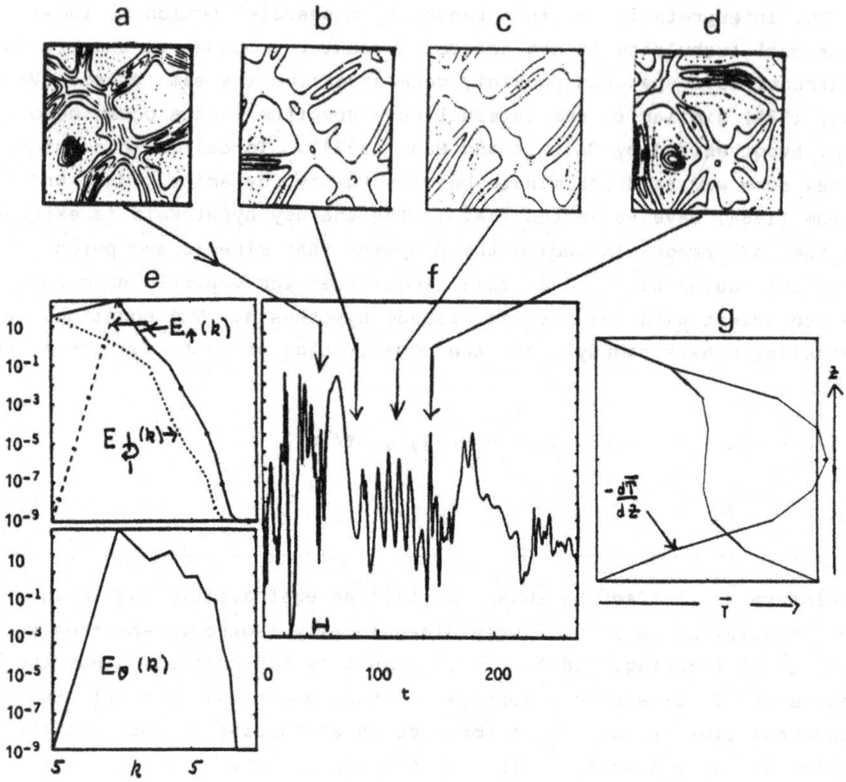

Fig. 3. Thermal convection for Ra = 7Rc (=654.5), Prandtl number = 10, and aspect ratio = 4√2. (a), (b), (c), (d) show mid-plane plan-forms of vertical velocity field at times as indicated by arrows; (e) spectra of polodial and toroidial energy; (f) time series for a point in mid-plane (vertical velocity); (g) profile of horizontally averaged temperature field, and mean temperature gradient.

The flow is reasonably chaotic; although there is a fair degree of spatial order, structures continuously deform and change their topology. At mid-time of the run--which was set up initially as simply Gaussian fields--the flow develops an ephemeral quasi-periodic state, which consists of deformed rectangles. This state is only metastable and, after a few eddy circulation times, gives way to more chaotic spatial states. Figure 3(b) shows a rough measure of energy spectra--subdivided into poloidal and toroidal components. The former is directly driven by buoyancy; the latter is horizontal swirling (vertical vorticity) driven here entirely by inertial terms. The toroidal component is strongly peaked at the largest available scale, but contributes negligibly to total energy. Its presence is very important in the planforms (a)-(d). Dynamically, it is responsible for destabilizing the roll structure, thought earlier to be stable at this Ra (see, e.g., Zippelius and Siggia, 1982).

The resolution of the numerical code for this run is (Nx,Ny,Nz = (64,64,8), which fits nicely within a Cray 1 core. However, if we are interested in stable statistics for an horizontally homogeneous system, this resolution is inadequate, since the peak toroidal energy is near the largest available scale. It seems entirely possible that (Nx,Ny) = (256,256) would be required, which is out of core. What is worse, the calculation becomes prohibitive (the time shown in Figure 3(d) represents 5 Cray hours!). What's needed here is a "super-grid" parameterization, but such is logically inconsistent.

Two-point closures, on the other hand, may be able to produce useful statistics for this problem much more readily. First, we point out that the flow field shown here is computed to be close to Gaussian (i.e., for the runs of Figure 3 [(3.11) $< \langle T'^4 \rangle / \langle T'^2 \rangle^2 \lesssim$ (3.12)]. On the other hand, implementing the closure for this problem is not easy; the flow is inhomogeneous in the vertical. Recently Dannevik (1984) has numerically integrated the Direct Interaction Approximation (DIA) for this problem. His results--although still at low Ra ($\lesssim$ 4Rc, Pr = .7)--appear promising, and extendable to larger Reynolds number.

We next discuss the application of the DIA to this problem. Rather than giving a systematic development of the formalism, we simply touch upon some essential physics not present in simpler procedures, such as the EDQNM (2.4). Readers interested in details are refered to Dannevik's thesis and to the earlier works of Kraichnan, especially Kraichnan (1964).

The DIA is formulated in terms of $\langle u_i(\underline{x},t)\, u_j(\underline{x}',t')\rangle$, $\langle T(\underline{x},t)\, T(\underline{x}',t')\rangle$, and $\langle u_i(\underline{x},t)\, T(\underline{x}',t')\rangle$. It further introduces response functions $G_{ij}(\underline{x},\underline{x}';t,t') = \langle \delta X_i(\underline{x},t) / \delta X_j(\underline{x}',t')\rangle$, $(i = 0,1,2,3)$, $X_i = u_i$, $i > 0$; $X_o = T_o$. No assumption is made concerning homogeneity or isotropy. The closure consists of a system of integro-differential equations for $\langle X_i(x,t)\, X_j(\underline{x}',t')\rangle$ and $G_{ij}(x,x';t,t')$. This set contains time integrals over the past history of the flow--extending from the initial spatial state of multivariate Gaussianity to the current time. It reduces to a simple set like (2.4) for homogeneous flow, if in addition these historical integrals are parameterized simply. Such a parameterization shows up in (2.4) as the $\theta(k,p,q)$ factor, and represents physically the length of time the energy transfer process remains coherent. In (2.4), the way the historical effects have been eliminated is to simulate them with a (hopefully) near-equivalent Markov system. However, for inhomogeneous flows, important instability effects are incorporated in G (for example, the buoyant instability), and it would seem unwise to Markovianize. Further, the interactions of G with the mean field [<T> (z,t)] are very important in establishing the correct thermal boundary layering (see, e.g., Herring, 1969).

Nevertheless, without some economizing of historical integrals, the DIA is simply impractical for inhomogeneous problems. What Dannevik proposed was to use Padé tables to render these integrals into algebraic form, nearly as simple as the Markovianized equations. To illustrate how this is done, let us consider a prototype of the calculation of the DIA calculation of the relaxation factor $\theta(k,p,q)$, as needed in (2.4), and as crudely specified by (2.6) in the EDQNM. In this prototype calculation we suppress all indexing superfluous to the historical integrals. In the DIA we must compute Green's functions from an integral equation whose time structure is,

$$[\partial/\partial t + L]\, G\,(t,t') = B\int_{t'}^{t} ds U\,(t,s)\; G\,(t,s)\; G\,(s,t') \quad . \tag{3.1}$$

Here, U(t,s) is the totality of Reynolds stress, temperature variance and heat flux, while the matrix L stands for all linear (and quasi-linear) aspects of the problem. We have suppressed all matrix (tensor two-point and (u,T)-cross-coupling) aspects of the problem, for simplicity. The two-time U was further approximated by utilizing the "fluctuation-dissipation theorem,"

$$U\,(t,t') = U\,(t',t')\; G\,(t,t') \quad , \tag{3.2}$$

although this step is not essential. The θ-relaxation time scale is computed from

$$\theta\,(t) = \int_0^{t} G\,(t,s)\; G\,(t,s)\; G\,(t,s)\, ds \quad . \tag{3.3}$$

Finally, we should note the generic form of the U (t,t') equation:

$$[\partial/\partial t + L\,]\; U\,(t,t') = B\int_0^{t'} G\,(t',s)\; U\,(t,s)\; U\,(t,s)\, ds$$

$$- B\int_0^{t} U\,(t',s)\; G\,(t,s)\; U\,(t,s)\, ds \quad . \tag{3.4}$$

Dannevik first introduces a two-time scale analysis, in which $\tau = (t-t')$ and $T = (t+t') / 2$ are the basic variables. It is thus assumed that the dependence on τ is much sharper than T, so that the system (3.1)-(3.4) becomes an ODE in terms of T; historical effects then only involve τ. This is an approximation only out of the statistically steady state. The key problem is then to simplify (3.1) so that [using (3.2)] (3.4) becomes an ODE (in independent variable T) to be solved for its equilibrium value. To effect this, we first write the Taylor series for $G(\tau)$, successively using (3.1),

$$G(\tau) = 1 - L\tau - (B-L^2)\tau^2/2! - (4BL + L^3)\tau^3/3! + (3B^2 + 11B\,L^2 + L^4)\tau^4/4! + \ldots. \tag{3.5}$$

The expression (3.5) is then matched against the Padé approximant:

$$G^{[L/M]}(\tau) = [P_0 + P_1\tau + P_2\tau^2 + \ldots P_L\tau^L] / [1+Q_1\tau + \ldots Q_M\tau^M]\ . \tag{3.6}$$

By examining the prototype in which [L,B] are simply scalars, Dannevik was able to select appropriate [L/M] values. A simple rule is that $M > L+2$, for convergence of the ds integrals. Notice that although [L,B] are in reality matrices that depend on T, no problem stands in the way of implementing (3.6) for this general problem.

There remains to formulate the (x,x') integration. We give no details on this point except to point out that a wave-number decomposition was used. The horizontal spectra were divided into wave-number bins (6), and the vertical wave-number pair (trig-functions to match the slip-boundary conditions) were treated explicitly, without interpolation. In discussing Dannevik's results, comparison cases arenecessary in order to assess accuracy. Two reference runs are (1) direct numerical simulations (DNS), and (2) the quasi-linear (QL) [equivalent in some sense to the rapid distortion theory] (Herring, 1963). For the latter, there are no inertial effects in the overall flux balance.

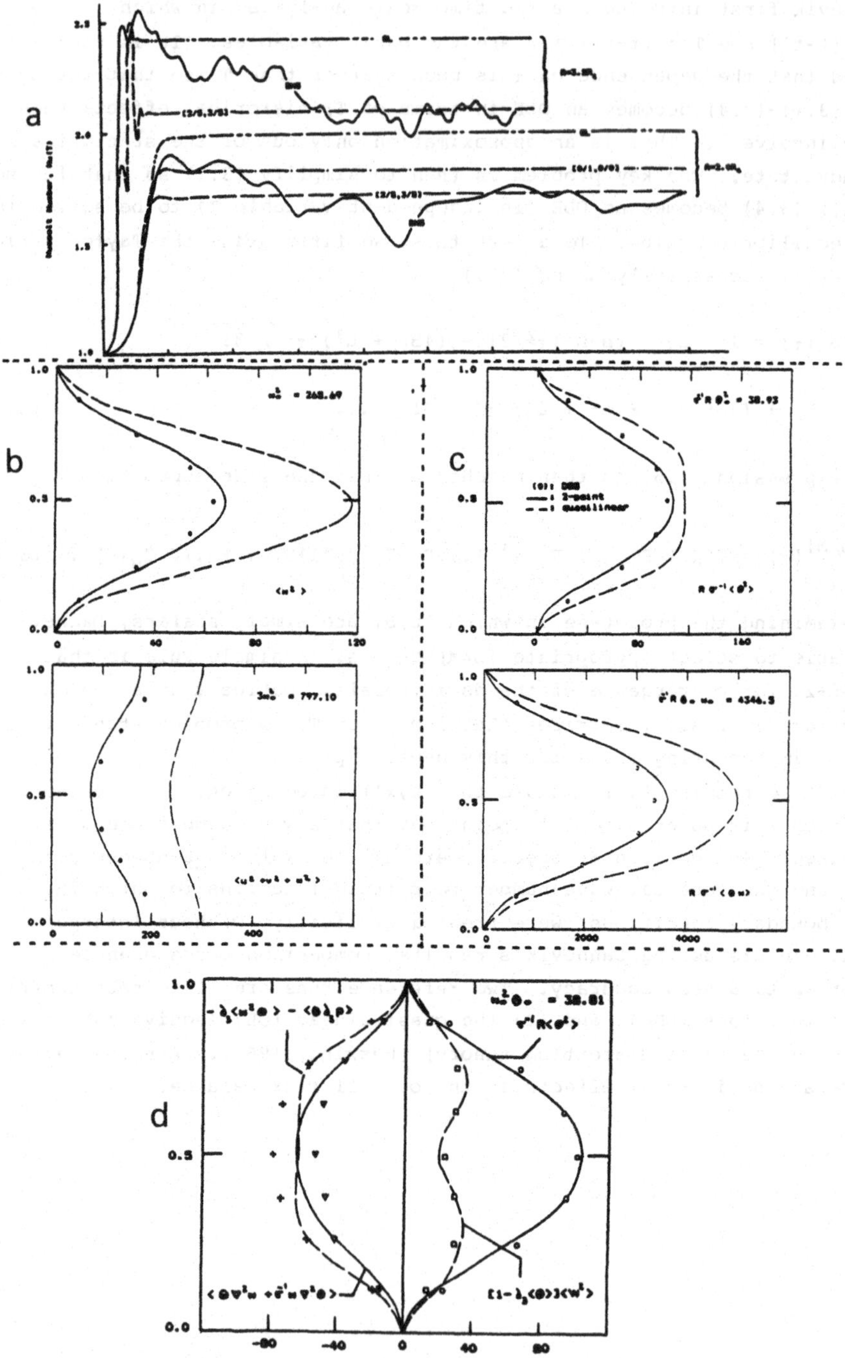
a
b
c
d
-80
-40
0
40
80

Fig. 4. Summary of Dannevik's (1984) comparison of Direct Numerical Simulations (DNS) to DIA and to quasilinear (QL) approximation for Pr. = .70, Ra = 2Rc and 3.5Rc. (a) Evolution of Nusselt number: DNS is solid curve, QL dash-dot-dot, DIA dot-dash. (b) $\langle w^2 \rangle$ (top) and $\langle u^2 + v^2 + w^2 \rangle$ (bottom). (c) Thermal variance (top), and convective heat flux. (d) Various terms in the heat budget, as indicated. Calculations are at Ra = 2Rc. Curves (b), (c), and (d) are for DIA only.

We note a pronounced improvement in the agreement between DIA and DNS as compared to that between DSN and QL. The two-time scale approximation in the DIA is undoubtedly responsible for the large DIA errors during the early transient phase of 4(a), for the Nusselt number.

The present calculations are rather preliminary in that Ra < 3.5Rc, not what one usually calls vigorous turbulence at large Reynolds numbers. Two points should be made here. First, as noted above, the case of low Ra, slip-boundary convection is not so trivial if one wants valid statistical information at large aspect ratios. It is clear that if the aspect ratio becomes much larger than 5, for example, the DIA has a computational advantage. Secondly, Dannevik's calculation may be augmented by using FFT's to accelerate the convolution sums, so as to extend the calculation to the strongly boundary-layered case.

We should note that as Ra increases, the DIA's accuracy will deteriorate because of its lack of Galilean invariance. The seriousness of this problem is not known. It is entirely possible that single-point profiles are accurate, despite large spectral errors (recall that the DIA gives correct energy decay laws even at large Reynolds numbers). If such errors are large, one has recourse to the more elaborate Lagrangian closures, but with considerably more labor.

4. COMMENTS ON SUB-GRID METHODS: RELATION TO CLOSURE

It is possible to combine closure with direct numerical simulation to obtain a rational sub-grid scale procedure. This approach was initiated by Kraichnan (1976) and by Leslie and Quarini (1979). The starting point is to write the closure for the energy spectrum, and then divide E (k,t) into two parts: (1) $E_<(k)$, $k < k_c$; and (2) $E_>(k)$, $k > k_c$. Since we have the full transfer in terms of E (k), through an equation like (2.4), it is possible to completely eliminate $E_>(k)$ from consideration. When this is done, it turns out that the $E_>(k)$ equation may be cast into a form identical to the original, but with a modified viscosity law. The entire procedure has the effect of mapping a large Reynolds number computation onto one at much lower (and more manageable) Reynolds number. The modified viscosity may be written approximately as

$$\nu(k) = \nu\,(k,k_c)\,[E\,(k_c)\,/\,k_c]^{1/2} \quad . \tag{4.1}$$

Here, $\nu\,(k,k_c)$ is derived from the closure. It generally has a cusp at the cutoff k_c (see, e.g., Chollet, 1983). The above prescription presumes that k_c is in the inertial range. Chollet (1983) has made an extensive examination of such procedures, comparing their predictions to closure without sub-grid scale approximations, using the EDQNM. The results are encouraging. Chollet also formulates the procedure for the more general case in which k is not necessarily in an inertial range.

The closure-derived $\nu\,(k,k_c)$ may be used in connection with amplitude equations; the solution to such a system then has an optimal ν (k) such that the truncated system matches the untruncated in some least square sense. One advantage of a formalism like (4.1) is that the eddy dissipation is present only to the degree needed; if $E\,(k_c) = 0$, at some stage, the eddy dissipation is zero. This is sensible in that no dissipation is required at this phase of the calculation, for the numerics to converge.

As noted above, the sub-grid scale method assures only that the spectra computed through its amplitude equations will be optimal--in the least square sense. The verisimilitude of its evolving structures in terms of velocity or vorticity amplitudes is not certain and is as yet an unsolved issue. The question may be conveniently put in terms of an initial value problem: how long are we to imagine the structures that evolve from the initial state to be real features of the flow, and at what point must we regard them as only typical aspects? May we regard the large scales more securely predicted than the small?

The difficulty is that this procedure seems to prohibit an exchange of phase information between sub-grid and grid scales. The problem has been discussed by Fox and Lilly (1972) and by Herring (1973). It seems inconsistent to treat a part of the system deterministically, and a part (the small) statistically, if there is no clear separation in their sizes. It may be that a properly posed renormalization group method could avoid this dilemma entirely.

REFERENCES

Brachet, M. E., and P.-L. Sulem, 1984: Direct numerical simulation of two-dimensional turbulence. Fourth Beer Sheva Seminar on MHD flows and turbulence. To appear in Progress in Astronautics and Aeronautics.

Charney, J. G., 1971: Geostrophic turbulence. J. Atmos. Sci., 28, 1087-1095.

Chollet, J.-P., 1983: Turbulence tridimensionelle isotrope: modelisation statistique des petites schelles et simulation numerique des grandes echelles. These de doctorat en Sciences, Institut de Mécanique de Grenoble, France.

Dannevik, W. P., 1984: Two-point closure study of covariance budgets for turbulent Rayleigh-Benard convection. PhD. Thesis, St. Louis University, St. Louis Mo. 166 pp.

Fornberg, B., 1977: A numerical study of 2-D turbulence. J. Comp. Phys., 25, 1-31.

Fox, D. G., and D. K. Lilly, 1972: Numerical simulation of turbulent flow. Rev. Geophys. Space Phys., 10/1, 51-72.

Gage, K. S., 1979: Evidence for a k-5/3 inertial range in mesoscale two-dimensional turbulence. J. Atmos. Sci., 36, 1950-1954.

-----, and G. D. Nastrom, 1984: The second workshop on technical aspects of MST radar, Univ. of Illinois, Urbana, 21-25 May.

Garrett, C., and W. Munk, 1979: Inertial waves in the ocean. Annual Review of Fluid Mechanics, Vol. 11, Annual Reviews, 339-369.

Herring, J. R., 1963: Investigation of problems in thermal convection, J. Atmos. Sci., 20, 325-338.

-----, 1969: Statistical theory of thermal convection at large Prandtl number. Phys. Fluids, 12, 39-52.

-----, 1973: Statistical turbulence theory and turbulence phenomenology. Proceedings of the Langley Working Conference on Free Turbulent Shear Flows, NASA SP 321, Langley Research Center, VA, 41-66.

-----, 1980: Statistical theory of quasi-geostrophic turbulence. J. Atmos. Sci., 37, 969-977.

-----, and J. C. McWilliams, 1984: Comparison of direct numerical simulation of two-dimensional turbulence with two-point closure: effects of intermittency. To appear in J. Fluid Mech.

Kraichnan, R. H., 1964: Direct interaction for shear and thermally driven turbulence. Phys. Fluids, 7, 1048-1062.

-----, 1971: An almost-Markovian Galilean-invariant turbulence model. J. Fluid Mech., 47, 513-524.

-----, 1976: Eddy viscosity in two and three dimensions. J. Atmos. Sci., 33, 1521-1536.

Leslie, D. C., and G. L. Quarini, 1979: The application of turbulence theory to the formulation of sub-grid scale modeling procedures. J. Fluid Mech., 91, 65-91.

Lilly, D. K., and E. L. Petersen, 1983: Aircraft measurements of atmospheric kinetic energy spectrum. Tellus, 35 (in press).

McWilliams, J. C., 1984: The emergence of isolated, coherent vortices in turbulent flow. To appear in J. Fluid Mech.

Ogura, Y., 1963: A consequence of the zero-fourth-order cumulant approximation in the decay of isotropic turbulence, J. Fluid Mech., 16, 33-40.

Van Zandt, T. E., 1982: A universal spectrum of buoyancy waves in the atmosphere. Geophys. Res. Lett., 9, 575-578.

Weiss, J., 1981: The dynamics of enstrophy transfer in two-dimensional hydrodynamics. La Jolla Institute Report, La Jolla, CA. 123 pp.

Zippelius, A., and E. D. Siggia, 1982: Disappearance of stable convection between free-slip boundaries. Phys. Rev., A26, 1788-1790.

CHAPTER V. Decimated Amplitude Equations in Turbulence Dynamics

Robert H. Kraichnan

ABSTRACT

A scheme of decimation, or statistical interpolation, is developed for stochastic systems which exhibit dynamical and statistical symmetries among groups of modes. The end product is generalized Langevin equations for a sample set of explicitly followed modes. All the other modes are represented by the random forcing amplitudes in the Langevin equations. The random forcing is constrained by realizability inequalities and by appeal to the statistical symmetries. The latter yield expressions for moments of the joint probability distribution of forcing amplitudes and sample-set amplitudes in terms of moments of the sample-set amplitudes alone. This permits integration of the dynamical equations for an ensemble of sample-set amplitudes. Converging approximation sequences may be generated by systematically enlarging the set of constraints.

The decimation scheme (DS) bridges among several kinds of attack on the turbulence problem: moment hierarchy equations, renormalized perturbation theory (RPT), renormalization group methods, and direct modeling of large systems by systems with fewer modes. The DS does not appeal to perturbation theory, but in the limit of strong decimation it can be analyzed perturbatively. The direct-interaction approximation and other RPT approximations may thereby be obtained from appropriate moment constraints. One feature of the DS is that invariance to random Galilean transformation, a severe problem in the RPT treatment of turbulence, can be directly assured by constraints which relate 3rd- and 4th-order moments.

1. INTRODUCTION

Turbulence in an incompressible Navier-Stokes fluid exhibits an interplay of randomness and order which challenges both measurement and theoretical treatment.[1] There are characteristic structures such as shear layers and wall vortices, but these objects are plastic in form. There is randomness in shape and size of the typical structures as well as in position, orientation, and time of appearance. Already at modest Reynolds numbers the dynamics are strongly nonlinear. At high Reynolds numbers, huge numbers of modes contribute to the dynamics and must be represented in direct numerical simulations of the flow. Randomness arises from sensitivity to initial conditions and to perturbations; from sometimes violent instabilities; and from the interplay of irreducibly large numbers of degrees of freedom.

Statistical description of some sort seems called for by the complicated structure of turbulence, the large number of degrees of freedom, and the impossibility of reproducing or predicting the full details of the velocity field. It is known empirically that well-chosen averages over ensemble, space, or time can be stable and smooth where the detailed velocity field is not. Turbulence statistics are commonly described in terms of moments of the velocity field because moments seem relatively simple to manipulate experimentally and theoretically. Moreover, low-order moments yield directly some quantities of physical interest such as the kinetic energy spectrum. However, the typically intermittent and plastic structures of turbulence probably cannot be well specified by low-order moments, and this suggests that moments may not be the most appropriate descriptors of the nonlinear dynamics. Moreover, moments provide a mathematically incomplete description of highly intermittent probability distributions.[2] The success of a theoretical treatment of turbulence may be intimately tied to the use of apt statistical descriptors.

A variety of methods have been described for the construction of approximations to turbulence statistics. Two questions are: (i) does a given method provide convergent sequences of approximants; (ii) are any of the approximants physically faithful enough to be acceptable and at the same time feasible to compute? Two kinds of approximation which have received attention are renormalized perturbation theory (RPT) for moments, and truncations or other closures of the hierarchy of moment equations built from the Navier-Stokes equation. Perturbation expansions in powers of nonlinearity (or Reynolds number) probably are divergent at all Reynolds numbers, and the divergence is not cured by renormalization.[3] Closure of the moment hierarchy by discard of cumulants above a given order is also probably divergent, but there are alternative, convergent truncations (see Sec. 2).

The labor of computing any of the named approximation methods rises very rapidly with order. Fourth-order moments already are very complicated to handle. In the case of RPT, there exists a truncation with unique properties at the level of 2nd-order moments, the direct-interaction approximation (DIA). The DIA can be represented as the exact dynamics of a stochastic model, and thereby has some basic physical consistency properties, including conservation of energy and positivity of power spectra.[4] Higher-order truncations of RPT do not guarantee positivity of power spectra and have the potential for explosive divergence when integrated in time. The approximation of setting 4th-order cumulants to zero in the moment hierarchy also has this potential; this is not a consequence of ultimate divergence of the cumulant expansion at high orders but is due to a failure to impose conditions on moments which assure an underlying positive-definite probability distribution.

The present paper describes a decimation scheme (DS) designed to exploit statistical redundancy among the modes of turbulent flows too large for full direct simulation. It is applicable generally to dynamical systems which contain large numbers of modes and which exhibit dynamical and statistical symmetries among groups of modes. The end product is a class of generalized Langevin equations for selected sets of explicitly followed modes. Constrained stochastic forcing terms in the Langevin equations represent the effects of all the other modes. The statistical constraints on the forcing terms typically are moment constraints, but other statistical descriptors may be used instead.

The following example illustrates the scheme. Consider the system

$$dx/dt = A_x \sum_{n=1}^{N} y_n z_n, \quad dy_n/dt = A_y x z_n, \quad dz_n/dt = A_z x y_n, \tag{1.1}$$

where $A_x + A_y + A_z = 0$. There are two quadratic constants of motion,

$$E = x^2 + \sum (y_n{}^2 + z_n{}^2), \quad F = A_y x^2 - A_x \sum y_n{}^2,$$

and a Liouville property which yields absolute equilibrium distributions of the form $\exp(-\beta E - \alpha F)$. Suppose N is large and economical modeling is sought. Consider the three-mode model

$$dx/dt = A_x' yz, \quad dy/dt = A_y' xz, \quad dz/dt = A_z' xy. \tag{1.2}$$

There is no choice of the A' which can reproduce the equipartition behavior of the original system in equilibrium. Whether the A' are chosen to conserve $E' = x^2 + y^2 + z^2$ or $E' = x^2 + N(y^2 + z^2)$, a canonical distribution with $\alpha = 0$ gives $\langle x^2 \rangle = \langle E' \rangle / 3$, while for the original system $\langle x^2 \rangle = \langle E \rangle / (1 + 2N)$.

In the DS, statistical symmetry is assumed with respect to n, and the modeling is

$$dx/dt = A_x \sum_{n=1}^{S} y_n z_n + q,$$

$$dy_n/dt = A_y x z_n, \quad dz_n/dt = A_z x y_n \quad (n \le S), \tag{1.3}$$

where x, y_n, z_n $(n \le S)$ are a sample set of modes and $q = A_x \sum_{n=S+1}^{N} y_n z_n$. The amplitudes y_n, z_n $(n > S)$ are not treated explicitly. Instead, the stochastic forcing q is constrained by statistical relations which express the symmetry with respect to n. In particular, the constraint

$$\langle q(t) x(t) \rangle = [(N-S)/S] \sum_{n=1}^{S} A_x \langle x(t) y_n(t) z_n(t) \rangle \tag{1.4}$$

gives conservation of

$$\langle E \rangle = \langle x^2 + (N/S) \sum_{n=1}^{S} (y_n{}^2 + z_n{}^2) \rangle, \quad \langle F \rangle = \langle A_y x^2 - (N/S) A_x \sum_{n=1}^{S} y_n{}^2 \rangle.$$

Solutions to (1.3) are sought by starting with an ensemble of initial x, y_n, z_n, q values $(n \le S)$ with symmetrical statistics; integrating to find x, y_n, z_n $(n \le S)$ amplitudes in each realization at the next time step; and then using an appropriate algorithm to construct an explicit ensemble of q amplitudes at the new time step which satisfy (1.4) and whatever other constraints are imposed. The process is then repeated for successive time steps. Alternatively, iteration methods may be used to solve for the amplitudes at a number of time steps simultaneously.

Now consider the conservation and equilibrium properties of the model (1.3), with (1.4) among the imposed constraints. A contribution $\langle x^2 + \sum_{n=1}^{S} (y_n{}^2 + z_n{}^2) \rangle$ to the conserved $\langle E \rangle$ may be regarded as coming from the conservatively interacting subsystem comprising x, y_n, z_n $(n \le S)$, and a remainder $[(N-S)/S] \sum_{n=1}^{S} \langle y_n{}^2 + z_n{}^2 \rangle$ from an effective external reservoir system, coupled via q. The subsystem in the absence of q has a Liouville property which yields equipartition of $\langle E \rangle$ in equilibrium $(\alpha = 0)$. If the reservoir system has any dynamics whatever which give a Liouville property to the entire system, then the exact equipartition ratio $\langle x^2 \rangle = \langle E \rangle / (1 + 2N)$ associated with (1.1) is preserved. If the constraints imposed on q do not give an effective Liouville property then the exact equipartition ratio is not assured, but exotic q statistics are needed to cause much error in the ratio. The essential difference from the model (1.2) is that the effective size of the system is still $2N + 1$, although only $2S + 1$ modes are treated explicitly.

The difference between (1.2) and (1.3) is already apparent in the case $N = 2$, $S = 1$. Fig. 1 shows the results of numerical integrations of (1.1), (1.2), and (1.3) in this case, with values $A_x = .5$, $A_y = .75$, $A_z = -1.25$, $A' = A$, and initial values $\langle x^2 \rangle = 0$, $\langle y_n{}^2 \rangle = 1$, $\langle z_n{}^2 \rangle = 1$. The constraint (1.4) was augmented by

$$\langle q(t)q(t')\rangle = A_x{}^2\langle y_1(t)z_1(t)y_1(t')z_1(t')\rangle, \tag{1.5}$$

an obvious consequence of symmetry for N = 2, which serves to fix the covariance of q at each time step. With the chosen initial values, (1.1) goes to an equilibrium with $\alpha \neq 0$, more complex than simple equipartition of $\langle E\rangle$. Further details are given in Sec. 6.

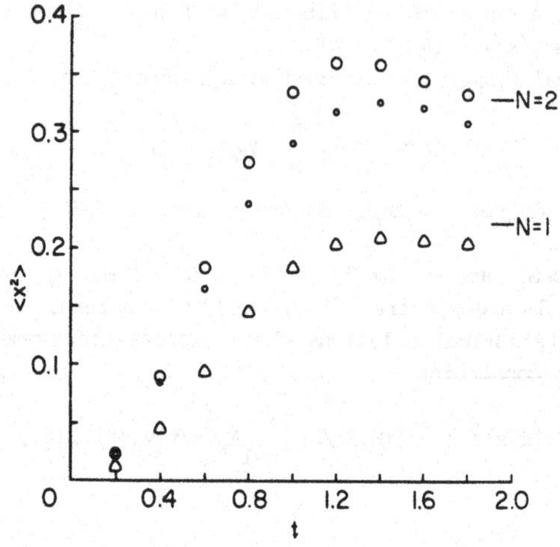

FIG. 1. Evolution of $\langle x^2\rangle$ for (i) N = 2 under (1.1) [small circles]; (ii) N = 2, S = 1 under (1.3) [large circles]; (iii) N = 1 under (1.1), identical with (1.2) [triangles]. The horizontal line segments give the absolute equilibrium levels for N = 2 and N = 1.

The DS may be viewed as an economizing technique for numerical computation. At the same time, it extends and illuminates some existing theoretical approaches to stochastic dynamical systems and provides bridges among them. In particular, the DS links RPT and the moment-equation hierarchy. Appeal to perturbation theory is not intrinsic to the DS, but when the decimation is strong $(N\to\infty)$, the equations can be analyzed by treating explicit terms (yz in the example) as small perturbations on the stochastic forcing terms (q in the example. The result is that the solution under a given set of moment constraints is expressed in terms of a particular set of renormalized perturbation terms. This will be discussed in detail in Secs. 9 and 10.

There is a obvious kinship between the DS and and renormalization-group ideas. Modes are weeded out in the DS, rather than removed from one end or the other of the wavenumber spectrum. The weeding can be performed all in one step rather than by small incremental transformations, and no appeal need be made to perturbation treatments. The incremental decimation typical of applications of renormalization-group ideas may be handled as specialization of the DS.

The theory and implementation of the DS lean heavily on systematic use of the conditions that the probability density be nowhere negative, particularly in the form of realizability inequalities for moments. If these inequalities are ignored, formal solutions of equations for moments may be physically meaningless. The common requirement that power spectra be non-negative is an example of realizability inequalities but can be insufficient to insure physically acceptable solutions. Positive-definite power spectra and certain other realizability conditions are automatically built into the DS. The complete set of realizability conditions is the key to obtaining converging sequences of approximation by the systematic imposition of increasing numbers of statistical constraints.

Sections 2 and 3 of this paper collect and review some facts about the approximate representation of probability distributions and about realizability inequalities. Section 4 describes some general methods of constructing convergent approximant sequences. The material in Sec. 2-4 is mostly not new, but it is developed in some detail because these matters have not been recognized to the extent they should in the turbulence literature. Section 5 discusses and illustrates numerical methods for explicitly constructing an ensemble of amplitudes which satisfy given statistical constraints. With this background in place, the development of the DS itself starts in Sec. 6. The worked examples of the DS are confined to simple, idealized systems. No attempt is made in the present paper to compute the Navier-Stokes system itself, but an outline is given of one way the DS may be applied to homogeneous turbulence.

2. FINITE REPRESENTATIONS OF A PROBABILITY DISTRIBUTION

The probability distribution (PD) of a set of N random variables obeying N independent algebraic equations consists of some number of singular spikes. The one-time PD of N functions obeying N independent first-order differential equations in time will be continuous or singular depending on the initial PD. If it is continuous, it may exhibit ever-finer small-scale structure (fine-graining) as time progresses. Finally, consider the joint PD of the NT random variables defined by the values, at T instants, of the N time-functions. This PD is NT-dimensional, but is confined to an N-dimensional surface, weighted according to the initial conditions. In any of the three cases, practicable computability may require that the PD be represented by a relatively small set of numbers. Such a representation may not distinguish between PD's which are continuous and PD's which consist of a sufficient number of spikes, and does not resolve fine-graining.

Let x be a single real random variable in $(-\infty,\infty)$ with the nowhere negative probability density $\rho(x)$. Let

$$\langle f(x)\rangle \equiv \int f(x)\rho(x)\,dx.$$

Two possible representations of $\rho(x)$ are by its moments $\langle x^n\rangle$ or by values at a discrete set of points x_n. These two descriptions come together in a natural way in the theory of Gaussian integration with weight $\rho(x)$. Moment representation can also be stated in terms of complete sets of continuous orthogonal functions, and this is the simplest starting point.

Let $w(x)$ be a positive-definite weight $[\int w(x)dx = 1]$ and define the orthonormal polynomials $p_n(x)$ of degree n by

$$\int p_n(x)p_m(x)w(x)\,dx = \delta_{nm}. \tag{2.1}$$

Then $\rho(x)$ has the formal approximants

$$\rho_M(x) = w(x)\sum_{n=0}^{M} b_n p_n(x), \tag{2.2}$$

where $b_n = \langle p_n(x)\rangle$. Each $\langle p_n(x)\rangle$ is constructed from the $\langle x^m\rangle$ for $0 \le m \le n$. The approximants converge in mean square if $\int [\rho(x)]^2 [w(x)]^{-1} dx$ exists. If, also, the orthonormal functions $[w(x)]^{1/2} p_n(x)$ are complete, then[5]

$$\int [\rho(x) - \rho_M(x)]^2 \frac{dx}{w(x)} \to 0 \quad (M \to \infty) \tag{2.3}$$

and, for any $f(x)$, a simple use of Schwarz' inequality yields

$$[\langle f(x)\rangle - \langle f(x)\rangle_M]^2 \le \int [\rho(x) - \rho_M(x)]^2 \frac{dx}{w(x)} \int [f(x')]^2 w(x')\,dx'. \tag{2.4}$$

The sufficient conditions for completeness of the orthonormal functions are that $w(x)$ vanish exponentially or faster at ∞ but not vanish at finite x. The moments $\langle x^n\rangle$ uniquely specify $\rho(x)$ provided $\rho(x)$ vanishes exponentially or faster at ∞ [$\langle x^n\rangle$ grows no faster than $n!$ as $n\to\infty$], so that completeness of the orthogonal functions is consistent with the existence of the integrals (2.3). This is essentially Carleman's criterion.[2] If $\rho(x)$ falls off more slowly, the weighted moments $\langle w^{1/2}(x) x^n\rangle$ uniquely specify $\rho(x)$. This follows from repeating the analysis starting at (2.2) but with $\rho'(x) = w^{1/2}(x)\rho(x)$. Eq. (2.4) can then be replaced by

$$[\langle f(x)\rangle - \langle f(x)\rangle_M]^2 \le \int [\rho(x) - \rho_M(x)]^2 \frac{dx}{\xi(x)} \int [f(x')]^2 \xi(x')\,dx', \tag{2.5}$$

where $\xi(x)$ is any positive weight which vanishes weakly enough at ∞ to make the first integral on the right-hand side exist. Again, that integral goes to zero as $M \to \infty$. An example of a PD which violates Carleman's criterion is the log-normal distribution, for which $\ln\langle x^n\rangle \propto n^2$.

The representation of $\langle f(x)\rangle$ by Gaussian integration can be illuminatingly formulated in terms of the orthogonal polynomials $Q_r(x)$, with leading term x^r, which satisfy

$$\langle Q_r(x) Q_s(x)\rangle = W_r \delta_{rs}, \tag{2.6}$$

where W_r is a normalization factor defined by (2.6). The Q_r also satisfy the discrete orthogonality relations[6]

$$\sum_{i=1}^{M+1} Q_r(x_i) Q_s(x_i)\,\rho_i = W_r \delta_{rs} \quad (r,s \le M), \tag{2.7}$$

where ρ_i is given by

$$\rho_i = 1/\sum_{r=0}^{M} [Q_r(x_i)]^2 / W_r \tag{2.8}$$

and the x_i $(i = 1,2,\ldots M+1)$ are the zero's of $Q_{M+1}(x)$. A standard recursion relation among the Q_r leads to the extended orthogonality condition

$$\sum_{i=1}^{M+1} Q_r(x_i) Q_s(x_i)\,\rho_i = 0 \quad (r+s \le 2M+1,\ r \ne s) \tag{2.9}$$

It is easy to show from (2.6)-(2.9) that

$$\langle P_r(x)\rangle = \sum_{i=1}^{M+1} P_r(x_i)\,\rho_i \tag{2.10}$$

for all polynomials $P_r(x)$ of degree $r \le 2M+1$. The equality in (2.10) of averages over the arbitrary original $\rho(x)$ and an $(M+1)$-spike singular PD shows that the values of moments to finite order do not determine whether a PD is continuous.

Eq. (2.10) and a use of Schwarz's inequality, noting $\int \rho(x)dx = 1$, yield

$$[\langle f(x)\rangle - \langle f(x)\rangle_M]^2 \le \int\{\rho^{1/2}(x)f(x) - [\rho^{1/2}(x)f(x)]_M\}^2 dx, \tag{2.11}$$

where $\langle f(x)\rangle_M = \sum_{i=1}^M f(x_i)\rho_i$ and $[g(x)]_M$ denotes a truncation of the expansion of $g(x)$ in the orthogonal functions $\rho^{1/2}(x)Q_n(x)$. The right-hand side of (2.11) goes to zero as $M \to \infty$ (completeness of the orthogonal functions with respect to $\langle\rangle$ operations) if the moments $\langle x^n\rangle$ increase no more rapidly than $n!$ as $n \to \infty$. If $\rho(x)$ is continuous this corresponds to the condition that $\rho(x)$ vanish exponentially or faster at ∞.

If $\rho(x)$ consists of spikes, the completeness property again follows from the stated growth condition on moments. This may be seen by writing $f(x)$ as a Fourier integral, noting completeness of the Fourier functions, and noting that the Taylor expansion of $\langle \exp(ikx)\rangle$ has a nonzero radius of convergence in x when the growth condition is satisfied. If $\rho(x)$ consists exclusively of S spikes, then W_n vanishes for $n > S$, and the $\rho^{1/2}(x)Q_n(x)$ for $n \le S$ form a complete set for expansions within $\langle\rangle$.

If the moments increase with n faster than n!, a complete description of $\rho(x)$ via Gaussian integration can be constructed from the weighted moments $\langle w(x)x^n\rangle$, where $w(x)$ is a positive weight such that $\rho'(x) = \rho(x)w(x)$ vanishes exponentially or faster at ∞, or the weighted moments grow no faster than n!. The construction proceeds via the polynomials $Q_n'(x)$ orthogonal with respect to $\rho'(x)$.

Representation by continuous orthogonal functions [(2.1) - (2.5)] is directly extensible to the multidimensional PD $\rho(x_1,x_2,\dots x_N)$ and corresponding weight w, with associated vector-indexed orthogonal polynomials. The investigation of completeness[5] is extended by performing analytic continuation along rays in the space of the x_n. It leads to the sufficient condition that w vanish exponentially or faster at ∞ along any ray.

There are difficulties in extending the Gaussian integration procedure to many dimensions, both because of problems in classifying the intersections of the nodal surfaces of orthogonal polynomials in many variables and because of practical considerations of the number of spikes involved. Heuristic schemes for constructing general finite-ensemble approximations to multi-dimensional PD's will be illustrated in Section 5.

The expansion (2.2) is distinct from cumulant expansion, even if $w(x)$ is Gaussian. The cumulant expansion of $\rho(x)$ is defined by

$$\phi(u) = \exp[\sum_{n=0}^{\infty} C_n(iu)^n/n!], \tag{2.12}$$

where

$$\phi(u) = \langle e^{iux}\rangle, \quad \rho(x) = \int_{-\infty}^{\infty} \phi(u)e^{-iux}du. \tag{2.13}$$

The moments are given by

$$\langle x^n\rangle = (-i)^n[d^n\phi(u)/du^n]_{u=0}. \tag{2.14}$$

The principal qualititative difference is that the approximants $\rho_M(x)$ defined by (2.2) converge to $\rho(x)$ in mean square and yield approximants to the characteristic function $\phi(u)$, defined by (2.13), which converge uniformly.[5] In contrast, the approximants to $\phi(u)$, obtained by truncating the sum in (2.12), in general diverge at ∞, and the $\rho(x)$ approximants then obtained from (2.13) in general do not exist. Moreover, the sequence C_n itself can diverge when $\rho(x)$ is healthy. An example is

$$\rho(x) = (2\pi)^{-1/2}x^2\exp(-x^2/2),$$

for which

$$C_0 = 1, \quad C_{2n+1} = 0, \quad C_2 = 3/2, \quad C_{2n} = -(-1)^n[(2n)!/n] \quad (n > 1).$$

3. REALIZABILITY INEQUALITIES

The necessary and sufficient conditions that $\rho(x)$ be non-negative everywhere are $\langle F(x)\rangle \geq 0$, where $F(x)$ is any positive-definite test function such that the integral exists. Define $f(x)$ by $[f(x)]^2 = F(x)$. If the moments $\langle x^n\rangle$ grow no faster than n! as $n\to\infty$, then the expansion of $f(x)$ in the orthogonal functions $[\rho(x)]^{1/2}Q_n(x)$ defined by (2.6) is complete with respect to the operation $\langle\rangle$. That is, if $r(x)$ is the remainder in the expansion of $f(x)$, $r(x)$ does not contribute to $\langle F(x)\rangle$. If, also, $\rho(x) > 0$ everywhere, then $r(x) = 0$. If the expansion of $f(x)$ is carried out, the necessary and sufficient conditions for non-negativity take the form

$$W_n \geq 0 \quad (n=0,1,2,\dots\infty), \tag{3.1}$$

where W_n is defined by (2.6). These relations constitute the realizability conditions for the $\langle x^n\rangle$. They are equivalent to the conditions that the matrices $M_{rs} =\langle x^{r+s}\rangle$ $(0\leq r,s\leq n)$ have no negative eigenvalues for any n.

Suppose $\rho(x)$ is expanded in the form (2.2). $[Q_n(x)]^2$ is a polynomial of degree 2n and therefore is orthogonal to the $p_r(x)$ for $r > 2n$. Thus (3.1) for given n is a constraint on the b_r only for $r\leq 2n$. If $\rho(x)$ contains a number of spikes, (3.1) remains the necessary and sufficient condition for non-negativity.

The first three $Q_n(x)$ defined by (2.6) are

$$Q_0(x) = 1, \quad Q_1(x) = x - \langle x\rangle,$$

$$Q_2 = x^2 - \langle x^2\rangle - [x-\langle x\rangle][\langle x^3\rangle - \langle x\rangle\langle x^2\rangle]/[\langle x^2\rangle - \langle x\rangle^2].$$

For n = 1 and n = 2, (3.1) then gives $x^2 - \langle x\rangle^2 \geq 0$ and

$$\langle x^4\rangle - \langle x^2\rangle^2 \geq [\langle x^3\rangle - \langle x\rangle\langle x^2\rangle]^2/[\langle x^2\rangle - \langle x\rangle^2]^2.$$

In the case where the moments rise more rapidly than n! [$\rho(x)$ vanishes more weakly than exponentially at ∞], (3.1) is replaced by the necessary and sufficient conditions for non-negativity

$$\langle w(x)\,[Q_n'(x)]^2\rangle \geq 0 \quad (n=0,1,2,\dots\infty), \tag{3.2}$$

where the Q_n' are orthogonal with respect to $\rho'(x) = \rho(x)w(x)$.

Eqs. (3.1) and (3.2) are easily extended to vector-valued x: if $\rho(x)$ vanishes exponentially or faster at ∞ in all directions [$\langle(k\cdot x)^n\rangle$ rises no faster than n! for all vectors k as $n\to\infty$], then positivity in mean square of all multi-dimensional orthogonal or weighted orthogonal polynomials is the necessary and sufficient condition that $\rho(x)$ be positive-definite.

The orthogonal polynomials Q defined by (2.6) are nonunique for vector-valued x. In the case of two scalar variables x and y, the first three Q(x,y) may be taken as

$$Q_{00}(x,y) = 1, \quad Q_{10}(x,y) = x', \quad Q_{01}(x,y) = y' - x'\langle x'y'\rangle/\langle x'^2\rangle,$$

where $x' = x-\langle x\rangle$, $y' = y-\langle y\rangle$. Equation (3.1) for $Q_{01}(x,y)$ then gives the Schwarz inequality $\langle x'y'\rangle^2 \leq \langle x'^2\rangle\langle y'^2\rangle$.

A problem of practical interest is the extent to which truncated sets of the inequalities (3.1) or (3.2) assure realizability of finite sets of moments or weighted moments. Here a set of moment values is said to be realizable if there is some positive-definite $\rho(x)$ which yields the given values. In the case of a single variable x, the Gaussian-integration construction gives a simple answer: if (3.1) is satisfied for $n \le M$, then the Gaussian weights ρ_i are all non-negative; the spike distribution is positive-definite and yields moments of all orders which satisfy (3.1) to $n \to \infty$.

The Gaussian-integration construction does not extend simply to vector x. If (3.1) is satisfied, with inequality signs only, up to some order of polynomial, it is possible to construct moments of higher order to extend (3.1) to indefinitely higher orders.[7] However, in general there is the possibility that the moments so constructed will grow faster than n!.

The general positive-definite polynomial P_{2M} of degree 2M in m real variables cannot be written as a sum of squares of real polynomials for $M > 2$ if $m = 3$ or for $M > 1$ if $m > 3$.[8] Therefore in these cases satisfaction of (3.1) for all Q of degree $n \ge M$ does not assure non-negativity of $\langle P_{2M}(x) \rangle$. It should be noted also that truncations of a complete expansion of a positive-definite $\rho(x)$ in orthogonal functions will usually not be positive-definite.

Here is a simple counter-example which shows that moments to order 4 which satisfy (3.1) to n = 2 need not be realizable. Consider the equations

$$x - yz = 0, \quad y - zx = 0, \quad z - xy = 0, \tag{3.3}$$

of which the solutions are (x,y,z) = (0,0,0), (1,1,1), (1,-1,-1), (-1,1,-1), or (-1,-1,1). Consider the moment values

$$\langle x^2 \rangle = \langle y^2 \rangle = \langle z^2 \rangle = \langle xyz \rangle = \langle x^2y^2 \rangle = \langle y^2z^2 \rangle = \langle z^2x^2 \rangle = 2,$$
$$\langle x^4 \rangle = \langle y^4 \rangle = \langle z^4 \rangle = 8, \tag{3.4}$$

with all other moments zero to 4th order. To degree 2, the orthogonal Q may be taken as

$$Q_{000} = 1, \quad Q_{100} = x, \quad Q_{010} = y, \quad Q_{001} = z,$$
$$Q_{011} = yz - x, \quad Q_{101} = zx - y, \quad Q_{110} = xy - z, \tag{3.5}$$
$$Q_{200} = x^2 + x^2 + y^2 - 6, \quad Q_{020} = y^2 - x^2, \quad Q_{002} = z^2 - (x^2 + y^2)/2.$$

These Q satisfy (3.1) and the moments satisfy (3.3) in mean square [$\langle (x - yz)^2 \rangle = \langle (Q_{011})^2 \rangle = 0$, etc]. But the moments clearly are not realizable since they cannot be constructed by averaging of the solutions of (3.3). It is easy to show that there are no finite values for 5th and 6th order moments which, when adjoined to the values (3.4), extend (3.1) to Q of degree 3. Also, if an auxiliary variable s is added, and $s - (x^2 + y^2 + z^2) = 0$ is included in (3.3), there are no finite values for all moments to 4th order which include the values (3.4) and satisfy (3.1) and the mean square of (3.3). With s added, the only moment values to 4th order which satisfy (3.1) and the mean square of (3.3) represent positive-weight averages of the 5 solutions of (3.3).

4. HIERARCHY EQUATIONS AND APPROXIMANT SEQUENCES

Suppose it is desired to integrate equations of the form

$$L_i(y,t) \equiv dy_i/dt + F_i(y,t) = 0 \tag{4.1}$$

under statistically specified initial conditions or other statistically described constraints. Here the y_i are a set of real variables and each F_i is a function of all the y's. In the case of the Navier-Stokes equation, the y_i can be the amplitudes of a modal expansion in space and the F_i are then polynomials of 2nd degree. The evolution of the many-time moments of an ensemble of solutions of (4.1) is described by the moment hierarchy equations

$$\langle P_n(y)L_i(y,t)\rangle = 0, \tag{4.2}$$

where $P_n(y)$ is a general nth-degree many-time polynomial in the y's. If the F's are polynomials of degree D, then (4.2) gives the time derivatives of a many-time y moment of degree n as linear expressions in y moments of degree $n+D-1$ and lower. Thus the infinite sequence of many-time y moments can be generated by integrating the infinite set of equations (4.2) forward in time from initial values of the moments of all orders.

A finite approximation to (4.1) can be written in the form

$$L_{is}(y) \equiv y_i(t_s) - y_i(t_{s-1}) + \Delta t F_i[y(t_{s-1})] = 0, \tag{4.3}$$

where $t_s = s\Delta t$. The finite approximation to (4.2) is then

$$\langle P_n(y)L_{is}(y)\rangle = 0. \tag{4.4}$$

Alternatively, more elaborate differencing schemes can be taken, or (4.1) can be replaced by equations for the amplitudes in a modal expansion of the y_i in time.

Suppose that the initial moment values represent a $\rho[(y(t_0)]$ which vanishes exponentially or faster at ∞ so that the initial moment representation is complete. Then, if (4.3) preserves the exponential decay property, (4.4) provides a formal construction of all the many-time moments. Unless also the realizability inequalities are satisfied to all orders, moments which satisfy (4.4) do not represent an average with positive weights over solutions of (4.3). If the realizability inequalities are satisfied to all orders, then the entire hierarchy (4.4) follows from the mean-square equation of motion[9]

$$\langle [L_{is}(y)]^2\rangle = 0, \tag{4.5}$$

by virtue of the Schwarz inequality

$$\langle P_n(y)L_{is}(y)\rangle^2 \le \langle [P_n(y)]^2\rangle\langle [L_{is}(y)]^2\rangle.$$

Conversely, if (4.5) is satisfied and any equation in the hierarchy (4.4) is not, then the moments represent a PD which is not positive-definite. Any approximation which satisfies (4.4) for all P_n of degree $\le D$ satisfies (4.5). Therefore, such an approximation either is an exact solution of the full hierarchy or violates realizability.

The hierarchy (4.4) is in fact a special subset of the realizability inequalities. Equation (4.3) implies that some of the eigenvalues of the moment matrices of the form introduced after (3.1) vanish, and (4.4) expresses the extremal property of the associated eigenvector polynomials $L_{is}(y)$. Thus, for any variation $\delta L_{is}(y) = \varepsilon P_n(y)$, where ε is an infinitesimal, the realizability inequality

$$\langle [L_{is}(y) + \delta L_{is}(y)]^2 \rangle \geq 0 \tag{4.6}$$

immediately yields (4.4). The full set of realizability inequalities, which are compactly expressed by (3.1), covers all positive test functions, not just those in the neighborhood of $[L_{is}(y)]^2$.

If (4.3) leads to PD's which vanish more weakly than exponentially at ∞, the unweighted moment hierarchy (4.4) does not provide a complete description. In this case, the analysis can be repeated with weighted moments. Eq. (4.4) is replaced by

$$\langle w(y) P_n(y) L_{is}(y) \rangle = 0, \tag{4.7}$$

where w is such that $\rho(y)w(y)$ falls off with exponential strength as ∞ is approached in any direction in the space of the $y_i(t_s)$.

Finite (but not necessarily practicable) approximants to the realizable solutions of (4.4) can be constructed by first expanding the many-time probability density $\rho[y(t_0), y(t_1), \ldots]$ in orthogonal functions with respect to a weight $w[(y(t_0), y(t_1), \ldots]$, as in (2.2). Each coefficient amplitude is an expression in a finite set of moments. Eq. (4.4) thereby is transformed into an infinite hierarchy of linear equations in the amplitudes of the expansion. Approximants may then be constructed, in more than one way, by (i) truncating ρ to a finite set of amplitudes; (ii) truncating the hierarchy equations; (iii) imposing all or some of those realizability inequalities which project only on retained amplitudes [see the remarks after (3.1].

One way of constructing the approximants is a straightforward Galerkin procedure. Equation (4.4) is imposed only for P which belong to some chosen set of polynomials orthogonal on w. In the resulting equations, all terms involving a coefficient amplitude which does not correspond to the chosen set are omitted. The result is a set of equations which uniquely determine the surviving many-time amplitudes from the surviving initial-value amplitudes. The realizability inequalities of the form (3.1) are imposed for all Q which involve only initial values of the y's and are linear combinations of the chosen set of orthogonal polynomials on w. The resulting solutions may violate (3.1) for Q which are linear combinations of the chosen set of many-time orthogonal polynomials.

An alternative way starts also with a chosen set of polynomials orthogonal on w, and the corresponding chosen set of amplitudes. But now only those equations (4.4) are retained which contain no amplitude outside the chosen set. These equations are insufficient in number to determine the chosen amplitudes uniquely from the chosen initial amplitudes. Among their solutions are truncations (to the chosen set of amplitudes) of the exact solutions of the infinite hierarchy (4.4). The solutions set is reduced in size by imposing realizability inequalities and some appropriate optimization criterion. The realizability inequalities are imposed now on the many-time amplitudes by asserting (3.1) for all Q which are linear combinations of the entire chosen set of polynomials. If the optimization criterion is independent of the particular choice of weight w, then the final solution set will also be independent of this choice.

The convergence of approximant sequences formed in either of these ways requires first that there exist ensembles of solutions of (4.3) which satisfy the conditions for completeness of the orthogonal expansion with respect to w. A second requirement is that these ensembles be stable in the sense that small perturbations of the equations for the expansion amplitudes produce small changes in the amplitudes. If both requirements are met, then, for initial conditions corresponding to a stable ensemble, successive enlargements of the chosen set yield a sequence of approximants whose limit is the complete moment specification of a positive-definite PD of solutions of (4.3). Similar convergence arguments cannot be made for truncations of an expansion in cumulants, for the reasons noted at the end of Sec. 2.

Converging approximant sequences (again not necessarily practicable) can also be set up in which the PD is exactly positive-definite at each stage and successive approximants reduce the magnitude of the mean-square errors $\langle [L_{is}(y)]^2 \rangle$. One way is to take

$$\rho_M(y) = |\psi_M(y)|^2,$$

$$\psi_M(y) = \sum_{n=0}^{M} [w(y)]^{1/2} c_n p_n(y), \tag{4.8}$$

where the $[w(y)]^{1/2} p_n(y)$ are a complete orthonormal set in the space of the $y(t_s)$ and n is a vector index. The moments of the $y(t_s)$ are then quadratic forms in the c's, and the latter are determined variationally to minimize some linear combination, with positive coefficients, of the left-hand sides of (4.5). Another kind of approximant sequence with full realizability at each stage can be constructed from truncations of an expansion of the $y(t_s)$ in powers of Gaussian random processes.[7]

The idealized schemes for constructing approximant sequences outlined above all express the approximant $\rho_M(y)$ in terms of continuous orthogonal functions. The rate of convergence therefore depends on how smooth is the exact $\rho(y)$. The solutions of (4.3) all lie on subspaces of the full y space so that the exact $\rho(y)$ cannot be continuous. The rate of convergence at a given stage of expansion in continuous orthogonal functions depends on how many singular surfaces of the exact $\rho(y)$ lie between zeros of the highest orthogonal functions retained. To take a single variable, the expansion of one δ-function in continuous orthogonal functions converges slowly, but the expansion of a sequence of δ-functions with smoothly varying amplitude envelope can converge rapidly until an order of expansion is reached in which the zero-spacing of the orthogonal functions becomes as small as the spacing of the δ-functions.

5. EXPLICIT REALIZATION UNDER CONSTRAINTS

Implementation of the DS illustrated in Sec. 1 involves the general problem of explicitly realizing ensembles of stochastic variables which satisfy moment constraints or other statistical constraints. If the ensemble contains R realizations of a number- or vector-valued variable x, the constraints can be written in the form

$$\langle f_i(x) \rangle = 0 \quad (i = 1, 2, \ldots C), \tag{5.1}$$

Here C is the number of constraints and

$$\langle f(x) \rangle \equiv \sum_{\nu=1}^{R} \rho_\nu f(x_\nu),$$

where ρ_ν is a weight for the νth realization x_ν. The present Section discusses some algorithms for the explicit construction of ensembles satisfying constraints of the form (5.1).

R is finite in an explicit construction, and an immediate problem is how to handle the fluctuation effects associated with finite sized samples. For example, suppose (5.1) consists of four constraints $\langle x\rangle = 0$, $\langle x^2\rangle = 1$, $\langle x^3\rangle = 0$, $\langle x^4\rangle = 3$, appropriate to a Gaussian ensemble for a number-valued variable x. A typical R-sized sample from an infinite Gaussian distribution will show fluctuations from these values. In particular, the variance in the expectation of the variance is

$$\langle(\langle x^2\rangle_R - \langle x^2\rangle_\infty)^2\rangle_\infty = R^{-1}(\langle x^4\rangle_\infty - \langle x^2\rangle_\infty^2),$$

if $\rho_\nu = 1/R$ (all ν).

Thus the constraints on a typical finite sample taken from an infinite distribution which obeys constraints of the form (5.1) are stochastic. In this Section, the problem is resolved by requiring the R-sized sample to satisfy precisely the constraints for the infinite distribution. The sample therefore is no longer typical but is a compact embodiment of chosen properties of the infinite ensemble. This is important in the application to decimated systems like the example of Sec. 1. In that example, the conservation property of the original system has an exact expression in terms of the decimated system, provided that the ensemble is infinite so the variances of all the statistically symmetric modes are exactly equal. If representation is by a finite ensemble, the conservation property can be exactly preserved by requiring that the (nontypical) finite sample of realizations obey the constraints (1.4) for the infinite ensemble.

A straightforward algorithm for satisfying (5.1) is a stochastic Newton-Raphson iteration method in which the successive linear equations

$$\langle\Delta x_{n+1} f_i'(x_n)\rangle = -\langle f_i(x_n)\rangle \qquad (i = 1, 2, \ldots C) \tag{5.2}$$

are solved for $\Delta x_{n+1} = x_{n+1} - x_n$ until the $\langle f_i(x_n)\rangle$ converge. Here n is an iteration label (the realization labels are suppressed), and $f_i'(x) = \partial f_i(x)/\partial x$. A variety of related, more sophisticated iteration schemes may be formulated. A simple and useful one includes an implicit step, replacing (5.2) by the pair of relations

$$\langle\Delta x_{n+1}^E f_i'(x_n)\rangle = -\langle f_i(x_n)\rangle, \tag{5.3}$$

$$\langle\Delta x_{n+1}(f_i'(x_{n+1}^E) + f_i'(x_n))\rangle = -2\langle f_i(x_n)\rangle, \tag{5.4}$$

$(i = 1, 2, \ldots C)$, where $x_{n+1}^E = \Delta x_{n+1}^E + x_n$. An iteration cycle consists of one use of (5.3) and one or more uses of (5.4).

If $R \geq C$, (5.2)-(5.4) are usually underdetermined linear systems.[10] The algorithms are completed by requiring that $\langle(\Delta x_{n+1})^2\rangle$ or $\langle(\Delta x_{n+1}^E)^2\rangle$ be minimized. This is accomplished by regarding the $f_i'(x_n)$ as vectors in an R-dimensional space and discarding any part of Δx_{n+1} orthogonal to these vectors.

The algorithm (5.3)-(5.4), with one use of (5.4), is rapidly convergent in many problems. Here are some elementary illustrations. In all of them the initial ensemble consists of a sample of 64 realizations of a Gaussian ensemble.

1. Single variable x with constraints $\langle x\rangle = 0$, $\langle x^2\rangle = 1$, $\langle x^3\rangle = 0$, $\langle x^4\rangle = 3$ (exact Gaussian values). The initial errors in these moments, due to finite sample size, were .1470, -.1640, .4019, -1.5829 respectively. After 4 iterations the errors were

$.93\times10^{-8}$, $.12\times10^{-6}$, $.76\times10^{-7}$, $.72\times10^{-6}$.

2. Single variable x with constraints $\langle x\rangle = 1$, $\langle x^{3/2}\rangle = 1.32934$, $\langle x^2\rangle = 2$, $\langle x^{5/2}\rangle = 3.32335$ [exact values for distribution $\rho(x) = .5\exp(|x|)$]. The initial errors in these constraints were $-.8526$, $-.5437$, $-.1164$, -2.4019. The errors after 5 iterations were all less than 10^{-6}. In this example two of the constraints are nonmoment constraints.

3. Single variable with constraints $\langle x^2\rangle = 1$, $\langle x^4\rangle = 1$. These constraints restrict x to the values ±1. The initial errors were .1474, $-.1640$, .4019, .4171. The errors after 5 iterations were all less than 10^{-4} and after 10 iterations were all less than 10^{-5}. The slower convergence in this example is directly associated with the sharpness of the value of x^2 in the target distribution.

4. Two variables x and y with initial values drawn from independent Gaussian distributions. Constraints $\langle x^2\rangle = 1$, $\langle y^2\rangle = 1$, $\langle x^4\rangle = 2$, $\langle y^4\rangle = 2$, $\langle x^2y^2\rangle = 0$. These constraints require that x and y never both be nonzero in any realization and restrict x^2 and y^2 to the sharp values 0 or 2. The initial errors were $-.1640$, .1556, $-.5829$, 2.203, .7231. The errors after 5 iterations were all less than .0015. After 10 iterations, the errors were all less than $.2\times10^{-5}$. In this example also, slower convergence is associated with the restriction to sharp values.

Now consider the problem of realizing a process $q(t)$ in a time interval $0\le t\le T$ under many-time constraints of the form (5.1). Some kind of discretizing is needed in order to compute. It can be of the equidistant form $t_s = s\Delta t$, or, alternatively, $q(t)$ can be expanded in some appropriate set of orthogonal functions in $(0,T)$ which is then truncated. In either event, algorithms like (5.2)-(5.4) can be applied to realize the constraints on the resulting set of variables. It may be possible to carry out a time-stepping process, in which constraints are applied in successive sub-intervals of $(0,T)$, each of which either adjoins or overlaps the previous sub-interval. If successful, the time-stepping can reduce the total labor. Again, discretizing within each subinterval can be equidistant or by expansion in orthogonal functions.

As an example, take scalar $q(t)$ and the covariance constraint $\langle q(t)q(t')\rangle = Q(t,t')$, where $Q(t,t')$ is a prescribed function. Here the constraint becomes trivial to satisfy if $q(t)$ is expanded in the orthogonal eigenfunctions of $Q(t,t')$;[11] the amplitudes of the eigenfunctions can be taken statistically independent and the covariance constraint reduces to the condition that the variance of each amplitude equal the corresponding eigenvalue. But it is illuminating to do the problem by time-stepping with equidistant discretizing. The constraints on the rth amplitude $q(t_r)$ are

$$\langle q(t_r)q(t_s)\rangle = Q(r\Delta t, s\Delta t) \quad (s = 0, 1, 2, \ldots, r-1), \tag{5.5a}$$

$$\langle [q(t_r)]^2\rangle = 1. \tag{5.5b}$$

These constraints may be satisfied as follows for an ensemble of R realizations: Take $q(t_0)$ as R samples from (say) a Gaussian distribution, normalized to give $\langle [q(t_0)]^2\rangle = 1$. For each $r > 0$, use (5.2) to find amplitudes $[q(t_r)]_\uparrow$ which satisfy (5.5a), considering $q(t_s)$ given. Since these constraints are linear, one iteration of (5.2) gives exact satisfaction. The amplitudes $[q(t_r)]_\uparrow$ so found are the least-squares solution of (5.5a). To satisfy (5.5b), write

$$q(t_r) = [q(t_r)]_\uparrow + [q(t_r)]_\rightarrow, \tag{5.6}$$

where $[q(t_r)]_\rightarrow$ obeys

$$\langle [q(t_r)]_\rightarrow q(t_s)\rangle = 0 \quad (s = 0, 1, \ldots r-1),$$

$$\langle [q(t_r)]_{\perp}^2\rangle = 1 - \langle [q(t_r)]_{\parallel}^2\rangle. \tag{5.7}$$

If $Q(t,t') = \exp(-|t-t'|)$, q can be realized as a Markov process and the solution to (5.6) and (5.7) can be immediately written down. It is

$$[q(t_r)]_{\parallel} = e^{-\Delta t}q(t_{r-1}), \quad [q(t_r)]_{\perp} = (1-e^{-2\Delta t})^{1/2}\xi(t_r), \tag{5.8}$$

where the $\xi(t_r)$ are any processes which obey $\langle\xi(t_r)\xi(t_s)\rangle = \delta_{rs}$. In this special example, only the $s = r-1$ constraint in (5.5a) is independent; the others are redundant. And, the part $[q(t_r)]_{\perp}$ fixed by (5.7) has variance independent of r.

The covariance $\exp(-|t-t'|)$ implies than none of the derivatives of q(t) exist in mean square. More interesting results are obtained for continuous processes. For example, consider

$$Q(t,t') = [1-(t-t')^2]\exp[-(t-t')^2/2].$$

Eq. (5.7) can be solved at each r by starting with a random variable $\xi(t_r)$, projecting out the part that correlates with the $q(t_s)$ for $s<r$, and normalizing to obtain $[x(t_r)]_{\perp}$. Thus,

$$[\xi(t_r)]_{\perp} = \xi(t_r) - \sum_{s=0}^{r-1}\zeta_s\langle\xi(t_r)\zeta_s\rangle, \tag{5.9}$$

$$[q(t_r)]_{\perp} = (1-\langle[q(t_r)]_{\parallel}^2\rangle)^{1/2}[\xi(t_r)]_{\perp}/\langle[\xi(t_r)]_{\perp}^2\rangle^{1/2}, \tag{5.10}$$

where the ζ_s are linear combinations of the $q(t_s)$ $(0\le s\le r-1)$ which satisfy $\langle\zeta_s\zeta_{s'}\rangle = \delta_{ss'}$. The results for $\langle[q(t_r)]_{\perp}^2\rangle$ are independent of the precise values of of $q(t_0)$ and $\xi(t_r)$ in the individual realizations, and independent of R if R exceeds the total number of constraints imposed at any time step. If $\Delta t = .1$, the numerical values of $\langle[q(t_r)]_{\perp}^2\rangle$ for $r = 1, 2, 3, 4, 5, 6$ are $.297\times10^{-1}$, $.587\times10^{-3}$, $.287\times10^{-4}$, $.113\times10^{-5}$, $.764\times10^{-7}$, $.445\times10^{-8}$. The rapid decrease with increasing r shows that the constraints (5.5a) strongly determine the future evolution of q(t) after a few steps. The q(t) obtained in individual realizations are typically smooth. If (5.5a) is truncated to $s\ge r-6$, and the time-stepping continued to $r>6$, the covariance continues to closely approximate the exact value. Thus Q(0,2) so obtained is in error by $<.005$ (exact value $-.406$).

Similar techniques can be used to obtain explicit realizations of ensembles of solutions of differential equations. Consider the simple illustration $dy/dt = q$. Discretizing can be done according to some chosen integration scheme. For example,

$$L(t_r) \equiv y(t_r) - y(t_{r-1}) - \tfrac{1}{2}\Delta t[q(t_r) + q(t_{r-1})] = 0. \tag{5.11}$$

Let the constraints be (5.5) plus the additional constraint

$$\langle q(t_r)y(t_r)\rangle = 0, \tag{5.12}$$

which gives the conservation property $d\langle y^2\rangle/dt = 0$. Again, solutions can be sought by iterating over an entire time interval (0,T) at once, or doing time-stepping over adjoining or overlapping subintervals. The simplest choice is to step one time step at a time, as above. The procedure is just as above except that (5.12) is added to the constraints (5.5a) which determine $[q(t_r)]_{\parallel}$. Eq. (5.11) is handled in standard fashion. An initial explicit step is performed in which $q(t_r)$ is replaced by the known $q(t_{r-1})$, $y(t_r)$ determined, and the constraints solved to yield $q(t_r)$. Then the procedure is repeated with (5.11) as written. The values obtained for $\langle [q(t_r)]_+^2 \rangle$ decrease rapidly with r, in close correspondence to the case where only the covariance constraint is imposed on q.

More general treatments can be carried out in which all amplitudes in a subinterval consisting of several time steps are solved for simultaneously: Eq. (5.11) is replaced by the equivalent statistical constraint

$$\langle [L(t_r)]^2 \rangle = 0. \tag{5.13}$$

The Newton-Raphson method can then be used to determine y and q values which satisfy (5.5), (5.12) and (5.13), subject to given values at the start of the subinterval.

It cannot be assumed in advance that stepping by single adjoining time steps will work. Even when it is clear that it would work with infinite ensembles, it may not with finite R because of the fact that the constraints imposed are actually those for an infinite ensemble, as discussed at the beginning of this Section. Evidence of failure would be a negative value of the righthand side of (5.7) at some time step.

Constraints in the form of realizability inequalities can be fitted into the framework of this Section by introducing auxiliary variables. Thus the constraint

$$\langle F_i(x) \rangle \geq 0 \tag{5.14}$$

can be written

$$\langle [\mu_i^2 - F_i(x)] \rangle = 0, \tag{5.15}$$

a constraint of the form (5.1) in the enlarged set of real variables x, μ_i.

The stochastic Newton-Raphson construction may be contrasted with the attempt to construct a "most probable distribution" by minimizing $\int \rho \ln \rho dx$ under moment constraints. Suppose the constraints first are $\langle x^2 \rangle = 1$, $\langle x^4 \rangle = 3$. The constrained minimization gives the Gaussian distribution $\rho(x) \propto \exp(-x^2/2)$. Let the second constraint be modified to $\langle x^4 \rangle = 3 + \varepsilon$, with $|\varepsilon| \ll 1$. The constrained minimization now gives

$$\rho(x) \propto \exp(-x^2/2 - \alpha x^2 + \beta x^4),$$

where, to $O(\varepsilon)$, $\alpha = \varepsilon/4$, $\beta = \varepsilon/24$. For $\varepsilon > 0$, this $\rho(x)$ cannot be normalized. The Newton-Raphson construction, starting from the initial Gaussian constraints, realizes the modified constraints by

$$\Delta x = (\varepsilon/24)(x^3 - 3x),$$

to $O(\varepsilon)$. There is no problem with $\varepsilon > 0$. The Newton-Raphson construction makes a least-squares shift of amplitudes in the individual realizations, while the attempt to construct a most probable distribution deals with the function $\rho(x)$.

6. CONVERGENCE AND REALIZATION UNDER DECIMATION

Consider N variables which obey first-order equations of motion and are sampled at T time steps. The integration of R realizations of the system then produces NTR numbers. If V operations must be performed to advance each equation one time step then the total number of operations required to generate this data is NTRV. Once the data exists, any desired moment can be constructed by direct averaging.

Suppose instead that a statistical approximation is constructed, by the methods of Sec. 4 or otherwise, such that equations of motion result for moments up to Mth order. If $N \gg M$, the number of such moments is of order $(NT)^M/M!$. The number of operations required to advance a moment equation of motion one step in time will not be less than V. Thus, for fixed R, the ratio of computing resources needed to integrate the moment approximation to those needed for direct integration of the ensemble goes up rapidly with NT and M. Moreover, the R required for good statistics may decrease as N becomes large because of statistical similarity among modes of a large system. This also increases the ratio.

Any computational usefulness of moment-based approximations for large systems, particularly approximations beyond the lowest possible order, depends on exploiting smoothness properties of moments, which permit description of their dependence on mode labels and time with drastically fewer numbers than required to describe unaveraged products of realization amplitudes. In particular, when the ratio of highest to lowest wavenumber in a turbulence calculation is large, the smoothness properties may permit major economies from specifying moments at logarithmic steps in wavenumber.

The DS treated in this paper is a particular way to exploit smoothness of moments (or other statistics). A sample set is chosen from all modes of a large system in such a way that moments involving any modes of the whole system can be expressed with sufficient accuracy by interpolating between moment values for the sample set alone. In order for such a procedure to be valuable, the whole system must exhibit statistical symmetry among large sets of modes, like the example in Sec. 1. The amplitudes of realizations of the sample set evolve according to equations of motion constructed to be consistent with constraints from the moment hierarchy (or other statistical constraints) and with sets of realizability inequalities. Some of the realizability inequalities are automatically built into the scheme by the fact that explicit realizations are constructed for the sample-set amplitudes.

The amplitudes of the sample set are intermediate quantities: a device for computing moments and other statistics economically. But at the same time, the equations of motion for the sample-set amplitudes provide physical insight with regard to the dynamical interaction between a set of explicitly followed modes and a sea of modes known only by their statistics. These equations of motion are in the spirit of renormalization-group theory, although the techniques are different.

Consider a set of modes y which obey equations of the form (4.1). Let u denote a subset of y called the sample set. Let v denote the rest of the modes y. Then (4.1) for the modes u can be written in the form

$$du_i(t)/dt + F_i^{<}(u,t) + q_i(t) = 0, \tag{6.1}$$

where $F_i^{<}$ contains all terms in F_i which involve only u modes and q_i represents the remainder of F_i. The objective of the DS is to use statistical symmetries among the y modes to form a closed dynamical system $\{u,q\}$ with the v modes eliminated except insofar as they are represented by the q_i. If there is to be no error, the joint PD of u and q must be unaltered by the elimination. This will be accomplished if the statistical symmetries among the y_i are strong enough to yield

$$\langle f(q,u)\rangle = \langle g(u)\rangle, \tag{6.2}$$

where f is any many-time functional of the u_i and q_i and g(u) is some many-time functional of the u_i alone. Here g(u) is found by writing q as an explicit function of the full set y and using the symmetries to reduce the resulting y averages. If $\langle g(u)\rangle$ can be formed for all f(q,u), then, from the considerations of Sec. 4, converging approximations to the exact statistics of the u_i and q_i can be obtained by solving (6.1) under constraints consisting of successively more relations (6.2), together with appropriate realizability inequalities.

If $\langle g(u)\rangle$ can be formed for all f(q,u), the complete set of realizability inequalities on $\{u,q\}$ is $\langle g(u)\rangle \geq 0$, for all positive-definite f(q,u). It is important to realize that positive-definite f(q,u) does not imply that g(u) is a positive-definite functional of the u's. A simple counter-example is the form

$$f(q,u) = [\textstyle\sum_{n=1}^{N} x_n]^2,$$

where the x_n are real variables statistically symmetric on n, the sample set u is x_1, x_2, and $q = \sum_{n=3}^{N} x_n$. The symmetry yields

$$\langle f(q,u)\rangle = \langle (N/2)(x_1^2 + x_2^2) + N(N-1)x_1x_2\rangle,$$

and the right-hand side is not the average of a positive-definite form in x_1, x_2 if $N > 2$.

There are systems for which the statistical symmetries are strong enough to yield (6.2) for all f(q,u). An important example is the collective representation of a collection of statistically identical systems, introduced in Sec. 7 and used in Secs. 8-10 to demonstrate the link between the DS and RPT. But for most systems, there will be f(q,u) for which (6.2) cannot be written without errors due to inexact interpolation of moments (or other statistics) or due to the lack of a qualitatively acceptable g(u) to represent certain moments of the full y system. An example of the latter kind of error will be given in the discussion, later in this Section, of the decimation example introduced in Sec. 1. If there are irreducible errors of either kind in writing (6.2) for moments above a certain order, then the construction of converging approximation sequences may require systematic enlargement of the sample set u as the numbers of constraints (6.2) and of appropriate realizability inequalities are increased.

Low-order moments usually do not carry very sharp information about a random process. But this may not be the case when one or more of the realizability inequalities involving the given moments are equalities. A PD that yields one or more such marginal inequalities will be termed 'taut'. Eq. (4.5) is a marginal realizability inequality, and therefore any PD of solutions of (4.3) is taut. The example to follow illustrates the decimation treatment of a system with taut PD. The example is a simple one, but it exhibits non-trivial features and manipulations of the DS.

Consider a set of N variables x_i with the moment values

$$\langle x_i^2\rangle = 1, \quad \langle x_i^4\rangle = A, \quad \langle x_i^2 x_j^2\rangle = 0 \quad (i \neq j), \tag{6.3}$$

which imply that only one of the variables is nonzero in any realization. The value

$A = N$ yields a taut PD, and the moment values are unrealizable for $A < N$. At the value $A = N$, (6.3) yields two equations of the form (4.5):

$$\langle(\sum x_i^2 - N)^2\rangle = 0, \quad \langle(x_i x_j)^2\rangle = 0 \quad (i \neq j), \tag{6.4a,b}$$

and consequently, following the arguments of Sec. 4, the realizability inequalities for the system include all the moment hierarchy equations which can be built on

$$\sum x_i^2 - N = 0, \quad x_i x_j = 0 \quad (i \neq j). \tag{6.5a,b}$$

If (6.5a) is multiplied by x_k^{2n} and averaged, then use of (6.5b) yields $\langle x_k^{2n+2}\rangle = N\langle x_k^{2n}\rangle$ and, in particular, $\langle x_k^6\rangle = N^2$. This last value, taken with (6.3), yields

$$\langle(x_k^3 - Nx_k)^2\rangle = 0, \tag{6.6}$$

which implies that x_k^2 takes the values 0 and N only. If explicit realization of amplitudes is carried out by the Newton-Raphson method under the constraints (6.3) alone, these sharp values are obtained (cf. example 4 at the end of Sec. 5).

Now consider what happens when this system is treated by decimation. Let the sample set consist of x_1 and x_2, and let $q = \sum_{i=3}^N x_i^2$ represent the rest of the variables, as they appear in (6.4a). The three variables x_1, x_2, q are now to represent the entire system, no matter how large N is. The constraints (6.3) yield

$$\langle x_1^2\rangle = \langle x_2^2\rangle = 1, \quad \langle x_1^4\rangle = \langle x_2^4\rangle = A, \quad \langle x_1^2 x_2^2\rangle = 0, \tag{6.7a}$$

$$\langle q\rangle = (N-2), \quad \langle q^2\rangle = (N-2)A, \quad \langle qx_1^2\rangle = \langle qx_2^2\rangle = 0. \tag{6.7b}$$

Equations (6.4) and (6.5) become

$$\langle(x_1^2 + x_2^2 + q - N)^2\rangle = 0, \tag{6.8a}$$

$$\langle x_1^2 x_2^2\rangle = 0, \quad \langle qx_1^2\rangle = 0, \quad \langle qx_2^2\rangle = 0, \tag{6.8b}$$

$$x_1^2 + x_2^2 + q - N = 0, \tag{6.9a}$$

$$x_1 x_2 = 0, \quad qx_1 = 0, \quad qx_2 = 0. \tag{6.9b}$$

It is clear from (6.9) that the only values of x_1^2, x_2^2, or q permitted by (6.7) for $A = N$ are 0 and N. If the Newton-Raphson method is used on the system x_1, x_2, q with only the constraints (6.7) and $A = N$, the joint PD found for these variables is the exact PD obtained from solving the undecimated system. It should be noted that the PD of x_1 and x_2 alone is not taut at $A = N$; the value $A = 2$ satisfies all realizability inequalities involving only these two variables. But when q is added to the system, the inequality

$$\langle(x_1^2 + x_2^2 + q - N)^2\rangle \geq 0 \tag{6.10}$$

becomes marginal at $A = N$ and violated for $A < N$, as in the undecimated sytem. Thus decimation gives exact results for the system (6.3), with only low-order constraints needed.

The decimation (1.3) is one in which the required size S of the sample set increases with the number of imposed moment constraints. The first few of those relations of the form (6.2) which are linear or quadratic in q can be written

$$\langle q(t)\rangle = [(N-S)/S]A_x\sum_{n=1}^{S}\langle y_n(t)z_n(t)\rangle, \tag{6.11}$$

$$\langle q(t)x(t')\rangle = [(N-S)/S]A_x\sum_{n=1}^{S}\langle x(t')y_n(t)z_n(t)\rangle, \tag{6.12}$$

$$\langle q(t)y_n(t')z_n(t')\rangle =$$

$$[(N-S)/(S-1)]A_x\sum_{m=1}^{S}(1-\delta_{nm})\langle y_m(t)z_m(t)y_n(t')z_n(t')\rangle \quad (n\le S), \tag{6.13}$$

$$\langle q(t)q(t')\rangle - [(N-S-1)/S]\sum_{n=1}^{S}A_x\langle q(t)y_n(t')z_n(t')\rangle =$$

$$[(N-S)/S]A_x^2\sum_{n=1}^{S}\langle y_n(t)z_n(t)y_n(t')z_n(t')\rangle. \tag{6.14}$$

These relations are obtained solely from the definition of q(t) and by appeal to statistical symmetry in n. Note that (6.11), (6.12), (6.14) can be evaluated with $S\ge 1$, while (6.13) requires $S\ge 2$. In order for $\langle [q(t)]^r\rangle$ to be expressible in terms of sample-set amplitudes alone, S must be at least equal to r. If S is fixed at a value, say 1, an infinite set of relations (6.2) like (6.14) can be evaluated, but they are not exhaustive.

S = 2 is special because it is the smallest S for which all the constraints (6.12)-(6.14) can be imposed, thereby ensuring that the mean-square of the equation of motion in the form (4.5) is satisfied. If all realizability inequalities are satisfied, then the moments found under these constraints represent exact solutions of (1.1).

The numerical results for S = 1 displayed in Fig. 1 show that a very small set of constraints can give accurate results when the system (1.1) is decimated. The computations were done with ensembles of 500 realizations and a time step $\Delta t = .2$. The integrations of the undecimated equations were done by 4th-order Runge-Kutta-Gill integration. The decimated equations were integrated under constraints (1.4) and (1.5), the latter applied at $t = n\Delta t$, $t' = m\Delta t$ ($m = n, n-1, n-2$), using 4th-order Runge-Kutta-Gill integration and essentially the scheme (5.9), (5.10), (5.12) to satisfy the constraints. The random variables before projection were drawn from Gaussian ensembles.

The general symmetry constraint (6.2), and the examples (6.11)-(6.14), are exact only for $R = \infty$. If R sufficiently exceeds C, where C is the total number of constraints imposed on discretized variables, it is plausible that the infinite-R constraints can be exactly satisfied by finite R, as illustrated by the examples in Sec. 5. As C increases, so generally does the minimum R that can satisfy the imposed constraints. An alternative, which can decrease the minimum needed R, is to include stochastic fluctuation corrections in (6.2) and add additional constraints to control the statistics of the corrections.

7. DECIMATED COLLECTIVE REPRESENTATION

Homogeneous turbulence commonly is described by spatial Fourier amplitudes of the velocity field. This representation is natural in that the Fourier components are the eigenfunctions of the velocity covariance, which is therefore diagonal in the Fourier representation. However, it can be dynamically unnatural in the sense that the Fourier amplitudes are collective coordinates for spatial structures that are only weakly interacting. This is the case when the size of the periodic box containing the turbulence is large compared to any order length of the turbulence.

The present Section is concerned with the DS for an especially simple Fourier representation: the collective description of a collection of dynamically and statistically identical systems which are dynamically uncoupled.[4] This may be considered an idealization of the spatial Fourier representation of turbulence, in which interpolation errors associated with decimation are zero. But also it is the key to the relationship between the DS and RPT.

Let $x_{;n}$ be a set of N (N odd) statistically independent (scalar or vector) random variables with identical univariate statistics. The semicolons are for clarity in later many-index expressions. The Fourier collective description is

$$x_\alpha = N^{-1/2}\sum_{;n}\exp(i2\pi\alpha n/N)x_{;n} \quad [\alpha = 0, \pm1, \pm2, \pm(N-1)/2], \tag{7.1}$$

$$x_{;n} = N^{-1/2}\sum_\alpha \exp(-i2\pi\alpha n/N)x_\alpha. \tag{7.2}$$

Greek indices are used to label the collective coordinates. Eqs. (7.1) and (7.2) are related by the identities

$$\sum_\alpha \exp[i2\pi\alpha(n-m)/N] = N\delta_{nm}, \quad \sum_n \exp[i2\pi(\alpha-\beta)n/N] = N\delta_{\alpha\beta}. \tag{7.3}$$

Consider the collective description of a collection of systems $(x_{;n}, y_{;n}, z_{;n})$ each of which obeys (3.3), with zero means, statistical independence of distinct systems, and statistics independent of n. Eqs. (7.1)-(7.3) applied to (3.3) yield

$$x_\alpha - N^{-1/2}\sum_\beta y_\beta z_{\alpha-\beta} = 0, \tag{7.4}$$

together with similar equations for y_α and z_α. Eqs. (7.1)-(7.3), together with the assumptions of zero means, statistical independence and statistical symmetry, give explicit expressions for any moment of the collective coordinates in terms of the cumulants of the individual systems. Some examples are

$$\begin{aligned}
\langle x_\alpha x_\beta\rangle &= \delta_{\alpha+\beta}\langle x^2\rangle,\\
\langle x_\alpha y_\beta z_\gamma\rangle &= \delta_{\alpha+\beta+\gamma}N^{-1/2}\langle xyz\rangle,\\
\langle x_\alpha x_\beta y_\gamma y_\varepsilon\rangle &= \delta_{\alpha+\beta}\delta_{\gamma+\varepsilon}\langle x^2\rangle\langle y^2\rangle\\
&\quad + \delta_{\alpha+\beta+\gamma+\varepsilon}N^{-1}(\langle x^2y^2\rangle - \langle x^2\rangle\langle y^2\rangle),\\
\langle x_\alpha x_\beta x_\gamma x_\varepsilon\rangle &= (\delta_{\alpha+\beta}\delta_{\gamma+\varepsilon} + \delta_{\alpha+\varepsilon}\delta_{\beta+\gamma} + \delta_{\alpha+\gamma}\delta_{\beta+\varepsilon})\langle x^2\rangle^2\\
&\quad + \delta_{\alpha+\beta+\gamma+\varepsilon}N^{-1}(\langle x^4\rangle - 3\langle x^2\rangle^2).
\end{aligned} \tag{7.5}$$

Here $\delta_{\alpha+\beta} \equiv \delta_{0,\alpha+\beta}$ and the averages on the right-hand sides are over any individual system; for example, $\langle x^2y^2\rangle = \langle x_{;n}^2 y_{;n}^2\rangle$ (any n). Moments are nonzero only if the labels add to zero.

Relations like (7.5) make it possible to write any moment in the collective representation in terms of moments of a sample set consisting of x_0, y_0, z_0 alone. Thus,

$$\langle x_\alpha x_\beta\rangle = \delta_{\alpha+\beta}\langle x_0^2\rangle,$$

$$\langle x_\alpha y_\beta z_\gamma\rangle = \delta_{\alpha+\beta+\gamma}\langle x_0 y_0 z_0\rangle,$$

$$\langle x_\alpha x_\beta y_\gamma y_\varepsilon\rangle = \delta_{\alpha+\beta}\delta_{\gamma+\varepsilon}\langle x_0^2\rangle\langle y_0^2\rangle + \delta_{\alpha+\beta+\gamma+\varepsilon}(\langle x_0^2 y_0^2\rangle - \langle x_0^2\rangle\langle y_0^2\rangle),$$

$$\langle x_\alpha x_\beta x_\gamma x_\varepsilon\rangle = (\delta_{\alpha+\beta}\delta_{\gamma+\varepsilon} + \delta_{\alpha+\varepsilon}\delta_{\beta+\gamma} + \delta_{\alpha+\gamma}\delta_{\beta+\varepsilon})\langle x_0^2\rangle^2 + \delta_{\alpha+\beta+\gamma+\varepsilon}(\langle x_0^4\rangle - 3\langle x_0^2\rangle^2). \tag{7.6}$$

If $x_{;n}$, $y_{;n}$, $z_{;n}$ do not have zero means, the label zero amplitudes play a special role, and (7.5), (7.6) become more complicated. Sample relations are

$$\langle x_\alpha\rangle = \delta_\alpha N^{1/2}\langle x\rangle, \quad \langle x_\alpha x_\beta\rangle = N\delta_\alpha\delta_\beta\langle x\rangle^2 + \delta_{\alpha+\beta}(\langle x^2\rangle - \langle x\rangle^2), \tag{7.5'}$$

$$\langle x_\alpha\rangle = \delta_\alpha\langle x_0\rangle, \quad \langle x_\alpha x_\beta\rangle = \delta_\alpha\delta_\beta\langle x_0\rangle^2 + \delta_{\alpha+\beta}(\langle x_0^2\rangle - \langle x_0\rangle^2), \tag{7.6'}$$

Eq. (7.4) for the sample set x_0, y_0, z_0 is

$$x_0 - N^{-1/2}y_0z_0 - q_x = 0, \quad q_x = N^{-1/2}\textstyle\sum_\beta' y_\beta z_{-\beta}, \tag{7.7}$$

with similar equations for y_0 and z_0. Here $\sum_\beta'$ denotes a sum with $\beta = 0$ omitted. As after (3.5), an auxiliary variable s may be included, giving the additional sample-set equation

$$s_0 - (x_0^2 + y_0^2 + z_0^2) - q_s = 0, \quad q_s = \textstyle\sum_\beta'(x_\beta x_{-\beta} + y_\beta y_{-\beta} + z_\beta z_{-\beta}). \tag{7.8}$$

As noted in Sec. 3, the realizability inequalities to 4th order, plus the mean squares of (3.1) as constraints, are sufficient to force exact moment values to 4th order for each individual system $x_{;n}$, $y_{;n}$, $z_{;n}$, $s_{;n}$. Since the passage to collective coordinates is a linear transformation, it follows that, in the collective representation, 4th-order realizability inequalities, plus the mean squares of (7.4) and an equation for s_α, are also sufficient. By relations like (7.6), all moments to 4th order in the collective representation can be expressed in terms of moments to 4th order of the sample set x_0, x_0, x_0, s_0. Each q factor is counted as a quadratic expression in assigning orders to moments containing q_x, q_y, q_z, q_s. Thus the decimated system (7.7), (7.8) yields the exact moments to 4th order for the entire collection of systems.

It is important to note that amplitudes of the individual systems cannot be obtained from amplitudes of an explicit realization of the sample set. According to (7.2), all the amplitudes x_α are needed to determine amplitudes for any $x_{;n}$. As noted in Sec. 3, $x_{;n}$ can take only the values 0, -1, or 1 in any realization. This is expressed by the

6th-order moment relation

$$\langle x^2(x^2-1)^2\rangle = 0, \tag{7.9}$$

which can be rewritten as a 4th order relation by using, instead of the s previously defined, three auxiliary variables λ, μ, ν defined by

$$\lambda - x^2 = 0, \quad \mu - y^2 = 0, \quad \nu - z^2 = 0. \tag{7.10}$$

Thereby the sharp values 0, -1, 1 for $x_{;n}$ can be deduced from moment values to 4th order constructed from the sample set alone.

If N is large, any small set of the collective variables has nearly Gaussian low-order moments, and marginal inequalities, like the mean-square of (7.4), are marginal because N appears explicitly in them, magnifying the departure from Gaussian values. The strongly nonGaussian statistics of the individual system variables express the fact that all N collective variables contribute to each $x_{;n}$, $y_{;n}$, or $z_{;n}$.

8. DECIMATED COLLECTION OF RANDOM OSCILLATORS

Renormalized perturbation theory (RPT) for equations of motion nonlinear in stochastic quantities can be formulated in a variety of ways. Diagram summation and reversion of formal perturbation expansions[12] are two of the more direct ways. Although useful approximants can be extracted from the formal expansions even when they do not converge, it is unsatisfying to do perturbation theory based on smallness of the nonlinear terms when in fact they may be very large.

The collective representation of a collection of N independent, statistically identical systems is a device for doing RPT on strongly nonlinear systems in such a way that there is something small on which to base the expansion. As $N\to\infty$, the interaction among any finite set of the collective modes becomes an infinitesimal perturbation on the interactions of the set with all the rest of the modes. The final formal expansion produced by doing RPT in this way is the same as if, say, series reversion were used. However, the collective representation suggests particular approximants to the infinite expansions whose significance may be less clear in other formulations. In particular, the direct-interaction approximation (DIA), a truncation of RPT with important consistency properties, can be derived in a natural way using the Fourier collective representation.

In Sec. 9, the DIA will be constructed by decimation under particular moment constraints chosen from a systematic sequence of constraints. In Sec. 10, the decimation analysis will be compared with a formulation, given some years ago, which is based on the construction of stochastic models in the collective representation.[4] A simple system to demontrate the analysis is a collection of N statistically independent random oscillators. The DS for this system is developed in the present Section.

The equations of motion for the random oscillators are

$$(d/dt + \nu)y_{;n} + ia_{;n}y_{;n} = 0. \tag{8.1}$$

Here $y_{;n}$, $a_{;n}$, ν are the complex amplitude, real frequency, and real damping of the nth oscillator. Zero means are assumed to simplify the analysis, and statistical symmetry in n is assumed.

The collective representation in the form (7.1) yields

$$(d/dt + \nu)y_\alpha + iN^{-1/2}\sum_\beta a_\beta y_{\alpha-\beta} = 0. \tag{8.2}$$

As in Sec. 7, the sample set may be taken as y_0, a_0 with zero interpolation error in the expression of any moment of collective amplitudes. Eq. (8.2) for $\alpha = 0$ is rewritten as

$$(d/dt + \nu)y_0 + iN^{-1/2}a_0 y_0 + q = 0, \tag{8.3}$$

$$q = iN^{-1/2}\sum_\beta{}' a_\beta y_{-\beta}. \tag{8.4}$$

Let the real and imaginary parts of $y_{;n}(0)$ be statistically independent with identical univariate statistics. Then under (8.1), the average of any product of $y_{;n}$, $y^*_{;n}$ factors (arbitrary time arguments) vanishes unless the numbers of y and y^* factors are equal. The general moment in the collective representation

$$\langle y_\alpha \cdots y_\beta y^*_\gamma \cdots y^*_\varepsilon a_\delta \cdots a_\zeta \rangle$$

vanishes unless the indices add to zero (take the negative of the index of a y^* factor) and the numbers of y and y^* factors are equal. The general moment constraint of the form (6.2) can be written as

$$\langle q \ldots q q^* \ldots q^* y_0 \ldots y_0 y^*_0 \ldots y^*_0 a_0 \ldots a_0 \rangle = \langle g(y_0, y^*_0, a_0) \rangle, \tag{8.5}$$

where the time arguments are arbitrary. The function g can be found in explicit form from relations like (7.6); g vanishes unless the total number of q and y_0 factors equals the total number of q^* and y^*_0 factors.

Particular examples of (8.5) are

$$\langle q(t)q^*(t') \rangle = \frac{N-1}{N}\left(N\langle a_0^2 y_0(t)y^*_0(t')\rangle - (N-1)\langle a_0^2\rangle\langle y_0(t)y^*_0(t')\rangle\right), \tag{8.6}$$

$$\langle q(t)y^*_0(t') \rangle = i\frac{N-1}{N}N^{1/2}\langle a_0 y_0(t)y^*_0(t')\rangle, \tag{8.7}$$

$$\langle q(t)a_0 y^*_0(t') \rangle = i(N-1)\left(\langle a_0^2 y_0(t)y^*_0(t')\rangle - \langle a_0^2\rangle\langle y_0(t)y^*_0(t')\rangle\right). \tag{8.8}$$

The $y_{;n}(t)$ are bounded under (8.1). Therefore if the $a_{;n}$ and $y_{;n}(0)$ have PD's which fall off exponentially or faster, this is true also of the joint PD of $a_{;n}$ and $y_{;n}(t)$ and, therefore, of the a_α and $y_\beta(t)$. In this case, moments express completely the joint PD of y_0, y^*_0, q, q^*, and a_0 with arbitrary time arguments. By the arguments of Sec. 4, converging approximations to moments of an ensemble of solutions of (8.1) can be constructed by systematically imposing more and more of the moment hierarchy equations built on (8.3), together with realizability inequalities of corresponding order, and those constraints (8.5) needed to reduce the q-containing moments retained in a given approximation to moments of y_0, y^*_0, a_0 only.

Consider the form of $q(t)$ implied by the imposition of some particular subsets of the symmetry constraints (8.5). If only (8.6) is imposed, $q(t)$ can be an arbitrary random process whose covariance equals the right-hand side of (8.6). If (8.6) and (8.7) are both imposed, the result can be written

$$q(t) = q_\downarrow(t) + q_\uparrow(t). \tag{8.9}$$

In partial correspondence to the logic used in (5.6)-(5.9), $q_{\uparrow}(t)$ is now that linear functional of y_0 which is the least-squares solution of the linear constraints (8.7) and $q_{\downarrow}(t)$ is orthogonal to those constraints:

$$\langle q_{\downarrow}(t)y_0^*(t')\rangle = 0, \quad \langle q_{\downarrow}(t)q_{\uparrow}^*(t')\rangle = 0. \tag{8.10}$$

Eq. (8.6) also is a constraint on $q_{\downarrow}(t)$; it can be written

$$\langle q_{\downarrow}(t)q_{\downarrow}^*(t')\rangle = Q(t,t') - \langle q_{\uparrow}(t)q_{\uparrow}^*(t')\rangle, \tag{8.11}$$

where $Q(t,t')$ denotes the right-hand side of (8.6).

The decomposition (8.9) can be a starting point for computing by the stochastic Newton-Raphson methods of Sec. 5, or related algorithms. For later comparison with RPT, it is illuminating to write (8.9) in the form

$$q(t) = b(t) + \int_0^T \eta(t,s)y_0(s)ds, \tag{8.12}$$

where $(0,T)$ is the time interval in which (8.1) is to be solved. The linear constraints (8.7) then take the form of an integral equation for $\eta(t,s)$:

$$\int_0^T \eta(t,s)\langle y_0(s)y_0^*(t')\rangle ds = Z_0(t,t') - \langle b(t)y_0^*(t')\rangle, \tag{8.13}$$

where $Z_0(t,t')$ denotes the right-hand side of (8.7). It should be noted that y_0 is correlated with $b(t)$ according to (8.3). Therefore in general $b(t) \neq q_{\downarrow}(t)$, and instead $b(t)$ is a linear combination of $q_{\downarrow}(t)$ and $q_{\uparrow}(t)$.

An approximation satisfying the given constraints can be constructed by solving (8.3), (8.6), and (8.13) simultaneously, by the methods of Sec. 5 or otherwise, to give the function $\eta(t,s)$ and an ensemble of realizations of $b(t)$. Since all the constraints imposed are satisfied by exact solutions of (8.1), the full set of approximations so constructed includes exact-solution ensembles. The precise properties of the solution of (8.3), (8.6), and (8.13) actually constructed by a given algorithm depend on the properties of the algorithm. In particular, the realized $b(t)$ may or may not be statistically independent of a_0 and of $y_0(t')$ for $t' < t$.

If (8.8) is added to the imposed constraints, (8.12) is replaced by

$$q(t) = b(t) + \int_0^T \eta(t,s)y_0(s)ds + ia_0\int_0^T \gamma_1(t,s)y_0(s)ds, \tag{8.14}$$

and (8.14) is replaced by a pair of integral equations, for η and γ_1, involving the right-hand sides of both (8.7) and (8.8). Similarly, the addition of more constraints linear in q adds more terms to $q(t)$. If constraints of the form

$$\langle q(t)P_n(a_0)y_0^*(t')\rangle = Z_n(t,t') \tag{8.15}$$

are added, where $P_n(a_0)$ is an nth-degree polynomial in a_0 and $Z_n(t,t')$ denotes the appropriate right-hand side of (8.5), $q(t)$ can be written as

$$q(t) = b(t) + \int_0^T \eta(t,s)y_0(s)ds + \sum_{n=1}^{C} (i)^n P_n(a_0)\int_0^T \gamma_n(t,s)y_0(s)ds, \tag{8.16}$$

if $n = C$ is the highest such constraint added.

In addition to explicit moment constraints expressing the symmetries of the collective representation, for example (8.6)-(8.8), any solution for the random function $q(t)$ satisfies the hierarchy of moment equations built upon (8.3). If the stochastic Newton-Raphson method is used to construct the solution, an explicit constraint from this hierarchy will be that the mean square of (8.3) vanish. Whatever the method of solution, the moment constraints implied by (8.3) can have the effect of adding additional terms to (8.12), (8.14) or (8.16) and of inducing correlations of $b(t)$ with these and other terms in $q(t)$. In particular, it will be noted in Sec. 9 that an additional term of the form $ia_0\zeta(t)$, where $\zeta(t)$ has zero mean and is correlated with $b(t)$, is essential to the construction of a consistent solution embodying the vertex renormalization expressed by the γ_1 term in (8.14). Some of the higher terms which can arise when higher symmetry constraints are imposed can be expressed by making the kernels η and γ_n stochastic functions.

It should be emphasized that decompositions like (8.16) are made here for later comparison with RPT. They are not needed to apply the stochastic Newton-Raphson method under (8.3) and given symmetry constraints. The Newton-Raphson method works directly with $q(t)$ as a whole.

The constraints (8.5) so far considered for the random oscillator have been linear in q and in y^*. Constraints linear in q and polynomial in y and y^* add terms polynomial in y and y^* to the right-hand side of (8.16). However, the random oscillator equation of motion is linear in y, and a consequence can be seen to be that such constraints are redundant with the constraints linear in y^*. Thus the coefficients of the additional terms in (8.16) are zero.

The imposition of constraints nonlinear in q, in addition to (8.6), can have the effect of adding a stochastic part to $\eta(t,s)$ and $\gamma_n(t,s)$. Alternatively, constraints nonlinear in q can be handled by explicitly introducing functions of q on the right-hand side of (8.16). The following example illustrates this. Consider the constraint

$$\langle q(t)\{[q^*(t)q(t)]^n y_0^*(t')\}\rangle = X_n(t,t'), \tag{8.17}$$

where $X_n(t,t')$ is the appropriate right-hand side of (8.5). If the quantity in $\{\}$ brackets is treated temporarily as a known quantity, this constraint can be handled in the same way as (8.7). The result is a term of the form

$$[q^*(t)q(t)]^n \int_0^T \chi_n(t,s)y_0(s)ds, \tag{8.18}$$

where $\chi_n(t,s)$ is nonstochastic. Again, the methods of Sec. 5 can be used to obtain realized solutions.

Eq. (8.16) as written contains integrals over the entire interval $(0,T)$. If an initial value problem has been posed, elementary notions of causality suggest that the integrals should be reducible to the interval $(0,t)$. There is a subtle difficulty here. Full specification of initial conditions requires that moments of the initial PD be given to infinite order. Instead, the constraints imposed have involved explicit moments only of finite order. The retention of backward-looking constraints [e.g., (8.7) for $t' > t$] and realizability inequalities may help to assure maximum consistency by implicitly imposing realizibility conditions on the higher initial moments, which are never dealt with explicitly. For this reason, the integrals in (8.16) are written over the entire interval $(0,T)$. It may, of course, turn out that the kernels are nonzero only for causal ordering of time arguments.

The trial forms for $q(t)$ so far taken have all been the expressions of minimum complexity likely to be needed to satisfy given moment constraints. But an exact expression for $q(t)$, satisfying all moment constraints derived from the equations of motion and from symmetries, can be obtained by direct manipulation of (8.2). To do this, let $y_\alpha^0(t)$ be the solution of (8.2) with all the terms containing a_0 removed. Let the resulting equation for y_0^0 be written as

$$(d/dt + \nu)y_0^0(t) + q^0(t) = 0. \tag{8.19}$$

Because the dynamical equations are linear in the $y_\alpha^0(t)$, $q^0(t)$ may be decomposed in the form

$$q^0(t) = b^0(t) + \int_0^t \eta^0(t,s)y_0^0(s)ds, \tag{8.20}$$

where $b^0(t)$ represents the value of $iN^{-1/2}\sum_\beta' a_\beta y_{-\beta}^0$ with $y_0^0(t)$ clamped to zero and the η^0 term represents the response of $q^0(t)$ to the actual nonzero values of $y_0^0(t)$. The kernel $\eta^0(t,s)$ has both mean and fluctuating parts. By (8.2), with a_0 terms removed, it has the exact expression

$$\eta^0(t,s) = iN^{-1/2}\sum_\beta' a_\beta G_{-\beta,0}^0(t,s), \tag{8.21}$$

where $G_{-\beta,0}^0(t,s)$ is the unaveraged response $\delta y_{-\beta}^0(t)/\delta y_0^0(s)$ of $y_{-\beta}^0(t)$ to an impulsive perturbation $\delta y_0^0(s)$. Also from (8.2), $\eta^0(t,s)$ is related to the diagonal unaveraged response function by

$$(\partial/\partial t + \nu)G_{00}^0(t,s) + \int_0^t \eta^0(t,r)G_{00}^0(r,s)dr = 0. \tag{8.22}$$

If now the a_0 terms are reinstated, inspection of (8.2) shows that

$$y_\alpha(t) = \exp(-iN^{-1/2}a_0 t)y_\alpha^0(t), \tag{8.23}$$

$$q(t) = \exp(-iN^{-1/2}a_0 t)b^0(t) + \int_0^t \exp[-iN^{1/2}a_0(t-s)]\eta^0(t,s)y_0(s)ds. \tag{8.24}$$

Expansion of the exponentials in (8.24) yields an exact series expansion of $q(t)$ consistent with the forms inferred above from imposition of symmetry constraints. Several cautions should be stated. First, the $b(t)$ and $\eta(t,s)$ found by imposing partial constraints are approximate, and need not exactly equal the $b^0(t)$ and mean of $\eta^0(t,s)$ in (8.24). Moreover, $b^0(t)$ and $\eta^0(t,s)$ are functions of the a_β ($\beta \neq 0$) and are statistically dependent on a_0 unless the a_β are independent for distinct $|\beta|$. The latter will be true if the underlying $a_{;n}$ are Gaussian, but not in general.

The derivation of (8.24) can be extended to any decimated symmetrical collection with linear dynamics. Examples are convection of a passive scalar by turbulence and the Schrödinger equation with random potential. For systems with nonlinear dynamics, the exact $q(t)$ can still be analyzed into a part $b(t)$, corresponding to clamping of the sample-set variables at zero, and a remainder. The remainder can be expanded as a functional power series in the sample-set variables. The coefficient kernels of the expansion then are functional derivatives of $q(t)$ with respect to the sample-set variables, evaluated at the clamped values.

9. DECIMATED COLLECTION AT INFINITE N

Perturbation expansion was not used in Sec. 8. In particular, it was not used to obtain the form (8.16) for $q(t)$, inferred from symmetry constraints, or the exact form (8.24). But now let $N\to\infty$. In the limit, the n-variate PD of any n collective amplitudes u_α, u_β,...u_δ, where u stands for a, y, or y^*, is normal up to any finite order of moment. This is a consequence of the independence of the individual oscillators and is expressed by moment relations like (7.5). Further, if the $y_n(0)$ are statistically independent of the a_n, then $y_0(t)$ and a_0 are weakly dependent for all t as $N\to\infty$. Weak dependence means that the departure of any finite-order moment from its value in an independent PD goes to zero as some power of N in the limit.

Weak dependence is easily demonstrated by making the perturbation expansion

$$y_\alpha(t) = \sum_{n=0}^{\infty} y_{\alpha n}(t), \quad q(t) = \sum_{n=0}^{\infty} q_n(t), \tag{9.1}$$

where $y_{\alpha n}(t)$ and $q_n(t)$ are $O(N^{-n/2})$, and substituting into (8.2)-(8.4). It is readily found that the leading term in $y_0(t)$ proportional to a_0^n is $y_{0n}(t)$. Thus,

$$\langle y_0(t)P_n(a_0)y_0^*(t')\rangle = O(N^{-n/2}), \tag{9.2}$$

where $P_n(a_0)$ is the nth-degree polynomial defined by

$$\langle P_n(a_0)P_m(a_0)\rangle = \delta_{nm}.$$

Since the PD of a_0 approaches a Gaussian as $N\to\infty$, these $P_n(a_0)$ are Hermitean polynomials in the limit. The orthogonal P_n just defined are the appropriate polynomials to use in (8.16) for $N\to\infty$. Then (9.2) implies that $\gamma_n(t,s)$ is $O(N^{-n/2})$.

Consider the perturbation treatment ($N\to\infty$) of (8.3) with only the constraint (8.7) imposed and the ansatz (8.12) for $q(t)$. Make the further trial assumptions that $b(t)$ is statistically independent of a_0, that y_0 is statistically dependent on a_0 and $b(t)$ only to the extent induced by (8.3) and that $\eta(t,s)$ is causal,

$$\eta(t,s) = 0 \quad (s > t), \tag{9.3}$$

with zero variance. Then

$$y_{00}(t) = G(t,0)y_0(0) - \int_0^t G(t,s)b(s)ds,$$

$$y_{01}(t) = -ia_0N^{-1/2}\int_0^t G(t,s)y_{00}(s)ds, \tag{9.4}$$

where $G(t,s)$ is defined by

$$(\partial/\partial t + \nu)G(t,s) + \int_0^t \eta(t,s')G(s',s)ds' = 0, \quad G(s,s) = 1. \tag{9.5}$$

Eq. (9.4) gives the following values to $O(1)$ for the left and right sides of (8.7):

$$\langle q(t)y_0^*(t')\rangle =$$

$$\int_0^t \eta(t,s)\langle y_0(s)y_0^*(t')\rangle ds - \int_0^{t'} G^*(t',s)\langle b(t)b^*(s)\rangle ds, \tag{9.6}$$

$$iN^{1/2}\langle a_0y_0(t)y_0^*(t')\rangle =$$

$$\langle a_0^2\rangle\int_0^t G(t,s)\langle y_0(s)y_0^*(t')\rangle ds - \langle a_0^2\rangle\int_0^{t'} G^*(t',s)\langle y_0(t)y_0^*(s)\rangle ds. \tag{9.7}$$

Clearly the right sides of (9.6) and (9.7) are the same, and (8.7) is satisfied to O(1), if

$$\eta(t,s) = \langle a_0^2\rangle G(t,s) \quad (t \geq s), \tag{9.8}$$

$$\langle b(t)b^*(s)\rangle = \langle a_0^2\rangle\langle y_0(t)y_0^*(s)\rangle. \tag{9.9}$$

In the limit $N\to\infty$, (8.3), (8.12) and (9.3) give

$$(d/dt + \nu)y_0(t) + \int_0^t \eta(t,s)y_0(s)ds + b(t) = 0. \tag{9.10}$$

Eqs. (9.8)-(9.10), with the definition (9.5), are the Langevin form of the DIA[5,13] for the random oscillator. Eqs. (9.8) and (9.9) are unique consequences ($N\to\infty$) of the constraint (8.7), the form (8.12) for $q(t)$, the causality assumption (9.3), and the assumption that $b(t)$ is statistically independent of a_0. If (9.9) is altered by adding a term $\Delta B(t,s)$ to the right side, then the right sides of (9.6) and (9.7) could be made the same only if $\eta(t,s)$ were to depend on t' as well as t and s.

Note that the covariance of $b(t)$ is fixed by (9.9) even though the only symmetry constraint imposed in deriving the DIA is (8.7), which is linear in q. This is so because of the relationship between q and b induced by (8.3).

The uniqueness of the DIA under (8.7) is very much a consequence of the special assumptions made about the form of $q(t)$. If no such assumptions are made, (8.7) permits a variety of solutions, including ensembles of exact solutions of (8.3) for any given univariate statistics of the $a_{;n}$.

Now let (8.7) and (8.8) both be imposed. Clearly a linear dependence of $q(t)$ on a_0y_0, as in (8.14), is induced in the least-squares solution satisfying the two constraints. But it is easy to see that a consistent statistical solution cannot be formed unless $b(t)$ obeys additional constraints or, equivalently, an extra term is added to (8.14). Thus if (9.3)-(9.9) are rederived with (8.14) replacing (8.12), (9.6) is unchanged while (9.7) becomes

$$iN^{1/2}\langle a_0y_0(t)y_0^*(t')\rangle = \langle a_0^2\rangle\int_0^t ds\int_0^s ds' G(t,s)\Gamma(s,s')\langle y_0(s')y_0^*(t')\rangle$$
$$- \langle a_0^2\rangle\int_0^{t'} ds\int_0^s ds' G^*(t',s)\Gamma^*(s,s')\langle y_0(t)y_0^*(s')\rangle, \tag{9.11}$$

where

$$\Gamma(t,s) = \delta(t-s) + N^{1/2}\gamma_1(t,s) \tag{9.12}$$

and it is assumed that

$$\Gamma(t,s) = 0 \quad (t < s). \tag{9.13}$$

The relations which leave (8.7) satisfied, corresponding to (9.8) and (9.9), are now

$$\eta(t,s) = \langle a_0^2\rangle\int_s^t G(t,s')\Gamma(s',s)ds', \tag{9.14}$$

$$\langle b(t)b^*(t')\rangle = \langle a_0^2\rangle\int_0^s \Gamma(s,s')\langle y_0(t)y_0^*(s')\rangle ds'. \tag{9.15}$$

But (9.15) is impossible except for the DIA case $\Gamma(t,s) = \delta(t-s)$: Under the symmetry assumptions of Sec. 8, each covariance in (9.15) is real and symmetrical in its time arguments. This symmetry is in general inconsistent with (9.15) because $\Gamma(s,s')$ operates unsymmetrically on the two arguments of $\langle y_0(t)y_0^*(s')\rangle$.

Therefore, no ensemble of y_0, q amplitudes can be constructed which satisfy (8.7) and (8.8) such that (8.14) holds, η and γ_1 are causal, and b(t) is constrained only by the specification of its covariance. A solution to this problem is to relax (8.14) to a more general form. Suppose there is added to the right side a term of the form $-ia_0\zeta(t)$, where $\zeta(t)$ is a stochastic function with zero mean and nonzero correlation with b(t) [cf (8.24)]. This adds to the right side of (9.11) the additional term

$$\langle a_0^2\rangle\int_0^t ds\int_0^{t'} ds' G(t,s)G^*(t',s')\Lambda(s,s'), \tag{9.16}$$

where

$$\Lambda(t,s) = N^{1/2}[\langle b^*(t)\zeta(s)\rangle - \langle\zeta^*(t)b(s)\rangle] \tag{9.17}$$

If a $\Lambda(t,s)$ with the correct O(1) value can be realized, the additional term will make (9.11) consistent with the symmetry of the covariances.

If q(t) is expanded in a Taylor series about t = 0 by so expanding (8.2)-(8.4), it is found directly that the lowest-order contributions to $\zeta(t)$ are $\propto t$ and $O(N^{-1/2})$. The lowest-order contribution to (9.11) from the $ia_0\zeta(t)$ term is $\propto t^3$, while the leading terms in (9.11) are $\propto t$. It can be seen from this direct expansion, or from (8.24), that there are no additional contributions to q(t), other than the $ia_0\zeta(t)$ term, which can make O(1) additions to (9.11).

The DIA value for $\eta(t,s)$ and the values $\gamma_1(t,s) = \zeta(t) = 0$ are exact if $a_{;n}$ has a semicircle distribution.[4] This is a consequence of the particular dependence of $b^0(t)$ and $\eta^0(t,s)$ in (8.24) on a_0. Other $a_{;n}$ distributions give different $\eta(t,s)$ and nonzero $\gamma_1(t,s)$. Since the constraints (8.7) and (8.8) are exact consequences of the symmetry of the collection of systems, realizable $\Lambda(t,s)$ values must exist which make the modified (9.11) consistent and satisfy (8.7) and (8.8), with $\eta(t,s)$ and $\gamma_1(t,s)$ corresponding to any positive-definite $a_{;n}$ distribution. This is also clear from (8.24). The higher univariate statistics of $a_{;n}$ constrain $\eta(t,s)$ through higher members of the sequence that starts with (8.7) and (8.8).

It is important to note that although $\gamma_2(t,s)$ and other higher-order terms in (8.14) do not directly contribute to either $\eta(t,s)$ or $iN^{1/2}\langle a_0y_0(t)y_0^*(t')\rangle$ in the limit $N\to\infty$, they affect the values of these functions indirectly. Terms involving $\gamma_2(t,s)$ enter the right side of (8.8) to leading order and thereby affect the value of $\gamma_1(t,s)$. There is a chain effect, through the entire sequence that starts with (8.7) and (8.8), by which $\gamma_{n+1}(t,s)$ and $\gamma_n(t,s)$ are coupled. With the DIA values of $\eta(t,s)$ and $\gamma_1(t,s)$, (8.8) yields nonzero $\gamma_2(t,s)$.

At this point, it is useful to examine what changes when the random oscillator is replaced by the more general bilinear system

$$dx_{i;n}/dt + \sum_j \nu_{ij}x_{j;n} + \tfrac{1}{2}\sum_{jk} A_{ijk}x_{j;n}x_{k;n}, \tag{9.18}$$

where the ν_{ij} form a damping matrix, $A_{ijk} = A_{ikj}$ is a coupling coefficient, the $x_{i;n}$ are real variables and ;n labels systems in a collection of size N, as in (8.1). Assume that the $x_{i;n}$ have zero means, are statistically independent for distinct n, and have statistics independent of n. The collective representation $x_{i\alpha}$ may be formed just as for the random oscillator. Because the $x_{i;n}$ are real, the collective amplitudes satisfy $x_{i\alpha} = x^*_{i,-\alpha}$. The decimated set may be taken as x_{i0}, with equations of motion

$$dx_{i0}/dt + \sum_j \nu_{ij}x_{j0} + \tfrac{1}{2}N^{-1/2}\sum_{jk} A_{ijk}x_{j0}x_{k0} + q_i, \tag{9.19}$$

where

$$q_i = \tfrac{1}{2}N^{-1/2}\sum_{jk} A_{ijk}\sum_\beta{}' x_{j\beta}x_{k,-\beta} \tag{9.20}$$

and $\sum_\beta'$ omits $\beta = 0$.

The symmetry constraints corresponding to (8.6)-(8.8) can be worked out just as for the random oscillator. The results are

$$\langle q_i(t)q_n(s)\rangle = \frac{N-1}{N}\sum_{jkab}\{NA_{ijk}A_{nab}\langle x_{j0}(t)x_{k0}(t)x_{a0}(s)x_{b0}(s)\rangle$$

$$- (N-1)[\langle x_{j0}(t)x_{k0}(t)\rangle\langle x_{a0}(s)x_{b0}(s)\rangle$$

$$+ \langle x_{j0}(t)x_{a0}(s)\rangle\langle x_{k0}(t)x_{b0}(s)\rangle + \langle x_{j0}(t)x_{b0}(s)\rangle\langle x_{k0}(t)x_{a0}(s)\rangle]\}, \tag{9.21}$$

$$\langle q_i(t)x_n(s)\rangle = \frac{N-1}{N}N^{1/2}\sum_{jk} A_{ijk}\langle x_{j0}(t)x_{k0}(t)x_{n0}(s)\rangle, \tag{9.22}$$

$$\langle q_i(t)x_{a0}(s)x_{b0}(r)\rangle = (N-1)\sum_{ij} A_{ijk}(\langle x_{j0}(t)x_{k0}(t)x_{a0}(s)x_{b0}(r)\rangle$$

$$- \langle x_{j0}(t)x_{k0}(t)\rangle\langle x_{a0}(s)x_{b0}(r)\rangle - \langle x_{j0}(t)x_{a0}(s)\rangle\langle x_{k0}(t)x_{b0}(r)\rangle$$

$$- \langle x_{j0}(t)x_{b0}(r)\rangle\langle x_{k0}(t)x_{a0}(s)\rangle). \tag{9.23}$$

If the coupling coefficients obey

$$A_{ijk} + A_{jki} + A_{kij} = 0 \tag{9.24}$$

and the ν's vanish, then $\sum_i x_{i;n}^2$ is a constant of motion. Eq. (9.22) for $i = j$ and $t = s$ preserves this property in the mean, making $\sum_i \langle x_{i0}^2\rangle$ a constant of motion.

A general expression for $q_i(t)$ consistent with all the symmetry constraints linear in q_i and with (9.21) can be written

$$q_i(t) = b_i(t) + \sum_j \int_0^t \eta_{ij}(t,s)x_{j0}(s)ds + \sum_j \int_0^t \zeta_{ij}(t,s)x_{j0}(s)ds$$

$$+ \frac{1}{2}\sum_{jk} \iint_0^t \gamma_{ijk}(t,s,r)x_{j0}(s)x_{k0}(r)dsdr + \text{higher-order terms}. \tag{9.25}$$

Here $b_i(t)$ and $\zeta_{ij}(t,s)$ are stochastic with zero means while $\eta_{ij}(t,s)$, and $\gamma_{ijk}(t,s,r)$ are ordinary functions. The higher-order terms can include a bilinear functional of the x's with zero-mean kernel and higher-degree functionals. In the limit $N\to\infty$, $\eta_{ij}(t,s)$ and $\zeta_{ij}(t,s)$ are the mean and fluctuation of the impedance matrix of the x_{i0} and are $O(1)$ and $O(N^{-1/2})$ respectively. The mean vertex correction $\gamma_{ijk}(t,s,r) = \gamma_{ikj}(t,r,s)$ is $O(N^{-1/2})$. These magnitudes, and the general form of $q_i(t)$, may be verified by Taylor expansion about $t = 0$. However, they are independent of perturbation analysis and are implied also by symmetries and form of the transformation to collective variables. The trial assumption of causal form

$$\eta_{ij}(t,s) = 0 \ (t < s), \quad \zeta_{ij}(t,s) = 0 \ (t < s),$$

$$\gamma_{ijk}(t,s,r) = 0 \ (t < s \text{ or } t < r) \tag{9.26}$$

is made in writing (9.25). This assumption also is verifiable term by term by Taylor expansion about $t = 0$.

Basic 2-time and 3-time statistics can be expressed in terms of the kernels in (9.25) and the covariance

$$X_{ij}(t,s) = \langle x_{i0}(t)x_{j0}(s)\rangle$$

by a straightforward perturbation treatment of (9.19) in the limit $N\to\infty$. Detailed derivations of the following results will not be given here. The mean linear response matrix

$$G_{ik}(t,r) = \langle \delta x_{i0}(t)/\delta x_{k0}(r)\rangle,$$

where $\delta x_{k0}(r)$ is an infinitesimal disturbance, obeys

$$\partial G_{ik}(t,r)/\partial t + \sum_j \int_0^t [\nu_{ij}\delta(t-s) + \eta_{ij}(t,s)]G_{jk}(s,r)ds = 0,$$

$$G_{ik}(r,r) = \delta_{ik}. \tag{9.27}$$

There are two 3-time mean response functions. The '2 in 1 out' function is

$$N^{1/2}\langle \delta^2 x_{i0}(t)/[\delta x_{j0}(s)\delta x_{k0}(r)]\rangle = \sum_a \int_0^t G_{ia}(t,t')\Gamma_{ajk}(t',s,r)dt', \tag{9.28}$$

where

$$\Gamma_{ijk}(t,s,r) = A_{ijk}\delta(t-s)\delta(t-r) + N^{1/2}\gamma_{ijk}(t,s,r). \tag{9.29}$$

The '1 in 2 out' function is

$$N^{1/2}\langle \delta[x_{i0}(t)x_{j0}(s)]/\delta x_{k0}(r)\rangle = \sum_{ab}\int_0^t dt'\int_0^s ds'$$

$$\times [G_{ia}(t,t')G_{jb}(s,s')\Lambda_{abk}(t',s',r)$$

$$+ G_{ia}(t,t')X_{jb}(s,s')\Gamma_{abk}(t',s',r) + G_{jk}(s,s')X_{ia}(t,t')\Gamma_{bak}(s',t',r)], \tag{9.30}$$

where

$$\Lambda_{ijk}(t,s,r) = N^{1/2}(\langle \zeta_{ik}(t,r)b_j(s)\rangle + \langle \zeta_{jk}(s,r)b_i(t)\rangle). \tag{9.31}$$

Finally, the triple correlation is given by

$$N^{1/2}\langle x_{i0}(t)x_{j0}(s)x_{k0}(r)\rangle = \sum_{abc}\int_0^t dt'\int_0^s ds'\int_0^r dr'$$

$$\times [G_{ia}(t,t')G_{jb}(s,s')G_{kc}(r,r')\Pi_{abc}(t',s',r')$$

$$+ G_{ia}(t,t')G_{jb}(s,s')X_{kc}(r,r')\Lambda_{abc}(t',s',r')$$

$$+ G_{ia}(t,t')X_{jb}(s,s')G_{kc}(r,r')\Lambda_{cab}(r',t',s')$$

$$+ X_{ia}(t,t')G_{jb}(s,s')G_{kc}(r,r')\Lambda_{bca}(s',r',t')$$

$$+ G_{ia}(t,t')X_{jb}(s,s')X_{kc}(r,r')\Gamma_{abc}(t',s',r')$$

$$+ X_{ia}(t,t')X_{jb}(s,s')G_{kc}(r,r')\Gamma_{cab}(r',t',s')$$

$$+ X_{ia}(t,t')G_{jb}(s,s')X_{kc}(r,r')\Gamma_{bca}(s',r',t')], \tag{9.32}$$

where

$$\Pi_{ijk}(t,s,r) = N^{1/2}\langle b_i(t)b_j(s)b_k(r)\rangle. \tag{9.33}$$

Each side of (9.27)-(9.33) is O(1) and includes all O(1) contributions. The higher-order terms in (9.25) do not contribute. Thus the terms shown explicitly in (9.25) fully express the alterations of the DIA $q_i(t)$ which directly effect the 2-time and 3-time quantities evaluated above. The kernels $\eta_{ij}(t,s)$, $\zeta_{ij}(t,s)$, $\gamma_{ijk}(t,s,r)$ all have t^0 terms in a Taylor expansion about $t = 0$ (note that the ζ defined for the random oscillator corresponds to a time integral of a ζ_{ij}). The function $\Pi_{ijk}(t,s,r)$ was not represented in the analysis of the random oscillator, a special case of (9.18) $(y = x_1 + ix_2,\ a = x_3)$ in which a is a degenerate (external) variable. Π_{ijk} is nonzero only if i,j,k all denote nondegenerate variables. Zero means $\langle x_{i;n}(t)\rangle$ have been assumed in the derivations of (9.25)-(9.33).

The DIA for the system (9.18) will be discussed in Sec. 10.

10. DECIMATION AND RENORMALIZED DIAGRAM EXPANSIONS

Sec. 9 showed that decimation of the Fourier collective representation of an infinite collection of systems under appropriate symmetry constraints can yield approximations similar to those obtained by renormalized diagram expansions. In particular, the DIA was derived by decimation. The logics of approximation by the DS and by diagram expansion are different, however. In the DS, as applied to the Fourier collective representation, a subset of the exact symmetry constraints which relate statistics of the sample set and statistics of the other modes is imposed. The approximations so defined are all those (including exact statistical solutions) which satisfy the imposed constraints and imposed realizability conditions. A particular approximation, like the DIA, is obtained only if a particular algorithm is used to realize ensembles of amplitudes or, as in Sec. 9, a particular ansatz is used to restrict the form of the stochastic driving of the sample-set modes by the rest of the modes.

In diagram expansion methods, chosen infinite subsets of terms in a perturbation expansion are summed to provide closed equations for a limited set of statistics of the entire set of modes. A given summation recipe, together with the (usually Gaussian) initial statistics, fully defines the associated approximation. The subsets chosen for summation may be picked on the basis of summability, preservation of symmetry and invariance properties, computability, and realizability of the resulting moment approximations. Line-renormalized summations for the system (9.18) yield approximations for η_{ij}, G_{ij} and X_{ij} while vertex-renormalized summations yield approximations for η_{ij}, G_{ij}, X_{ij}, Γ_{ijk}, Λ_{ijk} and Π_{ijk}, all defined in Sec. 9.[4,14-16]

In this Section, the relationship between the DS and RPT in the collective Fourier representation will be described further. The discussion will be based on dynamical equations less general than (9.18), namely

$$dx_{i;n}/dt + \sum_j \nu_{ij}x_{j;n} + \sum_{jk} B_{ijk}x_{j;n}x_{k;n} = 0, \tag{10.1}$$

where the notation is that of (9.18) and the B coefficients satisfy

$$B_{ijk} + B_{kji} = 0,\quad B_{ijk} = 0 \quad (i = j). \tag{10.2}$$

Eqs. (10.1) and (10.2) are a specialization of (9.18) and (9.24) in which $A_{ijk} = B_{ijk} + B_{ikj}$. They include as particular cases the random oscillator, convection of a passive scalar by an incompressible velocity field, and incompressible Navier-Stokes flow, the latter two in an appropriate modal representation. Collective amplitudes formed as in (7.1) obey

$$dx_{i\alpha}/dt + \sum_j \nu_{ij}x_{j\alpha} + N^{-1/2}\sum_{jk} B_{ijk} \sum_\beta x_{j\beta}x_{k,\alpha-\beta} = 0. \tag{10.3}$$

The construction of renormalized diagram expansions is illuminated by working with an alteration of (10.3) to the form[4]

$$dx_{i\alpha}/dt + \sum_j \nu_{ij}x_{j\alpha} + N^{-1/2}\sum_{jk} B_{ijk} \sum_\beta \phi_{\alpha,\beta,\alpha-\beta}x_{j\beta}x_{k,\alpha-\beta} = 0. \tag{10.4}$$

Here the ϕ coefficients are the same in each realization but are random functions of their indices subject to to the conditions

$$\phi_{\alpha,\beta,\alpha-\beta} = \phi^*_{-\alpha,-\beta,-\beta+\alpha}, \tag{10.5a}$$

$$\phi_{\alpha,\beta,\alpha-\beta} = \phi^*_{\alpha-\beta,-\beta,\alpha}, \tag{10.5b}$$

$$\phi_{\alpha,\beta,\alpha-\beta} = 1 \quad (\alpha,\ \beta \text{ or } \alpha-\beta = 0). \tag{10.5c}$$

The ϕ factors permit classification, and selective manipulation and suppression of renormalized diagrams. But the primary motivation for their introduction[4] was to provide a tool, otherwise missing from RPT, for constructing realizable summations. Eq. (10.3) and hence (10.1) are recovered if all ϕ factors equal unity. Variation of the ϕ factors with their indices implies a coupling of the systems in the collection.[4] Eq. (10.5a) preserves the reality condition $x_{i\alpha} = x^*_{i,-\alpha}$. Eq. (10.5b) assures that

$$\sum_{in} x^2_{i;n} = \sum_{i\alpha} |x_{i\alpha}|^2$$

is a constant of motion if all the ν_{ij} vanish. Eq. (10.5c) is needed to give acceptable representation of cases in which the $x_{i;n}$ have nonzero means. A 4th condition

$$\phi_{\alpha,\beta,\alpha-\beta} = \phi_{\alpha,\alpha-\beta,\beta} \tag{10.5d}$$

may or may not be imposed. It affects qualitatively the evolution of nonzero means.[17,18]

Let the $x_{i;n}$ have zero means and Gaussian initial statistics, and let (10.4) be solved by iteration to give an expansion in powers of the B_{ijk}. Let the resulting infinite series be multiplied to give an infinite series expansion for

$$X_{ij}(t,s) = \langle x_{i;n}(t)x_{j;n}(s)\rangle = \langle x_{i\alpha}(t)x_{j,-\alpha}(s)\rangle \quad (\text{any } n \text{ or } \alpha). \tag{10.6}$$

The statistical symmetries of the collection can be used to reduce the expansion for $X_{ij}(t,s)$ to a sum of terms each multiplied by a cycle, or summed product, of ϕ factors.[4] Each ϕ cycle is associated with a perturbation-theory diagram.

By using perturbation theory to evaluate the effects of variations of individual ϕ's, it can be shown[4] that the infinite expansion for X_{ij} can be re-expressed as a line-renormalized expansion in which only a subset of ϕ cycles, called irreducible cycles, appear as prefactors. Each irreducible cycle multiplies a functional polynomial in X and G. There is a similar line-renormalized expansion for G_{ij}.

In Ref. 4, a program was proposed in which successive approximations would be constructed by assigning statistics to the ϕ's such that only summable sequences of ϕ cycles were nonzero. If the assigned ϕ statistics are realizable, this automatically assures partial realizability for the resulting approximations for moments of the $x_{i\alpha}$.

Realizability of the ϕ's, together with the constraints (10.5a) and (10.5b), assures that (i) $X_{ij}(t,s)$ is a realizable covariance; (ii) the quadratic constants of motion of the exact dynamics are preserved; (iii) a Liouville theorem, which can be demonstrated for (10.1), is preserved. All of this follows from the properties of the explicit $x_{i\alpha}$ amplitudes which can be constructed by integration of (10.4).

The DIA with zero means is obtained if each ϕ is assigned the value +1 or −1 at random, subject only to the constraints (10.5a), (10.5b) and, optionally, (10.5d). The result is closed equations for η_{ij}, X_{ij} and G_{ij}. If there are nonzero means and (10.5c) is imposed, the result is closed equations for η_{ij}, X_{ij}, G_{ij}, and $\langle x_{i0}\rangle$.

In Ref. 4, a higher approximation, including iterated vertex corrections, was proposed. This approximation assumes that, in addition to the irreducible ϕ cycles which survive in the DIA, infinite sequences survive which express iterated vertex corrections to the DIA. All other cycles are assumed to have value zero. In the case of zero means, the result is closed equations for η_{ij}, X_{ij}, G_{ij}, Γ_{ijk}, Λ_{ijk} and Π_{ijk}.[4,15,19]

The higher approximation was shown to give consistent results in some simple cases,[4,14] but it was never demonstrated that ϕ's with the assumed properties were realizable. Actually, it can be shown by simple applications of Schwarz inequalities that, if the assumed cycles survive with their values in the exact dynamics, then essentially all cycles also have the same values as in the exact dynamics. In fact, then, ϕ's yielding the higher approximation cannot be realized.[20] At the present time, the DIA family remain the only diagram summations for (10.1) in which realizability of X_{ij} is assured.

In contrast, the DS offers systematic sequences of approximations beyond the DIA level by adjoining further constraints to (8.7). Any approximation found by the DS gives realizable X_{ij} because an ensemble of amplitudes of x_{i0} is always constructed. In addition, higher realizability inequalities can be systematically imposed, as discussed previously. On the other hand, the Liouville property of (10.1) is only partially represented in the DS if the only constraints imposed are moment constraints of finite order.

The realizability of the DIA, which is clear from the ϕ construction above, is also easily seen from the Langevin representation obtained in the DS. In the case of the random oscillator, $y_0(t)$ is constructed by (8.3) and $q(t)$, in the form (8.13), is constructible because $b(t)$ amplitudes satisfying (9.9) clearly can be constructed.

A DIA in Langevin form for the system (9.19) with zero means is obtained by imposing the constraints (9.22) and assuming that $q_i(t)$ has the form (9.25), with only the b_i and η_{ij} terms retained. The equations corresponding to (9.8) and (9.9) can be written

$$\eta_{ij}(t,s) = \sum_{abmn} A_{iab}A_{mnj}G_{am}(t,s)X_{bn}(t,s), \tag{10.7}$$

$$b_i(t) = \frac{1}{2}\sum_{jk} A_{ijk}\xi_j(t)\xi_k(t), \tag{10.8}$$

where $\xi_i(t)$ is a Gaussian random variable with covariance

$$\langle \xi_i(t)\xi_j(s)\rangle = \langle x_{i0}(t)x_{j0}(s)\rangle. \tag{10.9}$$

The moment equations for this DIA are those obtained by RPT with (10.5d) imposed. Again the realizability in the Langevin representation is clear because $b_i(t)$ clearly is realizable.

It may be possible in similar fashion to demonstrate realizability (or lack thereof) for the $X_{ij}(t,s)$ obtained in the higher approximation, described above, with iterated vertex corrections. The corresponding Langevin representation is (9.25), with only the explicitly shown terms retained. Here the key question is whether realizable $b_i(t)$ and $\zeta_{ij}(t,s)$ exist that yield the closed equations for η_{ij}, X_{ij}, G_{ij}, Γ_{ijk}, Λ_{ijk} and Π_{ijk}.

If an approximation higher than DIA is sought directly in the DS, it is logical to impose both (9.22) and (9.23). The decomposition (9.25) need not be made to compute by the DS. But if it is made, the right side of (9.23) contains contributions not only from the explicitly shown terms in (9.25) but also from the next higher terms

$$\frac{1}{6}\sum_{jkm}\iiint_0^t \gamma_{ijkm}(t,s,r,u)x_{j0}(s)x_{k0}(r)x_{m0}(u)dsdrdu, \tag{10.10}$$

where $\gamma_{ijkn}(t,s,r,u)$ is $O(N^{-1})$.

The corresponding term for the random oscillator in (8.16) is

$$-(a_0^2 - \langle a_0^2\rangle)\int_0^t \gamma_2(t,s)y_0(s)ds. \tag{10.11}$$

If the underlying $a_{;n}$ for the random oscillator have a semicircle distribution, then the DIA $\eta(t,s)$ is exact and $\zeta(t)$ [see below (8.16)] and $\gamma_1(t,s)$ both vanish. In this case, (8.8) is satisfied, with the correct $\eta(t,s)$, only if the contribution of $\gamma_2(t,s)$ to the right side of (8.8) is included. The higher RPT approximation with iterated vertex corrections does not satisfy (8.8) (see Sec. 12).

So far, all the discussion has been for zero-mean statistics. The DIA with non-zero means contains some known inconsistencies.[17,18] If (10.5d) is imposed, the mean amplitudes in dynamically linear problems, like the random oscillator or convection of a passive scalar by turbulence, and infinitesimal perturbations of the means in dynamically nonlinear problems, like Navier-Stokes turbulence, do not decay like $G_{ij}(t,s)$ as they do in the exact dynamics. In the case of the random oscillator, the mean field decay exhibits unphysical persistent oscillations.[21] If (10.5d) is not imposed, this inconsistency is removed, but then important parts of the nonlinear interactions in dynamically nonlinear problems are not represented in the DIA. For example, in Navier-Stokes turbulence the correlation of the advecting velocity with the advected velocity in the $\mathbf{u}\cdot\nabla\mathbf{u}$ term is lost.[18] Since nonlinear systems can have solutions which are nearly linear, there appears to be no satisfactory resolution within the DIA framework.

The DS may offer some light here. When the x's in (9.18) have nonzero means, the x_{i0} defined by the transformation (7.1) have $O(N^{1/2})$ means. The result is that the terms (10.10) and (10.11) in the expansion of the exact $q_i(t)$ now make $O(1)$ contributions to the right sides of (9.22) and (8.7), respectively, whereas they make $O(N^{-1/2})$ contributions in the case of zero means. This suggests that the inconsistency of the DIA with nonzero means may be repairable by adding, to the DIA expressions for triple moments, suitable terms containing $\langle x_{i0}(t)\rangle$ as well as $G_{ij}(t,s)$ and the fluctuation covariance $X_{ij}(t,s)$.

For the random oscillator with semicircle distribution of the $a_{;n}$, the exact $\gamma_2(t,s)$ in (10.11) for $N\to\infty$ can be shown to equal $-N^{-1}G(t,s)/2$. Thus, to $O(1)$, (10.11) is

$$\frac{1}{2}N^{-1/2}(a_0^2 - \langle a_0^2\rangle)\int_0^t G(t,s)\bar{y}(s)ds, \tag{10.12}$$

where

$$\bar{y}(s) = N^{-1/2}\langle y_0(s)\rangle = \langle y_{;n}(s)\rangle \quad \text{(any n)}. \tag{10.13}$$

With the correction (10.12) added to (8.12), and the DIA values taken for $b(t)$ and $\eta(t,s)$, $q(t)$ satisfies all the constraints (8.6)-(8.8) for $N\to\infty$.

In the limit $N \to \infty$, the correction (10.12) to $q(t)$ does not affect the value of $N^{1/2}\langle a_0 y_0'(t) y_0'^*(s)\rangle$, where $y_0'(s)$ is the fluctuating part of $y_0(s)$, but it does have an effect on the value taken by $N^{1/2}\langle a_0^2 y_0'(t)\rangle$. The latter moment has the expression

$$\langle a_0^2 y_0'(t)\rangle = -2\langle a_0^2\rangle \int_0^t \langle a_0 y_0'(s)\rangle ds + \langle a_0^2\rangle^2 \iint_0^t G(t,s)G(s,r)\bar{y}(r)dr, \tag{10.14}$$

where, also in the limit,

$$\langle a_0 y_0'(t)\rangle = -\langle a_0^2\rangle \int_0^t G(t,s)\bar{y}(s)ds. \tag{10.15}$$

The first term on the right side of (10.14) is identical with the DIA value with (10.5d) imposed. The second term on the right side is the contribution from the correction (10.12) to the DIA $q(t)$. It exactly cancels one-half of the first term. What is left is the DIA value with (10.5d) omitted. The approximation (10.14) is exact if the $a_{;n}$ have semicircle distributions, as assumed.

The corresponding corrections for the system (9.18)-(9.20) will be discussed in a future paper.

11. DECIMATED NAVIER-STOKES EQUATION

The decimated collective representation of a collection of systems (9.18) has value for illustrating features of the DS, in particular, the relation to RPT. However, it has negative value as a tool for actual computation of decimated ensembles of amplitudes. This is because integration of the sample set equations (9.19) under a set of symmetry constraints is more labor than integrating an ensemble of solutions of one of the identical systems (9.18). In order for the DS to be directly useful for computation, the modes $x_i(t)$ of a single system must be decimated, leaving a sample set in which the index i runs over fewer values than in the original system.

Because all the systems considered are nonlinear in stochastic quantities, even when linear in dynamical variables, the results of mode decimation depend on just what modal representation is used; they are not invariant to linear transformations of variables. The representation of the Navier-Stokes (NS) equation (and other stochastic equations involving fields in space) which behaves under decimation most like the decimated collections treated in Secs. 7-10 is spatial Fourier decomposition in a cyclic box. If the statistics are spatially homogeneous (displacement invariant), the moments of the spatial Fourier amplitudes vanish unless the wavevectors add to zero, just as moments of the collective amplitudes treated in Secs. 7-10 vanish unless the collective indices add to zero. Also, field equations with quadratic nonlinearity like the NS equation exhibit a convolution structure in the Fourier representation like that in (8.2) and (10.3).

However, spatial Fourier representation is probably not the best start for decimating the Navier-Stokes equation, even in a cyclic box. This is fundamentally because individual Fourier amplitudes are not natural dynamic entities. Each of them extends over the whole cyclic box even when dynamical interactions are effectively local in the box and correlation lengths are small compared to box size. A more natural representation for homogeneous turbulence perhaps would be decomposition of the velocity field into wavenumber bands together with a spatial discretization of the velocity in each band. The same basic considerations hold for modes of high mode number in an orthogonal decomposition appropriate to arbitrary boundary conditions and domain shape. Spatially local elementary structures of some kind probably are physically more appropriate when modes of wavelength small compared to domain size are strongly excited. The Fourier representation of homogeneous turbulence nevertheless is what will be discussed here, partly because it is traditional and well-studied and partly because it yields close analogs to the development in Secs. 7-10, which, it is hoped, provided a simple and clear introduction to the DS.

The physically unnatural character of the Fourier representation results in difficulties in constructing a suitable sample set. In Secs. 7-10, the sample set consisted of the amplitudes with collective index zero, and all moments of all collective amplitudes could be simply expressed in terms of moments of sample-set amplitudes. This was because of the strong symmetry with respect to collective index α. With the spatial Fourier decomposition, statistics vary importantly with wavevector, and amplitudes with wavevector zero do not constitute an acceptable sample set.

Let $u_i(\mathbf{k},t)$ be a Fourier amplitude of a zero-mean velocity field which is cyclic in a box of side L and has a spatially homogeneous PD. Reality of the x-space velocity field implies $u_i(-\mathbf{k},t) = u_i^*(\mathbf{k},t)$. Homogeneity implies zero-sum conditions on moments:

$$\langle u_i(\mathbf{k},t)u_j(\mathbf{p},s)\rangle = \delta(\mathbf{k}+\mathbf{p})\langle u_i(\mathbf{k},t)u_j(-\mathbf{k},s\rangle, \tag{11.1a}$$

$$\langle u_i(\mathbf{k},t)u_j(\mathbf{p},s)u_m(\mathbf{q},r)\rangle = \delta(\mathbf{k}+\mathbf{p}+\mathbf{q})\langle u_i(\mathbf{k},t)u_j(\mathbf{p},s)u_m(-\mathbf{k}-\mathbf{p},r)\rangle, \tag{11.1b}$$

etc., where $\delta(\mathbf{k}) = 1$ $(\mathbf{k}=0)$, $= 0$ $(\mathbf{k}\neq 0)$. Suppose an attempt is made to construct a sample set of amplitudes $u_i(\mathbf{k}_n,t)$ $(n = 0,\pm1,\pm2,\ldots\pm S)$ by taking $\mathbf{k}_0 = 0$ and choosing the remaining $\mathbf{k}_n$ at random from the $\mathbf{k}$'s allowed by the cyclic condition, subject to $\mathbf{k}_n = -\mathbf{k}_{-n}$ and some weighting algorithm. All moments (11.1a) can be approximated in terms of sample-set moments by some chosen interpolation formula. But in general the 3rd-order moments (11.1b), or higher-order moments, cannot be acceptably approximated in terms of sample-set moments. This is because the randomly chosen $\mathbf{k}_n$ will only accidentally contain zero-sum triads $\mathbf{k}_a \pm \mathbf{k}_b \pm \mathbf{k}_c = 0$, and the non-vanishing sample-set triple moments will generally be of the form

$$\langle u_i(\mathbf{k}_n,t)u_j(-\mathbf{k}_n,s)u_m(\mathbf{0},r)\rangle.$$

If spatially local descriptors of the velocity field were used instead, the general triple moment would be nonzero, and this problem would not arise.

An acceptable random sample set can be constructed as follows. Choose at random, according to some weighting algorithm, a set of N_S wavevectors allowed by the cyclic condition. Take the sample set wavevectors $\mathbf{k}_n$ to be the $N_S(N_S-1)/2$ ± pairs formed by taking all pairwise differences of the randomly chosen set (Fig. 2). The set $\mathbf{k}_n$ is completed by $\mathbf{k}_0 = 0$, giving $S = N_S(N_S-1)/2$. Now $N_S(N_S-1)(N_S-2)/6$ distinct zero-sum triads can be formed from the nonzero $\mathbf{k}_n$, each sum corresponding to a triangle with three of the original random wavevectors as vertices. More generally, a zero-sum n-tuple of the $\mathbf{k}_n$ corresponds to each n-gon whose vertices are n of the original random wavevectors.

General moments of the $u_i(\mathbf{k},t)$ can now be written in terms of moments of sample-set amplitudes by interpolation. This must be done in a carefully symmetric fashion in order to keep conservation properties of the full equations. Thus,

$$\langle u_i(\mathbf{k},t)u_j(\mathbf{p},s)\rangle = \textstyle\sum_{ab} C_{ab}(\mathbf{k},\mathbf{p})\langle u_i(\mathbf{k}_a,t)u_j(\mathbf{k}_b,s)\rangle, \tag{11.2a}$$

$$\langle u_i(\mathbf{k},t)u_j(\mathbf{p},s)u_m(\mathbf{q},r)\rangle =$$

$$\sum_{abc} C_{abc}(\mathbf{k},\mathbf{p},\mathbf{q})\langle u_i(\mathbf{k}_a,t)u_j(\mathbf{k}_b,s)u_m(\mathbf{k}_c,r)\rangle, \tag{11.2b}$$

etc., where the the coefficients $C_{ab}(\mathbf{k},\mathbf{p})$, $C_{abc}(\mathbf{k},\mathbf{p},\mathbf{q})$,... which depend on the values of the $\mathbf{k}_n$ and the interpolation formulas chosen, are fully symmetric to simultaneous permutations of indices and corresponding arguments. The C's vanish unless the arguments add to zero and satisfy

$$C_{ab}(\mathbf{k}_a,\mathbf{k}_b) = \delta(\mathbf{k}_a+\mathbf{k}_b),\quad C_{abc}(\mathbf{k}_a,\mathbf{k}_b,\mathbf{k}_c) = \delta(\mathbf{k}_a+\mathbf{k}_b+\mathbf{k}_c),\ \text{etc.} \tag{11.3}$$

If Lk_{max} is large, where k_{max} is the largest wavenumber admitted in a calculation, the weighting chosen to form the sample set may be essentially uniform in $\ln(Lk)$. Then $N_S \propto \ln(Lk)$ and $S \propto [\ln(Lk)]^2$. Some kind of randomly chosen sample set probably is superior to one with regularities because the latter may artificially bias in some fashion the final approximations for moments obtained by the DS.

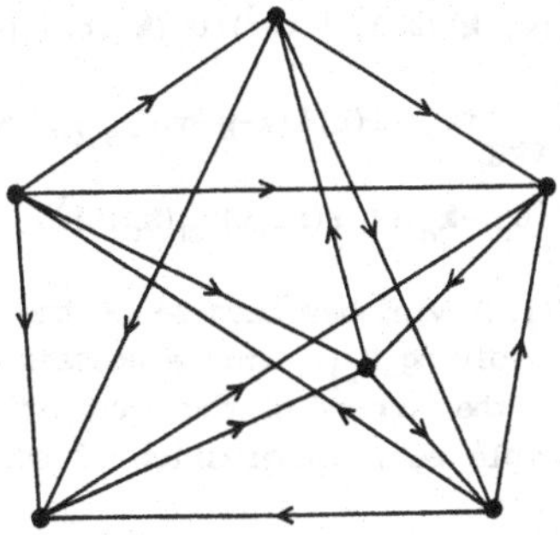

FIG. 2. Construction of sample-set wavevector network. The vertices are wavevectors picked by a stochastic algorithm from those allowed by the cyclic boundary conditions. The lines represent the $\mathbf{k}_n$ with the arrows (arbitrarily placed) indicating wavevector direction for positive n.

The incompressible NS equation in the cyclic box can be written

$$(\partial/\partial t + \nu k^2)u_i(\mathbf{k},t)$$
$$+ ik_m P_{ij}(\mathbf{k}) \sum_{\mathbf{pq}} \delta(\mathbf{k}-\mathbf{p}-\mathbf{q})u_j(\mathbf{p},t)u_m(\mathbf{q},t) = 0, \tag{11.4}$$

where ν is kinematic viscosity, the sum is over all $\mathbf{p},\mathbf{q}$ permitted by the cyclic condition, the tensor summation convention is used, and the projection operator

$$P_{ij}(\mathbf{k}) = \delta_{ij} - k_i k_j/k^2 \tag{11.5}$$

expresses the action of the pressure force in maintaining the incompressibility condition $k_i u_i(\mathbf{k},t) = 0$.

The dynamical equations for the sample-set amplitudes are then

$$(\partial/\partial t + \nu k_a^2)u_i(\mathbf{k}_a,t)$$

$$+ i(\mathbf{k}_a)_m P_{ij}(\mathbf{k}_a) \sum_{bc} \delta(\mathbf{k}_a-\mathbf{k}_b-\mathbf{k}_c) u_j(\mathbf{k}_b,t) u_m(\mathbf{k}_c,t) + q_i(\mathbf{k}_a,t) = 0, \tag{11.6}$$

where

$$q_i(\mathbf{k}_a,t) = i(\mathbf{k}_a)_m P_{ij}(\mathbf{k}_a) \sum'_{\mathbf{pq}} \delta(\mathbf{k}_a-\mathbf{p}-\mathbf{q}) u_j(\mathbf{p},t) u_m(\mathbf{q},t) \tag{11.7}$$

and the sum $\sum'$ is over all $\mathbf{p},\mathbf{q}$ pairs such that at least one of the pair is outside the sample set. Thus $q_i(\mathbf{k}_a,t)$ contains some terms with a sample-set amplitude as a factor.

The symmetry constraints on $q_i(\mathbf{k}_a,t)$ must be constructed carefully so that interpolation errors do not cause errors in energy conservation and other crucial integral properties. The sums over wavevector of the hierarchy equations constructed from (11.4) provide a guide. If left and right sides of each integrated hierarchy equation are expressed by the interpolation formulas, the resulting sums over sample-set wavevectors indicate forms, for the individual sample-set hierarchy equations, which keep the integral properties. Thus if (11.4) is multiplied by $u_n(-\mathbf{k},s)$, averaged, summed and expressed by (11.2), the result is

$$\sum_a \sum_{\mathbf{k}} C_{a,-a}(\mathbf{k},-\mathbf{k}) (\partial/\partial t + \nu k^2) \langle u_i(\mathbf{k}_a,t) u_n(-\mathbf{k}_a,s)\rangle$$

$$= - \sum_{abc} \sum_{\mathbf{kpq}} i k_m P_{ij}(\mathbf{k}) \delta(\mathbf{k}-\mathbf{p}-\mathbf{q}) C_{-a,bc}(-\mathbf{k},\mathbf{p},\mathbf{q})$$

$$\times \langle u_n(-\mathbf{k}_a,s) u_j(\mathbf{k}_b,t) u_m(\mathbf{k}_c,t)\rangle. \tag{11.8}$$

At $s = t$ the right side of (11.8) vanishes because of the symmetry of the C factors and trigonometric identities involving P_{ij}. This expresses energy conservation by the nonlinear interaction. Thus if the summand of the a-summation in (11.8) is taken as a hierarchy equation for the sample set, conservation of energy by the nonlinear term remains exact under interpolation.

Now let the corresponding hierarchy equation be constructed directly from (11.6); multiply (11.6) by $W_a u_n(-\mathbf{k}_a,s)$, average and sum over a, with

$$W_a = \sum_{\mathbf{k}} C_{a,-a}(\mathbf{k},-\mathbf{k}). \tag{11.9}$$

The nonlinear terms of the resulting equation are the same as in (11.8), and thereby give exact energy conservation, if $q_i(\mathbf{k}_a,t)$ is constrained by

$$W_a \langle q_i(\mathbf{k}_a,t) u_n(-\mathbf{k}_a,s)\rangle = S_a(t,s)$$

$$- i W_a (\mathbf{k}_a)_m P_{ij}(\mathbf{k}_a) \sum_{bc} \delta(\mathbf{k}_a-\mathbf{k}_b-\mathbf{k}_c) \langle u_n(-\mathbf{k}_a,s) u_j(\mathbf{k}_b,t) u_m(\mathbf{k}_c,t)\rangle, \tag{11.10}$$

where $-\sum_a S_a(t,s)$ is the right side of (11.8). Interpolation errors in the viscous terms have been ignored in the above discussion. If important, they can be reduced by using corrected damping factors in the sample-set equations.

The constraint (11.10) corresponds to (9.22). Higher constraints on $q_i(\mathbf{k}_a,t)$ that respect integral properties can be constructed by similar algorithms: approximating global sums by the moment interpolation formulas and expressing the result as explicit sums over the sample set. It should be noted that two things are involved in making interpolation approximations. First, the formulas chosen for moment approximation. Second, correction of dynamical equations for moments to preserve conservation and other invariance properties in the face of finite interpolation errors.

The direct-interaction approximation can be recovered in the present mode-decimation formulation by taking L very large compared to any correlation length, strongly decimating the modes in every neighborhood of **k** space, and reducing the interpolation formulas to simple identification of a zero-sum moment with the corresponding closest-neighbor moment of sample-set amplitudes.

12. RANDOM GALILEAN INVARIANCE

A long-recognized deficiency of the DIA is its failure to preserve the invariance of the NS equation to a random Galilean transformation: displacement of the flow by a spatially uniform velocity which varies randomly from realization to realization.[14] This failure is associated with spurious effects on spectral energy transfer at high wavenumbers in the presence of convection by turbulent excitation at low wavenumbers. In the exact dynamics, a random low-wavenumber excitation convects structures of small spatial scale without significant distortion and thereby has little effect on energy transfer at high wavenumbers. In the DIA, the phase mixing at high wavenumbers associated with the low-wavenumber excitation spuriously inhibits energy transfer at high wavenumbers. This problem with the DIA can be fixed by employing a Lagrangian formulation.[12] But it also can be cured in a direct way in the Eulerian framework by using the DS and imposing constraints, involving 4th-order moments, which are a subset of (9.23).

The random Galilean invariance problem is exhibited in its simplest form by a system of three random oscillators.[14] Consider the collection of N oscillator triads

$$dx_{;n}/dt + ia_{;n}kx_{;n} = 0, \quad dy_{;n}/dt + ia_{;n}py_{;n} = 0,$$

$$dz_{;n}/dt + ia_{;n}qz_{;n} = 0, \tag{12.1}$$

where

$$k + p + q = 0. \tag{12.2}$$

As in Sec. 8, statistical independence is assumed for distinct n and the statistics are independent of n. The $a_{;n}$ are assumed to have zero means, and $x_{;n}(0)$, $y_{;n}(0)$, $z_{;n}(0)$ are independent of $a_{;n}$. The collective representation is formed as in Sec. 8, the sample set is x_0, y_0, z_0, a_0, and the sample-set equations are

$$dx_0/dt + iN^{-1/2}a_0kx_0 + q_x = 0, \quad q_x = iN^{-1/2}k\sum_\beta{}' a_\beta x_{-\beta}, \text{ etc.} \tag{12.3}$$

In this model, $a_{;n}$ idealizes the role of the low-wavenumber excitation and x,y,z represent the high-wavenumber excitation.

Define S(t,r,s) by

$$S(t,r,s) = \langle x_{;n}(t)y_{;n}(s)z_{;n}(r)\rangle = N^{1/2}\langle x_0(t)y_0(s)z_0(r)\rangle, \tag{12.4}$$

and assume that the initial triple moment S(0,0,0) is nonzero. In the exact dynamics,

$$S(t,t,t) = \langle \exp[(-ia_{;n}(k + p + q)t]\rangle\langle x_{;n}(0)y_{;n}(0)z_{;n}(0)\rangle = S(0,0,0), \tag{12.5}$$

by (12.2) and the independence assumptions. According to (12.3), on the other hand,

$$dS(t,t,t)/dt$$

$$+ \langle q_x(t)y_0(t)z_0(t) + q_y(t)z_0(t)x_0(t) + q_z(t)x_0(t)y_0(t)\rangle = 0, \tag{12.6}$$

where (12.4) is used and the limit $N \to \infty$ is taken.

If the DIA form (8.12) is used for q_x, q_y, q_z, with $b_x(t)$, $b_y(t)$, $b_z(t)$ independent of $x_0(0)$, $y_0(0)$, $z_0(0)$, then (12.6) yields

$$dS(t,t,t)/dt + \int_0^t [\eta_x(t,s)S(s,t,t) + \eta_y(t,s)S(t,s,t) + \eta_z(t,s)S(t,t,s)]ds = 0. \tag{12.7}$$

Eq. (12.7) exhibits the spurious decay of triple moments in the DIA.

The DIA for this problem is obtained under the constraints

$$\langle q_x(t)x_0^*(s)\rangle = ikN^{1/2}\langle a_0x_0(t)x_0^*(s)\rangle, \text{ etc.} \tag{12.8}$$

Now let the additional symmetry constraints

$$\langle q_x(t)y_0(s)z_0(r)\rangle = ikN^{1/2}\langle a_0x_0(t)y_0(s)z_0(s)\rangle, \text{ etc.} \tag{12.9}$$

be imposed for $t = s = r$. Eq. (12.9) constitutes part of the constraints (9.23) for the present system. It follows immediately from (12.6) and (12.2) that $dS(t,t,t)/dt = 0$ for all t, as in the exact dynamics. Eq. (12.9) explicitly introduces correlation of q_x with y_0z_0, etc. when the triple moment $S(t,s,r)$ is nonzero, and this correlation just balances the spurious decay from the η terms.

In the case of the NS equation with wavevector decimation, the constraints corresponding to (12.9) are obtained, like (11.10), by forming an appropriate hierarchy equation first from (11.4) and then from (11.6) and expressing the results as sums over the sample-set wavevectors. The hierarchy equation corresponding to (12.9) is formed by multiplying (11.4) with $u_n(\mathbf{k}',s)u_\ell\mathbf{k}'',r)$ $(\mathbf{k}+\mathbf{k}'+\mathbf{k}''=0)$, averaging, and summing over $\mathbf{k},\mathbf{k}',\mathbf{k}''$.

It was shown in Ref. 14 that the approximation with iterated vertex corrections derived in Ref. 4 and discussed in Sec. 10 reduced the spurious decay of $S(t,t,t)$ compared with the DIA but did not eliminate it. This shows immediately that that approximation does not satisfy the constraints (9.23) [(8.8) for the random oscillator]. It should be noted that (8.6)-(8.8) or (9.21)-(9.23) constitute all the symmetry constraints associated with moments that appear in the mean square of the equation of motion. Any approximation that fully satisfies (8.6)-(8.8) or (9.21)-(9.23) and fully satisfies the realizability inequalities of all orders is exact.

13. REMARKS

Statistical mechanics usually treats large systems, traditionally systems which are spatially extensive. A system which is not spatially extensive can still be large if the mode density is high. Although statistical description of an ensemble is meaningful for a small system, statistical approximation as a device for computation makes sense only for systems large enough that direct integration of an ensemble of realizations is onerous or impossible. The potential value of statistical approximation derives from redundancy: statistical symmetry within large classes of modes of the large system. This symmetry can make statistics like moments, which are intrinsically more complicated than the underlying amplitudes, smooth functions of their arguments and thereby economically computable. Without statistical symmetries, a large system is no more amenable to statistical approximation than a small one.

The decimation scheme (DS) described in the present paper exploits statistical symmetries in a direct way. A relatively small sample set of modes is chosen to represent large sets of statistically redundant modes. The statistical symmetries are used to express statistics of the entire system in terms of statistics of the sample set alone. This permits integration of the equations of motion for an ensemble of amplitudes of the sample set. The accuracy of the resulting statistical approximations depends on how many constraints representing the symmetries are imposed, what realizability inequalities are imposed, and what interpolation errors arise from imperfect statistical symmetries. The systems treated in the present paper are idealizations of turbulence dynamics, but the DS is generally applicable to dynamical equations nonlinear in stochastic quantities. In particular, operator-valued variables can be treated, and approximations can be constructed involving decimated matrix representations of operators.

The DS can be used to form statistical approximations in two ways, not totally distinct. The first may be called hard closure. An example is the derivation of the direct interaction approximation (DIA) in Sec. 9 by imposing constraints associated with energy conservation and making a particular ansatz for the form of the random forcing in the Langevin equations for the sample-set amplitudes. Here the statistical approximation is wholly determined by the assumptions made, and the ansatz is an approximation justified by final results.

The second kind of statistical approximation may be called soft closure. Here a finite and incomplete set of constraints is imposed which are exact, satisfied by the exact dynamics. The constraints permit classes of approximations, including the exact dynamics. Particular results are determined by the algorithms used to construct ensembles of amplitudes satisfying the constraints. An example of soft closure is the decimation treatment in Secs. 1 and 6 of the system (1.1). Soft closure is safer because the only rigidly imposed assumptions are satisfied by the exact dynamics. Soft closures can be constructed with only desired constraints imposed. For example, instead of imposing (9.22) for all time arguments, which leads to the DIA in the limit of strong decimation, (9.22) may be imposed only for $t = s$ and $n = i$, which yields energy conservation, and (9.23) may be imposed only for $t = s = r$, which yields invariance to random Galilean transformations (see Sec. 12). The constraints could then be completed by partial imposition of (9.21), perhaps as time integrals over (9.21), giving a new approximation.

Approximations like the DIA are invariant to a linear change of basis. In contrast, the DS in general is sensitive to basis when used as a computational tool for systems nonlinear in stochastic quantities. This is illustrated by the potential use of the DS for subgrid-scale representation in turbulence calculations. If a straight Fourier representation is used, as in Sec. 11, each mode is coupled to essentially all modes of lower wavenumber and all modes of higher wavenumber. Under strong decimation, the explicit terms (involving only sample-set amplitudes) in the Langevin equation for a typical sample-set mode are a tiny part of the whole dynamics, and this makes harsh demands on numerical methods for the DS. On the other hand, if the velocity field is analyzed into wavenumber bands, and each band is represented by spatial discretizing, the dynamics exhibits a tree structure, roughly like the system (1.1) carried to multiple stages. In this case, the explicit terms make a qualitatively larger contribution to the dynamics of a typical mode.

The DS may be applied to the representation of a statistically stationary system by frequency amplitudes, or to other modal representations in time for nonstationary systems. This may yield useful approximations for the long-time statistics of chaotic maps and other low-order systems with stochastic behavior in time.

It was pointed out in Secs. 5 and 6 that in principle the DS leads to converging sequences of approximations as more symmetry constraints and more realizability inequalities are imposed. A more practical matter is estimation of the error in a given approximation. It is useful to think of two kinds of error. The first involves violation of a conservation, symmetry, or realizability condition and can be recognized and quantified directly from the final numerical products of an approximation. Examples are violation of energy conservation or negative regions in the mean density of ink convected by turbulence. The second kind, more difficult to quantify, is numerical error in quantities which satisfy demanded invariances and other qualitative properties.

A prime measure of quantitative error is the mean square of the equation of motion [e.g. (4.5)]. In an approximation where realizability of the probability density is guaranteed, this measure is immediately computable from the values of the moments which make up the mean square. In the case of bilinear dynamics, like incompressible turbulence, moments through 4th order are needed. In the DS, complete realizability is usually not guaranteed when low-order symmetry constraints and realizability inequalities are all that are imposed. If the moment constraints through 4th order [e.g. (9.21)-(9.23)] are satisfied, the mean square of the equation of motion is zero, but it is not guaranteed that the probability density is everywhere positive. In this case measures of error are provided by $\langle L^2F\rangle$, where L represents the equation of motion and F is some positive test function. Such measures can be calculated from the computed sample-set amplitudes.

The foundation for the DS is the existence of converging sequences of approximations in which the equations of motion are satisfied in mean square and successively more realizability inequalities are imposed. It was noted in Sec. 4 that the moment hierarchy equations are a subset of these inequalities. Nothing in this framework leads necessarily to decimation approaches. One alternative is to deal directly with moments, impose the mean square of the equation of motion as a moment constraint, and impose realizability inequalities by requiring that the eigenvalues of the associated moment matrices be nonnegative. Small systems like (3.3) can be successfully treated in this way by iteration methods which systematically correct moments to eliminate negative eigenvalues. The DS was originally suggested by the need for computational economies. It is perhaps surprising that such practical motivation leads to deep links between the moment hierarchy and renormalized perturbation theory.

ACKNOWLEDGMENTS

I have benefited from conversations with D. F. DuBois, B. Hasslacher, J. R. Herring, W. V. R. Malkus, S. A. Orszag, H. A. Rose, E. D. Siggia, and E. A. Spiegel. Drs. DuBois and Rose kindly commented on drafts of parts of the manuscript. This work was supported by the Global Atmospheric Research Program and Ocean Sciences Division of the National Science Foundation under Grant ATM-8008893 to Robert H. Kraichnan, Inc. and by the United States Department of Energy under Contract W-7405-Eng-36 with the University of California, Los Alamos National Laboratory.

REFERENCES

[1]M. Van Dyke, An Album of Fluid Motion (Parabolic, Stanford, 1982), Chs. 5 & 6.

[2]H. S. Wall, Analytical Theory of Continued Fractions (Chelsea, New York, 1967), p. 330.

[3]R. H. Kraichnan, Adv. Math. 16, 305 (1975).

[4]R. H. Kraichnan, J. Math. Phys. 2, 124 (1961); 3, 205 (1962).

[5]R. H. Kraichnan, J. Fluid Mech. 41, 189 (1970).

[6]C. Lanczos, Applied Analysis (Prentice Hall, Englewood Cliffs, New Jersey, 1956), p. 376.

[7]R. H. Kraichnan, Phys. Rev. Lett. 42, 1263 (1979).

[8]G. H. Hardy, J. E. Littlewood, G. Pólya, Inequalities (Cambridge University Press, Cambridge, England, 1967), p. 56.

[9]R. H. Kraichnan, in Nonlinear Dynamics, edited by H. G. Helleman (New York Academy of Sciences, New York, 1980), p. 37.

[10]But not always. For example, the two constraints $\langle x\rangle = 1$, $\langle x^2\rangle = 1$ force $x = 1$ in all realizations for any finite R.

[11]An example of realization of a velocity field with prescribed covariance is given by R. H. Kraichnan, Phys. Fluids 13, 22 (1970).

[12]R. H. Kraichnan, J. Fluid Mech. 83, 349 (1977).

[13]C. E. Leith, J. Atmos. Sci. 28, 145 (1971).

[14]R. H. Kraichnan, Phys. Fluids 11, 1723 (1964).

[15]P. C. Martin, E. D. Siggia & H. A. Rose, Phys. Rev. A 8, 423 (1973).

[16]R. Phythian, J. Phys. A 8, 1423 (1975); 9, 269 (1976).

[17]S. A. Orszag & R. H. Kraichnan, Phys. Fluids 10, 1720 (1967).

[18]J. R. Herring & R. H. Kraichnan, in Statistical Models and Turbulence, edited by M. Rosenblatt and C. Van Atta (Springer-Verlag, New York, 1972).

[19]R. H. Kraichnan in The Padé Approximant in Theoretical Physics, edited by G. A. Baker, Jr. and J. L. Gammel (Academic Press, New York, 1970), Ch. 4, Sec. V. The Γ defined in this ref. is η in the present notation.

[20]In the notation of Ref. 4, Schwarz inequalities show that, if $C_{2;1} = 1$ and $C_{4;3} = 1$, then infinite classes of cycles not reducible to these cycles by vertex contraction also have the value unity. In particular, the cycles represented by the diagrams of Ref. 4, Fig. 8(c) have this value.

[21]H. A. Rose, private communication.

CHAPTER VI. Flat-Eddy Model for Coherent Structures in Boundary Layer Turbulence

Marten T. Landahl

Abstract

An inviscid model for a "flat" eddy, i.e., a localized flow structure with large horizontal dimensions compared to its vertical extent (Landahl 1978, 1983, 1984; Russell and Landahl 1984), is explored further for the study of coherent structures in the wall region of a turbulent boundary layer or channel flow. Flat eddies may be excited by mixing in local instability regions, as may form, for example, in the thin internal shear layers produced by stretching of spanwise vorticity. Conditional sampling of the equations of motion establishes the relation beween the noncoherent motion due to local instability and the coherent structure. By making some reasonable assumptions about the statistical properties of the turbulent stresses produced by a region of local instability a model for a "typical" eddy is constructed. Comparisons with a linearized version of this model with VITA-educed sampled velocity signatures from measurements in a channel flow show good qualitative and quantitative agreement between theory and experiments for the later periods during the bursting cycle. The nonlinear version of the model applied to eddies that are highly elongated in the streamwise direction shows many of the qualitative features seen in experiments such as shear layer formation and the possible appearance of strong, localized ejection.

Introduction

The structure of turbulence is a problem of great scientific and engineering importance which has concerned many fluid dynamicists for a long time. One aspect which has attracted particular attention is the existence of large-scale coherent structures in turbulent shear flows. As an indication of the very strong interest in this field, a recent review paper by Hussain (1983) lists 278 references. This interest was inspired to a great extent by the reports from the Stanford group (Kline et al.,

1967, Kim et al. 1971) on their visualization experiments on boundary layers in a water channel revealing the existence of coherent flow structures in the near-wall showing intermittent turbulent bursting behavior. Coherent structures have also been found in free shear layers and in jets; the present paper will deal exclusively with those found in the near-wall region of a boundary layer.

The outstanding difficulty in both experimental and theoretical treatments of turbulence is that eddies of all possible scales, between the largest the flow geometry will allow down to the smallest that can be sustained against viscous dissipation, are present in the flow. Hence, even at modest Reynolds numbers the scale resolution problem becomes of almost insuperable difficulty both in laboratory investigations and in numerical simulation experiments. How the different scales interact is not yet well understood. It appears that there may be active eddies on practically all scales (in the sense that they have an important role in production of turbulence); the small scales are clearly responsible for most of the dissipation, as a simple scaling argument shows, but they also may initiate large scales through local instability, as will be illustrated below. Because of the complexity of the interaction between the various flow structures it is extremely difficult to establish cause-and-effect relationships in the occurring dynamical processes, especially since the flow is governed by nonlinear mechanisms (indeed, the random, "chaotic", behavior of turbulence is a consequence of its nonlinearity). This difficulty arises both in theoretical and experimental treatments of the turbulence problem. One possible way out of this difficulty is to use conditional sampling methods to single out a particular flow structure for detailed study. This technique has become popular in experimental studies of shear flow turbulence, and has also been used in recent theoretical studies by the present author (Landahl 1983, 1984). Of crucial importance for the success of such studies is the selection of a conditional sampling criterion that will bring out the essential mechanism. One such technique which has met with some success is the VITA-technique, first used for experimental investiation of the near-wall region by Blackwelder and Kaplan (1976), which also has turned out to have some some advantages for the theoretical treatment, as well (Landahl 1983, 1984).

Formulation of the flat-eddy model

The formulation of the flat-eddy model has been presented earlier (Russell an Landahl 1984, Landahl 1983) but will be repeated here for completeness.

Consider inviscid disturbances (of finite amplitude) in a constant density flow. The disturbances are characterised by large horizontal scales compared to the vertical one. Furthermore, we will here assume that the spanwise scale ℓ_3 is much smaller than the streamwise scale ℓ_1 but much bigger than the vertical scale ℓ_2. Hence, we will consider disturbances that have a "surfboard" character (Fig. 1) with

$$\ell_3/\ell_1 = O(\ell_2/\ell_3) = \varepsilon , \qquad \varepsilon << 1 . \tag{1}$$

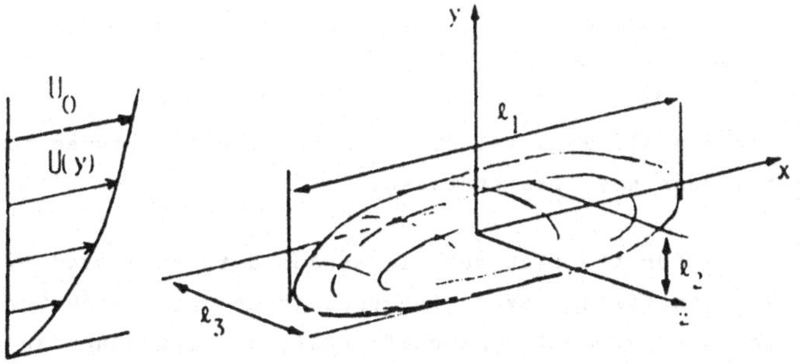

Fig. 1. A flat eddy of "surfboard" type with $\ell_3/\ell_1 = O(\ell_2/\ell_3) = \varepsilon$.

As we will see, this assumption allows for substantial simplifications in the problem, in particular for the calculation of the effects of pressure gradients.

We consider fluctuations of finite amplitude with velocities $u_1, u_2, u_3 = u, v, w$ and pressure p superimposed on a steady and incompressible shear flow, which will be assumed to be parallel and of the form $U_i = U(x_2)\delta_{i1}$, $x_1, x_2, x_3 = x, y, z$, y being the coordinate normal to the wall in the usual way. Although the assumption of a parallel mean flow holds strictly only for a channel flow it should give little errors for a boundary layer for flow structures which have scales of the same order as the boundary layer thickness or less. The fluctuations are governed by

$$\rho D[u_i + U(y)\delta_{i1}]/Dt = -\partial p/\partial x_i + \mu\nabla^2 u_i + \partial\tau_{ij}/\partial x_j \ , \qquad (2)$$

$$\partial u_i/\partial x_i = 0 \ , \qquad (3)$$

where τ_{ij} are the "fluctuating Reynolds stresses" defined by

$$\tau_{ij} = \rho(\overline{u_i u_j} - u_i u_j) \ , \qquad (4)$$

overbar denoting overall mean value. The mean flow is assumed to be in balance with the mean Reynolds stresses.

The velocity field for the flat eddy is asssumed to be educed by a (as yet unspecified) conditional averaging process through which the fluctuations are subdivided into conditionally averaged and random components as follows:

$$u_i = \tilde{u}_i + u_i' \ , \qquad (5)$$

where tilde denotes conditionally averaged quantities. Conditional averages over primed quantities are defined to be zero. By substitution into the equations of motion and application of conditional averaging one obtains

$$\rho\tilde{D}(\tilde{u}_i + U\delta_{1i})/Dt = -\partial\tilde{p}/\partial x_i + \mu\nabla^2\tilde{u}_i + \partial\tilde{\tau}_{ij} , \qquad (6)$$

$$\partial\tilde{u}_i/\partial x_i = 0 , \qquad (7)$$

where

$$\tilde{D}/Dt = \partial/\partial t + (\tilde{u}_j + U\delta_{1j})\partial/\partial x_j , \qquad (8)$$

is the time rate of change for a fluid element advected with the mean and the conditionally averaged flow field, and

$$\tilde{\tau}_{ij} = \rho(\overline{u_i u_j} - \widetilde{u_i' u_j'}) , \qquad (9)$$

are the conditionally averaged turbulent stresses due to the noncoherent motion.

We will be concerned with the evolution of the coherent structure over a time period of the order the eddy convection time

$$t_c = \ell_1/U_0 , \qquad (10)$$

where U_0 is the reference velocity (= a typical convection velocity), this being the time the eddy evolution can be traced by a fixed probe.

The turbulent stresses $\tilde{\tau}_{ij}$ will be modelled in a very simplified manner. We assume that they arise from local, small-scale instability which, as suggested by Kline et al. (1967), may arise on the thin internal shear layer produced by the large-scale coherent motion, and therefore have a vertical scale which is small compared to its horizontal ones. Secondly, since the time scale of such instabilities is set by the inverse of the mean shear, which is much shorter than the typical eddy evolution time, they may be expected to act only during a short time. Thirdly, they are assumed to produce complete mixing so as to restore the streamwise velocity to its mean value. One may therefore formulate the problem for the evolution of the coherent structure, after the cessation of small-scale instability activity, as an initial-value problem governed by (6) with $\tilde{\tau}_{ij}$=0. The initial $\tilde{v}$- and $\tilde{w}$-distributions will then be determined by the action of the stresses $\tilde{\tau}_{ij}$ as shown below.

Since we are here interested in isolating purely inviscid effects, we

Since we are here interested in isolating purely inviscid effects, we will at the outset neglect the effects of viscosity. One finds from the analysis of a simple model for a purely convected eddy (Landahl 1983) that, at least for eddies of not too small flatness parameter ε, the viscosity is likely to have only a small diffusive effect during a typical eddy evolution time scale. It is only for the long-time evolution of the eddy one needs to include viscosity (for example, in the determination of the mean velocity distribution).

For the inviscid initial-value problem we then have that $\tilde{v}_o(x_j)$ (and hence, because of continuity, also $\tilde{w}_o$) is determined by the action of the stresses $\tilde{\tau}_{ij}$, together with

$$\tilde{u}_o \equiv \tilde{u}_1(x_j,0) = 0 \ , \tag{11}$$

The boundary conditions are that u_i vanishes at large distances from the eddy and that, with the wall at $y=0$,

$$\tilde{v} \equiv \tilde{u}_2 = 0 \text{ at } y = 0 \ . \tag{12}$$

We seek a formal solution by introducing Lagrangian variables ξ_i ($\xi_1=\xi$, $\xi_2=\eta$, $\xi_3=\zeta$) which are equal to x_i at $t=0$. By straightforward integration of the first and third of (7) with $\mu=0$ and $\tilde{\tau}_{ij}=0$ we find

$$\tilde{u} + U(y) = U(\eta) + \tilde{u}_o(\xi,\eta,\zeta) - (1/\rho)\int_0^t (\partial\tilde{p}/\partial x)Dt' \ . \tag{13}$$

$$x = \xi + t[U(\eta) + \tilde{u}_o(\xi,\eta,\zeta)] - (1/\rho)\int_0^t (t - t')(\partial\tilde{p}/\partial x)Dt' \ , \tag{14}$$

$$\tilde{w} = \tilde{w}_o(\xi,\eta,\zeta) - (1/\rho)\int_0^t (\partial\tilde{p}/\partial z)Dt' \ , \tag{15}$$

$$z = \zeta + t\tilde{w}_o - (1/\rho)\int_0^t (t - t')(\partial\tilde{p}/\partial z)Dt' \ . \tag{16}$$

with $\tilde{u}_o=0$. Here, $\int \div Dt'$ means integration over time holding ξ_i fixed.

To determine the remaining vertical velocity component, $\tilde{v}$, and the coordinate y, for the fluid element ξ,η,ζ, we use the continuity equation, which in terms of the Lagrangian coordinates may be written (Lamb, 1932, art. 14)

$$J = \frac{\partial(x,y,z)}{\partial(\xi,\eta,\zeta)} = 1 \ . \tag{17}$$

where J is the Jacobian. Expansion of (17) leads to a set of first-order nonlinear partial differential equation for the Lagrangian variables, the solution of which may be found formally with the aid of the method of characteristics to be given by direct integration along the curve (Russell and Landahl 1984)

$$d\xi_i/dy = A_i \ , \tag{18}$$

where

$$A_1 = x_\zeta z_\eta - x_\eta z_\zeta \ , \tag{19}$$

$$A_2 = x_\xi z_\zeta - x_\zeta z_\xi \ , \tag{20}$$

$$A_3 = x_\eta z_\xi - x_\xi z_\eta \ . \tag{21}$$

The pressure, which enters into the expressions for x and z, may be found from the solution of the Poisson equation

$$\nabla^2(\tilde{p}/\rho) = \sigma = -2U'\partial\tilde{v}/\partial x - \partial^2(\tilde{u}_i\tilde{u}_j)/\partial x_i\partial x_j \tag{22}$$

which is obtained by taking the divergence of (7) (again with $\tilde{\tau}_{ij}=0$). It is in the handling of the pressure that the flat-eddy assumption produces considerable simplifications. Russell and Landahl (1984) showed that, to lowest order in ε, a solution consistent with the boundary condition (12) is

$$\tilde{p}/\rho = -\int_y^\infty (y-y')\sigma(x,y',z,t)dy' - (1/2\pi)\iint_{-\infty}^{\infty} \Sigma'(x',z',t)R^{-1}dx'dz' \ , \tag{23}$$

where

$$\Sigma'(x',z',t) = \int_0^\infty \sigma(x',y',z',t)dy'$$
$$= -\int_0^\infty \{[(U+\tilde{u})^2]_{xx} + 2[(U+\tilde{u})w]_{xz} + (\tilde{w}^2)_{zz}\}dy \ , \tag{24}$$

and

$$R = \{(x-x')^2 + y^2 + (z-z')^2\}^{1/2} . \tag{25}$$

For a "surfboard" type eddy with $\ell_3/\ell_1 = O(\varepsilon)$ a further simplification is possible in that, following the techniques used in slender-wing theory, the double integral in (24) may be replaced by a single one (Landahl 1984),

$$\iint_{-\infty}^{\infty} \Sigma(x',z',t)R^{-1}dx'dz' \approx -\int_{-\infty}^{\infty} \Sigma(x,z',t)\ln[y^2 + (z-z')^2]dz' . \tag{26}$$

One finds from the above that the pressure is of the order

$$p = O(\varepsilon\rho\tilde{v}_0 U_0) \tag{27}$$

and that, for times of the order t_c, its direct contribution to u through the streamwise pressure gradient is of order $\varepsilon\tilde{v}_0$ and hence negligible in comparisons with the effect of fluid element liftup, as expressed by the term $U(\eta)-U(y)$, which gives a contribution of order $\tilde{v}_0\varepsilon^{-2}$ during this time. The contribution from spanwise pressure gradient to $\tilde{w}$ is of order $\tilde{v}_0 = O(\varepsilon\tilde{w}_0)$, which will be retained as a correction, as will be the effect on the liftup of the fluid element due to the crossflow pressure gradient, which will be of order $\tilde{v}_0\varepsilon^{-1}$.

Since the pressure thus accounts for only a small correction, one may devise a simple iterative calculation procedure, in which one first ignores the effects of the pressure to determine from (18) a zeroth-order ("homobaric", Russell and Landahl 1984) approximation with

$$A_i = A_i^{(0)} , \tag{28}$$

where $A_i^{(0)}$ are calculated using

$$x = x^{(0)} = \xi + U(\eta)t, \; z = z^{(0)} = \zeta + \tilde{w}_0(\xi,\eta,\zeta)t \tag{29}$$

in (19). The results then obtained for the velocity components may then be used to determine the pressure and hence a first-order correction to the position and velocity of the fluid element. In Landahl (1978) the correction was obtained on basis of linearized theory and it was shown that the long-time development of the eddy contained a component in form of a damped wave with a spanwise wavelength of somewhat over a hundred measured in wall variables. In Russell and Landahl (1984) the pressure was extrapolated by a two-term Taylor series from t=0 (with the nonlinear terms included). In Landahl (1984) linearized slender-body theory was used to evaluate from $\tilde{u}_i^{(0)}$ the change in fluid element displacement due to pressure in a single time step. In the present calculations a multi-time-step procedure has been developed in which the velocity field adopted at a previous time step is used to determine the pressure effects during the next time step as described below.

Modeling of the initial velocity and stress fields

Under the assumption that the turbulent stresses due to the noncoherent flow field act during a time period which is very short compared to the eddy convection time the motion of the fluid elements will be small during this time and the solution of (6) may consequently be approximated by

$$\tilde{u} = \tilde{u}^{(i)} + (1/\rho)\int_{t_i}^{t} (\partial\tau_{12}/\partial y)Dt' \ , \tag{30}$$

$$\tilde{w} = \tilde{w}^{(i)} + (1/\rho)\int_{t_i}^{t} (\partial\tau_{23}/\partial y)Dt' \ . \tag{31}$$

Here, $t_i < 0$ is the time at which the stresses τ_{ij} commence their action and $\tilde{u}^{(i)},\tilde{v}^{(i)},\tilde{w}^{(i)}$ the initial velocity field at $t=t_i$. Because of the flatness of the eddy, only the stress terms involving y-derivatives have been retained. The pressure induced by the coherent motion is of order $\varepsilon\rho U_0\tilde{v}^{(i)}$, which may be shown to give a negligible correction of order $\varepsilon^2\tilde{v}^{(i)}=O(\varepsilon\tilde{u}^{(i)})$ during this time.

Application of continuity then gives for t=0

$$\partial\tilde{v}_0/\partial y = \partial\tilde{v}^{(i)}/\partial y - (1/\rho)(\partial/\partial y)\int_{t_i}^{0} (\partial\tilde{\tau}_{12}/\partial x + \partial\tilde{\tau}_{23}/\partial z)Dt' \ . \tag{32}$$

Since very little is known about the spanwise stress component $\tilde{\tau}_{23}$ we will assume it to be zero. Also, because the mean profile is stable to small $\tilde{v}$-disturbances (Landahl 1967) it is reasonable to set $\tilde{v}^{(i)}=0$ on the assumption that the events are well separated in time. Furthermore, we will assume that the mixing due to the local instability is complete so as to restore the velocity profile to its mean distribution. These assumptions, together with integration of (32), then gives the initial conditions

$$u_0 = 0\,,$$
$$v_0 = -(1/\rho)\int_{t_i}^{0}(\partial\tau_{12}/\partial\xi)Dt'\,, \qquad (33)$$
$$w_0 = -\int^{\zeta}(\partial v_0/\partial\eta)d\zeta'\,,$$

where we have replaced the Eulerian variables with the Lagrangian ones in view of the short time scale of the instability process. For the stress τ_{12} we select a modified "Gaussian hat" (Landahl 1983, 1984),

$$\tau_{12} = C\,\bar{\eta}^{\,3}(1-2\bar{\zeta}^{\,2})\exp(-\bar{\xi}^{\,2}-\bar{\eta}^{\,2}-\bar{\zeta}^{\,2})\,, \qquad (34)$$

where

$$\bar{\xi} = \xi/\ell_1,\ \bar{\eta} = \eta/\ell_2,\ \bar{\zeta} = \zeta/\ell_3 \qquad (35)$$

which, with the choice of $\ell_2^+=15$, gives reasonable fit to the y-distribution of the turbulent shear stresses during bursting obtained in the measurements by Kim et al. (1971). The scales ℓ_1 and ℓ_3, and the constant C measuring the disturbance amplitude, will be left unspecified for the time being. The factor $(1-2\bar{\zeta}^2)$ ensures that $w_0 \to 0$ for large $\bar{\zeta}$.

The initial velocity field thus selected has the form of two pairs of counterrotating streamwise vortex structures, as illustrated in Fig.2. In view of the prominent role longitudinal vortices have played in the discussion on wall-layer structures it is of interest to notice that longitudinal vorticity may be produced by the action of local instability.

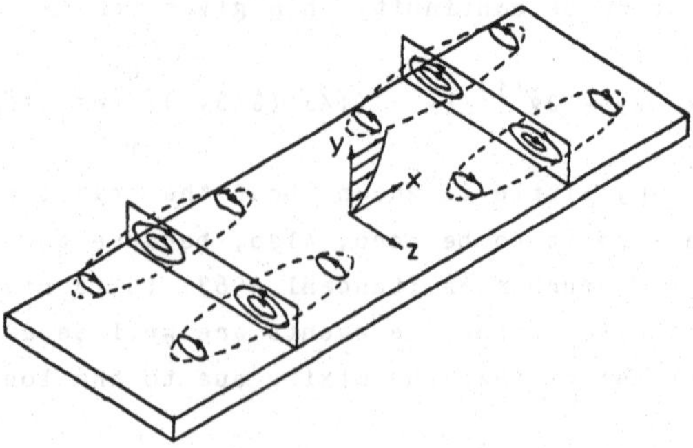

Fig. 2. Initial velocity field (schematic) for a "surfboard" type eddy (from Russell and Landahl, 1984).

Application to VITA-educed coherent structures

In the VITA-technique (Blackwelder and Kaplan, 1976) the sampling criterion is that the short-time variance for u exceeds a selected threshold k times the mean square fluctuations,

$$\mathrm{var}(u) = (1/T)\int_{-T/2}^{T/2} u^2 dt - [(1/T)\int_{-T/2}^{T/2}]^2 > k u_{rms}^2, \tag{36}$$

where T is the selected averaging time. As pointed out by Johansson and Alfredsson (1982) this serves as a filter bringing out structures of typical time scale of T. The threshold k is usually chosen between 0.5 and 2. The condition (36) provides a time reference time t_o for the detection of the eddy around which the sampling is carried out. Blackwelder and Kaplan (1976) discovered that the nondimensional conditional average

$$\langle u\rangle^* = u/(k u_{rms}^2)^{1/2} \tag{37}$$

collapses fairly well onto a single curve as function of the time $\tau = t - t_o$ for different values of the threshold parameter k. Johansson and Alfredsson (1982) found that the collapse becomes even better if one treats the events with accceleration and deceleration separately. In the present theoretical model we consider only the accelerating case.

In the theoretical model the threshold is assumed to be low so that the disturbances contributing to the conditional average are sufficiently weak to permit linerization. Expansion of the solution for small disturbances gives, for the "surfboard" eddy, after terms of order ε^2 and higher have been neglected,

$$u = -\ell U'(y) , \tag{38}$$

$$v = (\partial/\partial t + U\partial/\partial x)\ell , \tag{39}$$

where

$$\ell = y - \eta = \ell^{(0)} + \ell^{(1)} , \tag{40}$$

in which

$$\ell^{(o)} = t\int_0^y \tilde{v}_{o\eta}(\xi',y')dy' \ , \tag{41}$$

$$\ell^{(1)} = -(1/\rho)(\partial^2/\partial z^2)\int_0^y dy'\int_0^t (t-t')\tilde{p}(\xi'',y',z,t')dt' \ , \tag{42}$$

and

$$\xi' = x - U(y')t \ , \tag{43}$$

$$\xi'' = x - U(y')(t - t') \ , \tag{44}$$

Johansson and Alfredsson (1982) also noticed that the number of events detected is a very steep function of the threshold k, so that only a small interval of amplitudes contribute to the ensemble average. Therefore, the mean square of the conditionally averaged velocity may be approximated by the square of its conditional average, and in seeking the velocity signature satisfying the sampling condition (36), one may replace u^2_{rms} by $\overline{u^2}$ in (37).

The free parameters that still need to be determined in the model for the initial velocity field chosen are the horizontal scales ℓ_1, ℓ_3 and the coefficient C giving the amplitude of the conditionally sampled velocity field. Also, one needs to determine the time of detection relative to the time of the onset of the instability induced stresses as well as the distance x_o to the center of the eddy at the time of detection. Employing the iteration in the small parameter ε one finds that the lowest-order solution becomes independent of the spanwise scale ℓ_3 and also that it depends on the parameters x/ℓ_1, t/ℓ_1, only, not on x, t and ℓ_1, separately.

To determine the remaining free parameters for the lowest-order solution one searches for the combination of parameters that maximize

$$M = \mathrm{var}(u)/(\overline{u^2}\ )^{1/2} \quad , \tag{45}$$

for a given integration time T. An eddy with such a combination will give the maximum contribution to the threshold for a given amplitude C. The search for the two parameters x/ℓ_1 and t_0/ℓ_1 is easily accomplished with the aid of a computer. Because of (45), the nondimensional conditional average is then obtained as

$$\langle u\rangle^* = u/(M_{max}\overline{u^2})^{1/2} \; . \tag{46}$$

The nondimensional velocity becomes independent of C, and hence of k, in consistency with the experimental findings. A comparison of a $\langle u\rangle^*$-signature obtained in this manner with the experimental results of Johansson and Alfredsson (1984) is shown in Fig.3 (reproduced from Landahl 1983). Inclusion of the first-order terms in ε due to pressure (Landahl 1984) gives results that differ insignificantly from those obtained from the lowest-order solution. As seen, the theory agrees well with the experiments, except for the early time period, for which it gives a too rapid onset of the velocity defect. This discrepancy is perhaps not surprising in view of the very crude model employed for the stress initiation.

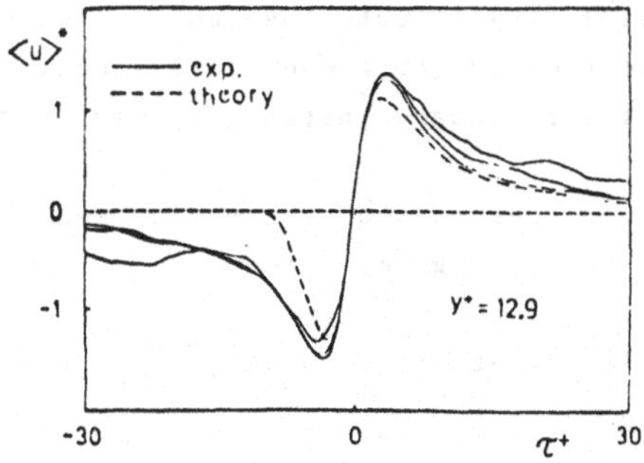

Fig. 3. Conditionally sampled and normalized streamwise velocity as function of time from detection $\tau^+=t^+-t_0^+$ educed by the VITA-technique with $T^+=10$ for accelerating events at $y^+=12.9$. Experiments by Johansson and Alfredsson (1982). (From Landahl, 1983).

It is particularly interesting that a basically linear theory is able to predict the amplitude of the nondimensional conditionally averaged velocity signature. The success of the theory is due in part to the fact that the quantity averaged, namely u, is also that used for the detection. Thus, the individual events contributing to the average are all lined up at $\tau=t-t_0=0$. For the conditional averaged v-component, on the other hand, the theory is found to give a maximum amplitude that exceeds the measured one by a factor of about 2.5 (see Landahl 1983), the reason being that the theory assumes perfect correlation between the two components in each realization, whereas in reality there is a certain phase jitter between u and v. This also shows up in the amplitude of the conditionally sampled uv-distribution, which the theory overpredicts by a similar factor. Nevertheless, the theory gives a good representation of the shape of the measured conditionally sampled uv-signature, as is seen in Fig.4 (reproduced from Landahl 1983).

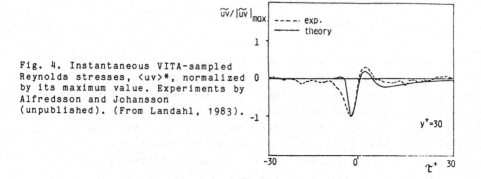

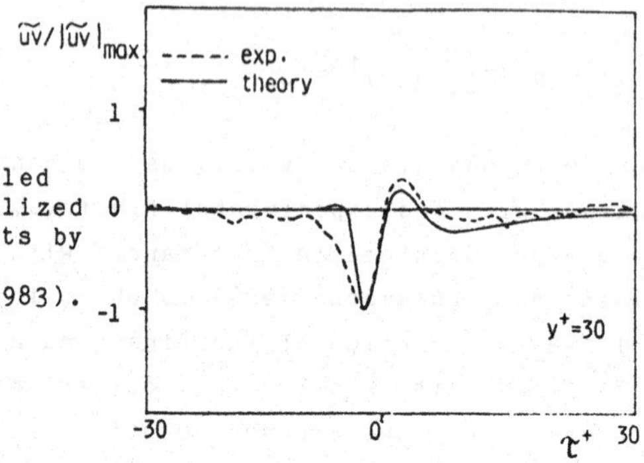

Fig. 4. Instantaneous VITA-sampled Reynolds stresses, <uv>*, normalized by its maximum value. Experiments by Alfredsson and Johansson (unpublished). (From Landahl, 1983).

Nonlinear eddy development

Nonlinearity effects the eddy in two main ways; i) it causes nonlinear distortion due to large cross-stream displacements of the fluid elements and ii) secondary instability may arise in the disturbed flow field.

The nonlinear distortion may be understood from (18) - (21). In the case of $\varepsilon \to 0$, (29) give, with $u_o = 0$

$$A_1^{(o)} = -tU'(1 - tv_{o\eta}) , \tag{47}$$

$$A_2^{(o)} = 1 - v_{o\eta}t , \tag{48}$$

$$A_3^{(o)} = t^2U'w_{o\xi} . \tag{49}$$

The solution becomes singular and breaks down (Stern and Paldor, 1983, Russell and Landahl 1984) if $A_i = 0$ at any point in the flow, since then all the material coordinates ξ_i=const possess a vertical tangent at the same point. Such a point was called a "tent-pole" singularity by Russell and Landahl (1984). For the initial velocity field selected here with spanwise symmetry with respect to $\zeta=0$, we have $w_o=0$ at the plane of symmetry. Hence $A_3=0$ at that plane, and the singularity for the $\varepsilon \to 0$ solution would then arise for points on the symmetry plane for which

$$1 - tv_{0\eta} = 0 \ . \tag{50}$$

For the chosen initial velocity field the maximum of $v_{0\eta}$ occurs for $(\overline{\xi},\overline{\eta})=(1/\sqrt{2},1/\sqrt{2})$ and leads to a tentpole singularity at

$$t = t_s = \ell_1 \ell_2 e/C\sqrt{2} \ . \tag{51}$$

For an illustration of the flow behavior near such a singularity, in Fig. 5 we have used the $\varepsilon=0$ solution to determine as function of time the location of a vertical line of fluid markers released at time $t=0$ and position $\overline{x} = x/\ell_1 = 0.5$. In this case the following parameters were chosen:

$$C = U_0 \ell_2, \ \ell_2^+ = 20 \ .$$

In the figure time is made dimensionless with "outer" variables ℓ_1 and U_0, where U_0 is taken as the velocity at y+=90. For this set of parameters,

$$\overline{t}_s = t_s U_0/\ell_1 \quad = 3.85$$

so that the curve for $\overline{t}=4$ is actually outside the range of validity of the theory. As seen, the marker patterns show a striking resemblance of those seen in visualization experiments by Kline et al. (1967) and Kim et al. (1971) indicating that the violent ejection phase of bursting may be associated with the approach to a singular conditions for the inviscid problem.

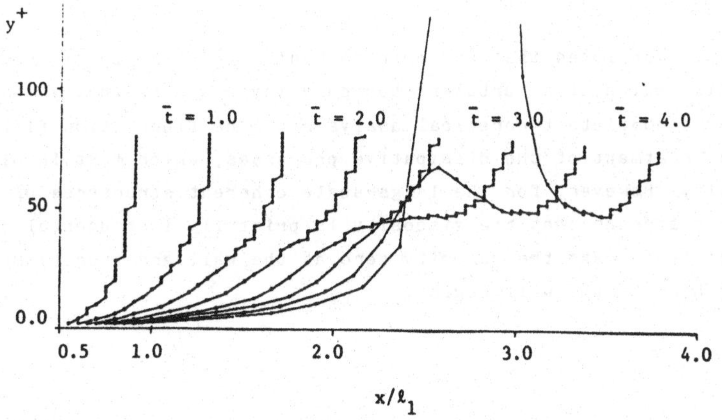

Fig. 5. Simulated fluid marker patterns obtained from homobaric solution. Markers are released at $t=0$ at $\overline{x}=x/\ell_1=0.5$.

However, it follows from (18) - (21) that in the neighborhood of the singularity for the homobaric ($\varepsilon=0$) approximation the pressure becomes increasingly important, and the contribution from it may actually prevent the singularity from appearing. This was the finding in the calculations by Russell and Landahl (1984), in which the pressure was approximated by its first two terms in a Taylor series expansion around $t=0$. Preliminary calculations employing a time-stepping procedure in which the pressure is updated at each time step indicate that the pressure may also introduce an oscillatory motion (see forthcoming MSc thesis by Henningson, 1985).

Conclusions

The flat-eddy model for coherent structures in the near-wall region of a turbulent boundary layer presented here shows surprisingly many of the qualitative features seen in the experiments. Crucial for the regeneration of new eddies from old ones in this model is the formation of thin shear layers due to stretching of spanwise mean vorticity introducing new regions of local instability. The nonlinear theory also shows the possible appearance of strong ejections arising because of horizontal convergence of fluid elements causing them to form jet-like eruptions from the surface. No corresponding strong jet-like inflows have been indicated by the theory. Conditional sampling treatment of weak, linear disturbances based on the VITA-technique gives velocity signatures in good qualitative and quantitative agreement with experimental data for events associated with accelerations.

It may be concluded that the main mechanisms of turbulence production in the wall region of a turbulent boundary layer are basically inviscid. Of course, a complete theoretical analysis of the fluctuation field requires also a treatment of the dissipative processes, which must involve viscosity. However, for the large-scale coherent structures of interest here, it appears that the viscosity is primarily instrumental on a long time scale to make the velocity zero at the wall and thereby maintain a large shear in the wall region.

Acknowledgements

This research was supported by the U.S. Air Force Office of Scientific Research under Contract F-49620-83C-0019.

References

Blackwelder, R.F. and Kaplan, R.E. 1976 On the wall structure of the turbulent boundary layer. J. Fluid Mech. 76, 89.

Henningson, D.S. 1985 Solution of the 3D Euler equations for the time evolution of a flat eddy in a boundary layer. MSc Thesis, Mass. Inst. Technology, Department of Aeronautics and Astronautics.

Hussain, A.K.M.F. 1983 Coherent structures - reality and myth. Phys. Fluids 26, 2763.

Johansson, A.V. and Alfredsson, P.H. 1982 On the structure of turbulent channel flow. J. Fluid Mech. 122, 295.

Kim, H.T., Kline, S.J. and Reynolds, W.C. 1971 The production of turbulence near a smooth wall in a turbulent boundary layer. J. Fluid Mech. 50, 133.

Kline, S.J., Reynolds, W,C,, Schraub, F.A. and Runstadler, P.W. 1967 The structure of turbulent boundary layers. J. Fluid Mech. 30, 741.

Lamb, H.E. 1932 Hydrodynamics, Sixth Edition. Dover Publications.

Landahl, M.T. 1967 A wave-guide model for turbulent shear flow. J. Fluid Mech. 29, 441.

Landahl, M.T. 1978 Modeling of coherent structures in boundary layer turbulence in: AFOSR/Lehigh University Workshop on Coherent Structure of Boundary Layers, 340. C.R. Smith & D.E. Abbott, (eds.).

Landahl, M.T. 1983 Theoretical modelling of coherent structures in wall bounded shear flows. in Proceedings, Eighth Biennial Symposium on Turbulence. (eds. G.K. Paterson and J.L. Zakin), Univ. of Missouri-Rolla.

Landahl, M.T. 1984 Coherent structures in turbulence and Prandtl's mixing length theory. Z. Flugwiss. and Weltraumforschung, 8(4), 233.

Reichardt, H. Vollständige Darstellung der Turbulenten Geschwindigkeitsverteilung in Glatten Leitungen. Z. ang. Math. & Mech. 31 (7) 208.

Russell, J.R., and Landahl, M.T. 1984 The evolution of a flat eddy near a wall in an inviscid shear flow. Phys. Fluids 27, 557.

CHAPTER VII. Progress and Prospects in Phenomenological Turbulence Models

B.E. Launder

1. INTRODUCTION

Despite what the dictionaries say, 'phenomenological' linked with 'turbulence model' means not so much that conformity with observed phenomena is an important factor in fixing the model's form, but rather that it is based on a *too superficial set* of observations to allow any very interesting or genuinely useful results to spring from it. In the present workshop, speakers have used the term in its essentially pejorative sense to dismiss models ranging from the mixing-length hypothesis to third moment closure.

What the *organizers* of the Workshop had in mind when requesting a paper on the title topic was, presumably, a personal view of where we stand in single-point closure. They recognized that in 1984 and for another decade at least the vast majority of large-scale calculations of flows through and around engineering equipment would be based on turbulence models of this type. My aim is to give some impression of the capabilities and limitations of representatives of this class. The territory is too vast, the model types too varied to cover in a brief presentation of this kind, but not too much will be lost by confining attention to the three types of model listed in Table 1 for it is at these levels that most current activity is directed.

Before proceeding to specific models and closure ideas, a few general observations are in order. First, single-point closures adopt such a simplified view of turbulence that nowadays no-one seriously supposes that a *universally valid* model can be developed with this approach. Nonetheless, the rationale underlying turbulence modelling at this or any other level must be to devise a set of formulae (differential and/or algebraic equations) that adequately describe a *broad* range of flows and phenomena. Some assert that a given single-point closure will only reliably handle flows with a single type of large-eddy structure (since such models take no explicit account of these large scale motions). That is patently a too pessimistic view to take. Even the simplest of the models to be discussed here [1] correctly predicts the turbulent shear stress in both the flat plate boundary layer and in the plane jet in stagnant surroundings - flows with very

	Model Type	Acronym	Example References	Turbulent Transport Equations
	2-equation eddy viscosity model	EVM	[1], [15]	k, ε
second-moment closures	Algebraic stress model	ASM	[15]	k, ε
second-moment closures	Differential stress model	DSM	[3], [4]	ε, $\overline{u_i u_j}$

Table 1 Types of single-point closures considered

Shear flow	$\dfrac{dy_{\frac{1}{2}}/dx \text{ - ASM}}{dy_{\frac{1}{2}}/dx \text{ - DSM}}$
Asymptotic plane wake	1.35
Round jet	1.20
Plane jet	1.05
Equilibrium wall jet in adverse pressure gradient (case 0261 [14])	1.0

Table 2 Relative rates of spread of equilibrium shear flows with ASM and DSM closures

different large-scale structures. The earth's gravitational field can provoke a bewildering, beautiful variety of large-scale structures in a turbulent shear flow depending upon the degree of stratification and orientation of the strain relative to the acceleration vector. Yet, while having nothing to say about the large-scale structures themselves, second-moment closures have nevertheless successfully reproduced the turbulent stress and heat flux levels under conditions ranging from strongly stable stratification to a purely buoyantly driven convection [2-4].

There are, nevertheless, several 'awkward' turbulent flows - perhaps notably the axisymmetric jet in stagnant surroundings - whose development is known to be seriously in disagreement with that predicted by models whose form and coefficients have been shaped by other shear flows. The aim of much current research is to determine how to adapt existing closures so that they mimic correctly the behaviour of the anomalous cases without destroying the satisfactory agreement generally achieved over a wide range of other flows.

Some turbulent flows are intrinsically unsuited to modelling by single-point closure. Some of the current studies in turbulence management outlined by Dr. Bushnell are of this type where amplification or suppression of particular frequencies or wave lengths in the spectrum is a crucial part of the phenomenology. Yet, this fact provides no encouragement to relegate single-point approaches to the closet - even supposing an alternative were available. The automobile has not replaced the bicycle, nor the aeroplane the car; rather, each new generation of transport machines has extended the range of journeys that could be undertaken. Thus, while the Workshop has presented several interesting prototypes and design concepts of advanced vehicles for turbulent flow exploration, even when some of these eventually reach the production stage there will remain a continuing need for models of the type described below.

2. A SUMMARY OF THE MODELS AND THEIR CAPABILITIES

Stress-Strain Connections

The transport equation describing the evolution of the Reynolds stresses $\overline{u_i u_j}$ is readily obtained by taking a turbulent velocity-weighted moment of the Navier-Stokes equations and averaging; we write this in symbolic form as:

$$\frac{D\overline{u_iu_j}}{Dt} - D_{ij} = P_{ij} - \phi_{ij} - \frac{2}{3}\varepsilon\delta_{ij} \tag{1}$$

where D_{ij}, P_{ij} and ϕ_{ij} denote respectively net diffusive gain or 'transport' of $\overline{u_iu_j}$; the rate of stress generation by mean velocity gradients, rotation, buoyancy, etc.; and the combined effect of non-dispersive pressure interactions and non-isotropic dissipation processes. The last of these terms is traceless, $\phi_{kk} = 0$. Finally, ε is the dissipation rate of turbulent kinetic energy (so that $\frac{2}{3}\varepsilon\delta_{ij}$ is the viscous dissipation rate of $\overline{u_iu_j}$ in high Reynolds-number, locally-isotropic turbulence). Of the above processes, only P_{ij} is exactly representable in terms of second moments and mean-field quantities. It is a very important term, however, and it is the fact that stress generation agencies *are* treated exactly in second-moment closures that gives them decisive advantages over simpler phenomenological models. In an orthodox second-moment treatment, approximations are supplied for ϕ_{ij}, ε_{ij} and D_{ij}. Alternatively, an approximation is often made for the *total* transport term, $\left\{ \frac{D\overline{u_iu_j}}{Dt} - D_{ij} \right\}$, which for brevity we write as T_{ij}. The usual assumption is that proposed by Rodi [5]

$$T_{ij} = \frac{\overline{u_iu_j}}{\overline{u_ku_k}}\, T_{mm} \tag{2}$$

Notice that since ϕ_{ij} is traceless, by contracting eq(1), eq(2) may be written

$$T_{ij} = \frac{\overline{u_iu_j}}{\overline{u_ku_k}}\,(P_{mm} - 2\varepsilon) \tag{3}$$

This approximation of T_{ij}, while broadly plausible, is motivated more by considerations of numerical simplicity than physical realism. The right-hand side of eq(3) contains no differential operations on the turbulent stresses; nor does P_{ij}. Thus, if (as is nearly always the practice) the approximations adopted for ϕ_{ij} also contain no stress differentials, the replacement of T_{ij} by the right side of eq(3) reduces eq(1) from a set of differential to a set of algebraic equations. Accordingly, this type of second-moment closure is often called an *algebraic* stress model (ASM). With models of this type only the trace of the Reynolds stress tensor (strictly the turbulent kinetic energy $k \equiv \overline{u_ku_k}/2$) is obtained from a transport equation. In all the schemes discussed here, the energy dissipation rate, ε, is also found from a rate equation (of which more later); so, ASMs require the numerical solution of two partial differential equations for turbulence quantities.

The alternative approach of treating convective transport, $D\overline{u_i u_j}/Dt$, exactly and modelling D_{ij} on its own (in a form that *does* habitually include spatial gradients of $\overline{u_i u_j}$) will here be called a *differential* stress model - or DSM for short. While several recent studies [6,7,8] have made a serious attempt to model the constituent parts of D_{ij} (involving contributions from both pressure and velocity fluctuations - and from fluctuating body forces, if present) in engineering computations the most frequently made approximation is:

$$D_{ij} = -c_s \frac{\partial}{\partial x_k}\left[\frac{k}{\varepsilon}\,\overline{u_k u_\ell}\,\frac{\partial \overline{u_i u_j}}{\partial x_\ell}\right] \tag{4}$$

which may be regarded as a particular application of the generalized gradient diffusion concept, perhaps first formally stated by Daly and Harlow [9]. The optimum value of the coefficient c_s is about 0.2. While eq(4) is a gross oversimplification of the interactions producing a net diffusive transport of $\overline{u_i u_j}$, there is a widely held view that replacing this model by one of the more elaborate, newer proposals will have little effect where it matters, i.e. on the mean velocity distribution, the shear-layer spreading and mixing rates, etc. For this reason, discussion of processes and their modelling is here directed mainly at ϕ_{ij} and ε.

The third model type represented in Table 1 shares with the second-moment closures the problem of obtaining the turbulence energy dissipation rate. Like an ASM it determines local values of the turbulence energy via a transport equation but instead of truncating eq(1) to obtain a constitutive equation linking stress and strain, it assumes the simple Newtonian stress-strain law

$$-\left(\overline{u_i u_j} - \frac{2}{3}\,\delta_{ij} k\right) = \underbrace{\frac{c_\mu k^2}{\varepsilon}}_{\text{effective viscosity}}\left(\frac{\partial U_i}{\partial x_j} + \frac{\partial U_j}{\partial x_i} - \frac{2}{3}\,\delta_{ij}\frac{\partial U_k}{\partial x_k}\right) \tag{5}$$

where U_i denotes the mean velocity component in direction x_i and the coefficient of the effective viscosity, c_μ, is usually given the value 0.09. In a simple shear, i.e. $U_1 = U_1(x_2)$ an Effective Viscosity Model (EVM) gives essentially the same connection between the mean strain, dU_1/dx_2, and the shear stress $\overline{u_1 u_2}$ as an ASM. Eq(5), however, gives *isotropic* normal stresses, a result that is strikingly different from experimental observation and, for that matter, from ASM predictions. This undeniable weakness is of no practical consequence in this case because the normal stresses do not affect the mean flow development. As the strain field becomes more complex, this decoupling no longer applies and the reliability of EVMs deteriorates faster than ASMs or DSMs. Nevertheless, even in highly complicated flows the $k \sim \varepsilon$ eddy viscosity model not infrequently generates results of sufficient accuracy for the purposes in question. The fact that EVMs adopt a diffusive-like stress-strain law means that the discretized momentum equations are easier to solve than those arising with either an ASM or a DSM since the equations are less stiff. There is a core saving too since instead of storing the Reynolds stresses at nodal points, just the scalar-effective viscosity has to be available. Partly for these reasons, EVMs have found very widespread use in the past and will be extensively employed for some years to come. In the writer's group, while general purpose codes are being written with ASM or DSM treatments for final calculations, they also contain the $k \sim \varepsilon$ EVM as an option to facilitate rapid initial explorations.

Some Examples of Model Performance

Before returning to the various closure questions for second-moment models, Figures 1-5 show some applications of the EVM discussed above to four near-wall flows and a free shear layer. The first example is the turbulent mixing between a wall jet and a turbulent wake, the shear flow being one developing in an adverse pressure gradient. This experiment by Irwin [10] is a somewhat simplified laboratory representation of the type of shear flow that arises with multi-element airfoils. Computed distributions of mean velocity are shown at several stations with the EVM and with two alternative ASM schemes. These results have been recently obtained at UMIST by Mr. Z. Nemouchi (personal communication). The computations were made with the parabolic solver PASSABLE (Leschziner [11]) and begin from assigned initial profiles at $x/b = 2.8$ (b being the height of the injection slot) and march downstream therefrom. Fifty cross-stream nodes were used and a forward step size of approximately 10% of the layer width was employed. Two things may be said of the computational results: firstly, that the EVM gives nearly the same behaviour as the ASMs, an observation that accords with remarks in the previous paragraph since $\partial U/\partial y$ (i.e. $\partial U_1/\partial x_2$) is the only significant component of strain; secondly, that the computed behaviour mimics the measurements well save that beyond 21 slot heights downstream the calculated wake fills

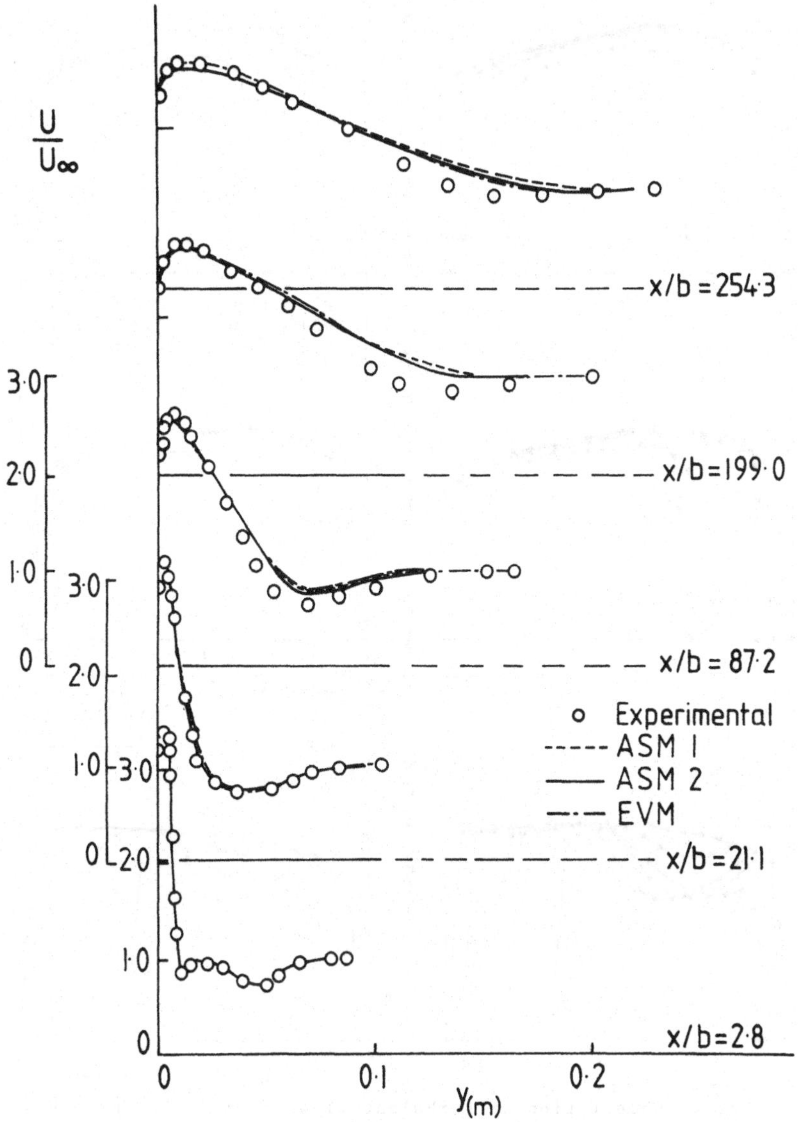

Figure 1 Merging of wake and boundary layer in adverse pressure gradient. Experiment Irwin [10]; Computations Nemouchi (personal communication)

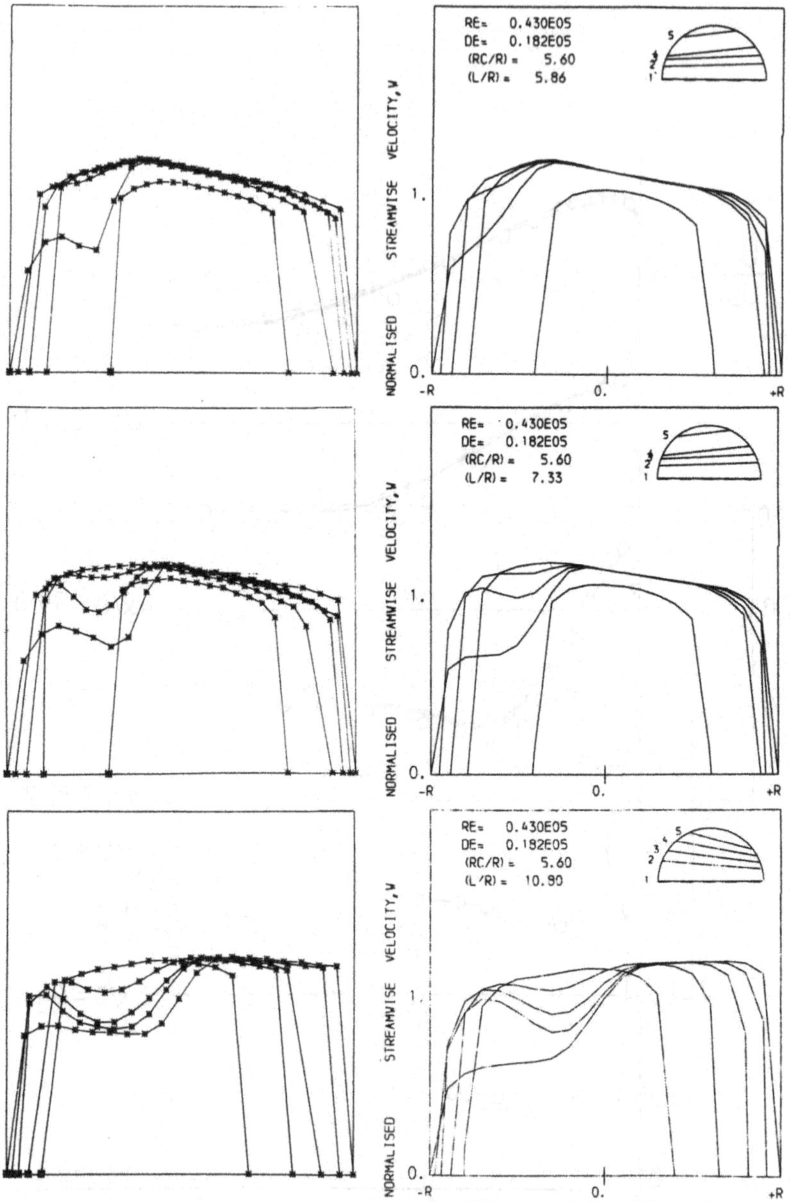

Figure 2 Computation of turbulent flow around a 90° pipe bend Experiment Enayet et al [12]; Computation Iacovides and Launder [13].
(Comparisons (top to bottom) at 60°, 75° and 1 diameter downstream)

in faster than the measurements. Even then, it seems that the differences owe at least as much to shortcomings in the experiment as in the turbulence model. Irwin, who also made computations of this flow, concluded that over the outer part of the shear flow there were significant effects of three-dimensionality.

Figure 2 considers the development of streamwise velocity profiles in the turbulent flow in a circular tube around a tight 90° bend. The flow, which is partly developed at entry to the bend, is strongly three-dimensional but symmetric about the bend's plane of symmetry. Due to the different refractive indices of the tube material and the working fluid, the measured laser-doppler traverses [12] on the left of the figure were made along the non-parallel lines shown in inset. The computors went to considerable lengths to achieve numerically grid-independent results for this flow [13]. The resultant computed profiles along the same lines shown on the right indicate a development in substantial agreement with experiment.

The above flow is really too complex to be able to test whether the turbulence model is functioning correctly - or whether it's a matter of cancelling errors. More appropriate for model evaluation purposes are the three curved *two*-dimensional shear flows whose development is recorded in Figures 3-5; these were all chosen as test cases for the 1980/81 Stanford Conference [14]. For each flow, comparison is drawn with either ASM or DSM computations; these higher level closures adopt the approximation for ϕ_{ij} recommended later. The self-preserving wall jet developing on a logarithmic spiral surface is notable for the very rapid growth of the shear flow as the spiral curvature (s/R) increases. For any given spiral $y_{\frac{1}{2}}/s$ reaches a constant value as s becomes large; for the plane wall jet ($R \to \infty$) this dimensionless growth rate is approximately 0.072. The solid line in Figure 3 shows the correlation of experimental data recommended as the target behaviour for the Stanford Conference. The EVM behaviour reported by Rodi and Scheuerer [15] and Khezzar [16] not only shows a far too weak sensitivity to curvature but also predicts a rate of spread of the wall jet on a flat surface that is too large by 30%. Khezzar's computations, with the same two variants of ASM tested by Nemouchi for Irwin's flow (Figure 1), show considerably better agreement. Closely similar behaviour had been obtained by Rodi and Scheuerer [15], again with an ASM.

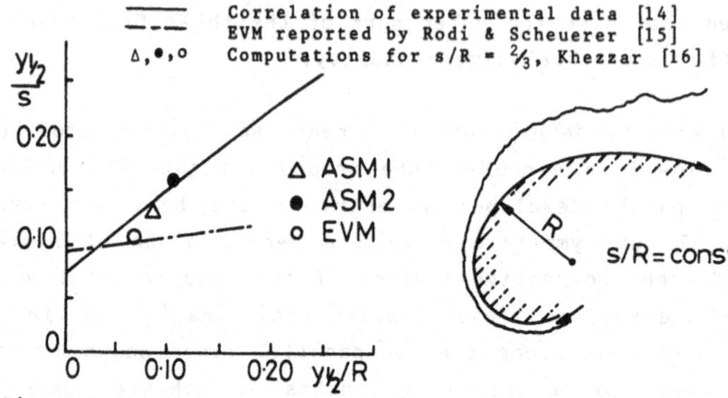

Figure 3 Development of self-preserving wall jet on logarithmic spiral surface for different degrees of curvature (s/R)

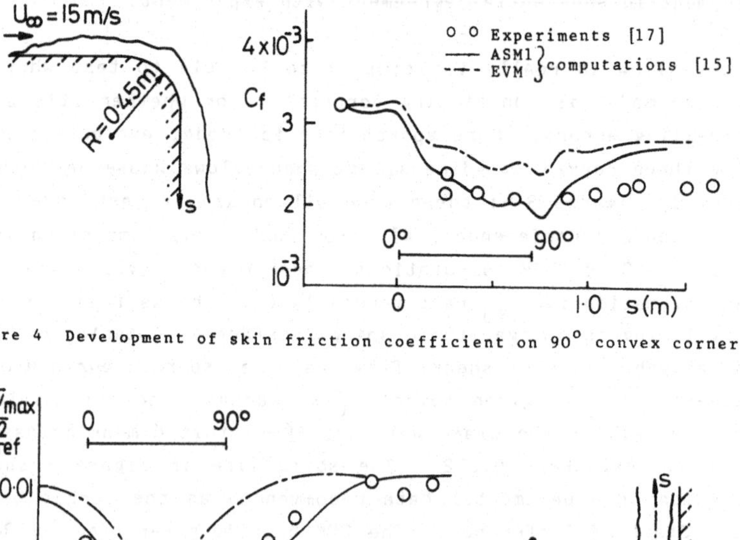

Figure 4 Development of skin friction coefficient on 90° convex corner

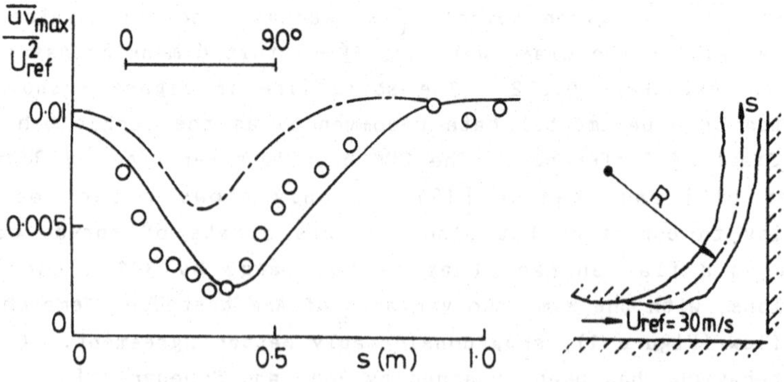

Figure 5 Curved mixing layer - development of maximum shear stress

o o Experiments [18]
—— DSM computations, Gibson & Younis [19]
—·— EVM computations [15]

The effect of convex curvature on the boundary-layer skin-friction coefficient is shown in Figure 4, the data being those of Gillis and Johnston [17]. Here the secondary strain due to curvature acts to damp the shear stress (since the angular momentum increases in the direction of increasing radius) producing a marked reduction in c_f. Rodi and Scheuerer's [15] results show that again the EVM misses this response to curvature while their ASM gives a satisfactory account of the skin-friction coefficient on the curved wall. Downstream from the bend, however, the ASM produces a too rapid rise of the length scale, an effect that we shall later trace to weaknesses in the ε equation in the bend region itself.

The final curved flow considered is the Castro-Bradshaw mixing layer [18]. Figure 5 shows the variation of the maximum turbulent shear stress in the layer as the flow is bent through a 90° arc. Besides the EVM results obtained by Rodi and Scheuerer [15], the figure includes the DSM computations reported by Gibson and Younis [19]. As on the two previous examples, the computations solved the thin shear flow equations rather than the complete Reynolds equations. Work in progress at UMIST by Mr. P.G. Huang suggests that for this flow such a numerical simplification is not entirely justified. Nevertheless, it is safe to conclude from Figure 5 that while the EVM underpredicts the damping of $\overline{uv}$ due to streamline curvature, the DSM provides a broadly satisfactory account.

The preceding examples are representative of the comparative behaviour to be expected between EVMs and DVMs or ASMs in simply-strained curved flows. The clear superiority of the latter types stems from the fact that they account *exactly* for the direct effects of mean strain on the Reynolds stress field.

3. THE APPROXIMATION OF ϕ_{ij}

Free Flows

Forty years ago Chou [20], in his proposals for a third-moment closure, showed that any approximation to the pressure-strain correlation, ϕ_{ij}, should contain two types of term, one containing purely turbulence interactions, ϕ_{ij1}, and another containing mean-strain terms, ϕ_{ij2}. Rotta [21], in respecting this distinction, proposed that the turbulence part of ϕ_{ij} should be modelled as:

$$\phi_{ij1} = - c_1 \frac{\varepsilon}{k} \left(\overline{u_i u_j} - \frac{\delta_{ij}}{3} \overline{u_k u_k} \right) \tag{6}$$

a representation that has remained popular until the present day. The coefficient c_1 is often termed the Rotta constant. In the absence of mean strain or inhomogeneity the coefficient c_1 can be determined by observing the return of a non-isotropic Reynolds stress field towards isotropy. Such studies indicate a non-unique value of c_1 - result that is to be expected since the actual process is non-linear whereas eq(6) approximates it by a linear model. Lumley and Newman [22] (see also Lumley [6,23]) have provided an extensive analysis and suggested adding non-linear terms to Rotta's model

$$\phi_{ij1} = - \left[c_1 \epsilon \, a_{ij} + c_1' \epsilon \left(a_{ik}a_{kj} - \frac{\delta_{ij}}{3} A_2 \right) \right]$$

(7)

where a_{ij} ($\equiv \overline{u_i u_j} - \frac{1}{3} \delta_{ij} \overline{u_k u_k}$)/k) is the dimensionless anisotropic part of the Reynolds stress and A_2 is the second invariant of the stress tensor, $A_2 \equiv a_{ik}a_{ik}$ - often termed the *anisotropy* . In fact, Lumley and Newman suggest that c_1' should be zero but retain a non-linear form by making c_1 a function of A_2 and A_3, the latter being the *third* invariant of the Reynolds stress, $a_{ik}a_{kj}a_{ji}$. Logically the argument for the third invariant is plain enough. Without it one cannot distinguish the rates of return to isotropy of nearly two-dimensional turbulence ($\overline{u_3^2} \ll \overline{u_1^2}$ or $\overline{u_2^2}$) from nearly one-dimensional turbulence ($\overline{u_1^2} \gg \overline{u_2^2}$ or $\overline{u_3^2}$) if each has the same value of A_2; in fast-food terms our model would be blind to differences between a hamburger and a frankfurter. Does it matter? "Yes" say Lumley and Newman from considering the decay behaviour of several sets of Reynolds stresses with differently anisotropic initial conditions. Yet, one might argue that in flows of practical interest, strong inhomogeneities in the stress field and the effects of mean strain on the eddy structure - what, in fast-food terms, we might call a SMOKE screen[†] - make such a distinction irrelevant. The writer's view is that, at the moment, the best practice is to use Rotta's proposal, either in its original form with $c_1 \simeq 3.0$ or with c_1 increasing with A_2 from unity in isotropic turbulence to about 4 in a strongly non-isotropic shear flow.

There is now a sufficient accumulation of experience in using second-moment closures in inhomogeneous flows to say unequivocally that the best available model of the mean-strain (or 'rapid') part of ϕ_{ij} is what we term the Isotropization of Production (IP) model:

$$\phi_{ij2} = - c_2 \left(P_{ij} - \frac{1}{3} \delta_{ij} P_{kk} \right)$$

(8)

[†] <u>S</u>mothered in <u>M</u>ustard <u>O</u>nions and <u>KE</u>tchup

This evidently produces a tendency towards isotropy of the production tensor equivalent to that which Rotta's model exerts on the Reynolds stresses. If, in fiscal terms, ϕ_{ij1} is seen as a wealth tax, the IP model of ϕ_{ij2} is an income tax. If, moreover, we consider turbulence as characterizing a state midway between chaos and organization not unlike Western civilizations, we should expect the tax coefficient c_2 to lie between about 0.3 and 0.6.

Before taking up the question of choosing c_2 it needs to be emphasized that the IP model, besides being simpler, offers wider applicability than the more theoretically based Quasi-Isotropic (QI) model (Launder et al [24,25], Naot et al [26], Lumley [23]). As an illustration of this, Figures 6 and 7 show predictions of Younis [27] (see also Gibson and Younis [28]) of the swirling jet in stagnant surroundings measured by Morse [29]. Launder and Morse [30] had earlier reported DSM results of this flow showing poor agreement with the data, a discrepancy they had traced to their use of the QI model. Younis' computations with the IP model display substantially higher stress levels than those of the QI computations (Figure 6) in agreement with experiment. As a result, the rate of spread of the jet and the rate of decay of the centre-line velocity shown in Figure 7 are markedly greater, again in very good agreement with experiment.

It may be wondered why the discovery that the IP model could correctly predict the swirling jet has taken so long to surface. The explanation lies in the $c_1 \sim c_2$ map shown in Figure 8. The points indicate values of the pairs of constants used by different workers over the past dozen years. There are also three lines:

$c_2 = 0.6$	-	the exact analytical result for suddenly strained isotropic turbulence
$c_1 = 1.8$	-	the value of c_1 indicated by a numerical simulation of grid-turbulence decay by Schumann and Patterson [31]
$\frac{(1 - c_2)}{c_1} = 0.23$	-	the interrelation between c_1 and c_2 that allows closest accord with stress levels in a free shear flow in local equilibrium, [32]

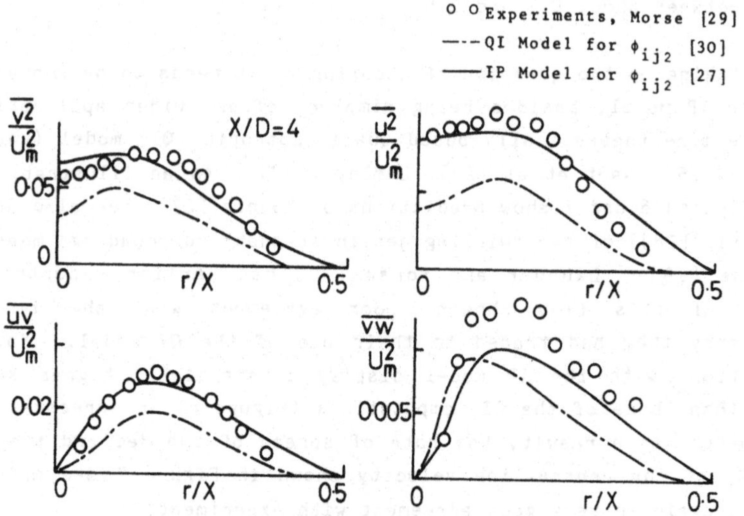

Figure 6 DSM computations and measurements of Reynolds stresses in swirling jet in stagnant surroundings; swirl number = 0.35

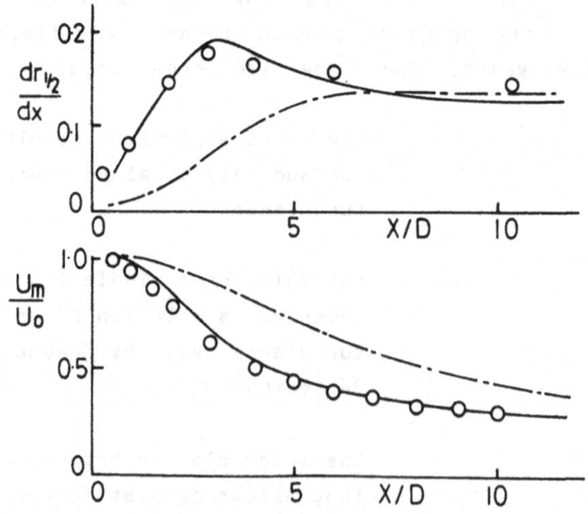

Figure 7 Rate of spread and decay of maximum velocity in swirling jet. Key as in Figure 6.

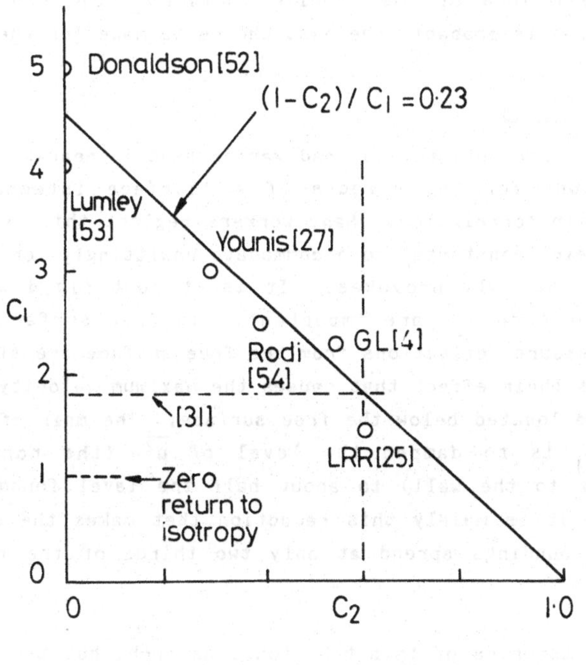

Figure 8 $c_1 \sim c_2$ indifference map

It will be noticed that these three lines nearly intersect at a point, suggesting that the optimum choice of c_1 and c_2 would lie in the vicinity of that intersection triangle. We note, however, that Donaldson's group and, in his pre-1974 papers, Lumley omit any contribution from ϕ_{ij2}. As long as one confines attention to flows near local equilibrium and doesn't stray too far from the $c_1 \sim c_2$ connection given above, this omission is of little consequence. Younis' calculations (which included, besides swirling jets, many plane and curved shear layers) adopted a larger value of c_1 (3.0) and a smaller value of c_2 (0.3) than has generally been used in recent years - but we notice that these have been chosen so that $(1 - c_2)/c_1$ takes the required value. It was only in calculating the swirling flows (for which the stress generation tensor is much more complex) that the advantages of reducing c_2 become apparent.

The indications are thus that if c_1 increases with the anisotropy of the stress field, c_2 should decrease. However, in the absence of any firm recommendations of the precise variation, the use of Younis' constant values is probably the best choice to make for the present.

Wall Effects on ϕ_{ij}

Chou's early contribution [20] had warned that in near-wall flows one had to account for the effects of a 'surface integral' on the pressure-strain correlation. Many workers neglect this 'echo' effect or adjust their 'constants' to accommodate unwittingly for the changes in ϕ_{ij} that the wall provokes. It isn't just rigid walls where surface effects on ϕ_{ij} are important. In free surface flows the resultant pressure reflections from the free surface are significant - indeed, it is their effect that causes the maximum velocity in a river usually to be located below the free surface. The main effect of the wall on ϕ_{ij} is to dampen the level of $\overline{u_2^2}$ (the normal stress perpendicular to the wall) to about half the level found in a free shear flow. It is mainly this reduction that makes the wall jet in stagnant surroundings spread at only two thirds of the rate of the free jet.

There are no theories of this behaviour, as such, but the effects are tolerably well correlated for a plane surface (for specific models see for example [4] or [25]). The strength of the wall reflection process is usually taken proportional to (ℓ/x_2), i.e. the size of a typical turbulent eddy divided by the distance from the surface in question.

Thus, far from a wall where the eddy size is determined by the thickness of the shear layer, the influence of the wall diminishes with height, while in the near-wall region, where ℓ is often proportional to x_2, the wall effect is fairly uniform. The length scale ℓ is habitually identified with $(k^{3/2} / \varepsilon)$ so any errors in predicting ε may lead to erroneous levels of length scale and thus to aberrant levels of the stresses. An example of this may be seen in Rodi and Scheuerer's [15] prediction of a boundary layer on a convex surface (the skin friction behaviour for this flow appeared in Figure 4). The ratio of $\overline{u_1^2}/\overline{u_2^2}$ along a streamline in the inner region of the boundary layer is shown in Figure 9. While the measured level of $\overline{u_1^2}/\overline{u_2^2}$ remains practically uniform in the region of convex curvature the predicted level rises sharply to more than twice the initial level. The reason for this relative augmentation is that while $\overline{u_1^2}$ and $\overline{u_2^2}$ are both damped by the convex curvature, the effect on $\overline{u_2^2}$ is direct (being proportional to the curvature strain), while $\overline{u_1^2}$ experiences a damping only at second hand via the effect on $\overline{u_1u_2}$. The growth of $\overline{u_1^2}/\overline{u_2^2}$ is thus an expected development. Why, one must ask, are these expectations not borne out by experiment? The explanation appears to be that in the experiment the near-wall length scales are reduced thereby reducing, at any height, the quantity (ℓ/x_2) and as a result diminishing the wall damping of $\overline{u_2^2}$. A precisely analogous phenomenon has been noted in the atmospheric boundary layer where under increasingly stable stratification $\overline{u_1^2}/\overline{u_2^2}$ is observed to decrease. Ref[4] shows that conventional wall-reflection models account for this behaviour provided that the correct length-scale dependence on stratification is supplied.

Before considering the procedure for determining ε (or ℓ) and its shortcomings, it needs to be said that the simple approaches currently used for correlating pressure reflections from a plane surface are not cast in a suitable form to permit their extension to surfaces of arbitrary topography. There is some suggestion that plane walls abutting at right angles (e.g. an interior corner) can be handled by a superposition procedure [33], but even if this is confirmed, a procedure for handling an arbitrary surface is still some way off.

A further weakness of present wall-reflection models (or perhaps another aspect of the same weakness) is that in the near wake of a thin plate the memory of the damping of $\overline{u_2^2}$ by wall effects lasts several boundary layer thicknesses downstream. The results seem to suggest that part of what is actually called a "wall reflection"

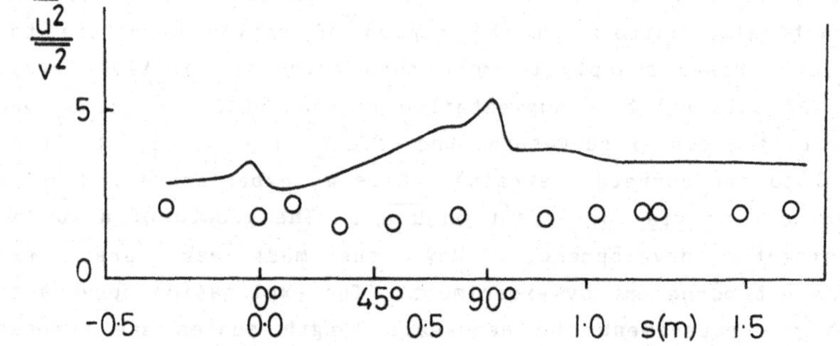

Figure 9 Development of streamwise:cross-stream normal stresses in a boundary layer on a convex surface
o Experiment [17] —— ASM computation [15]

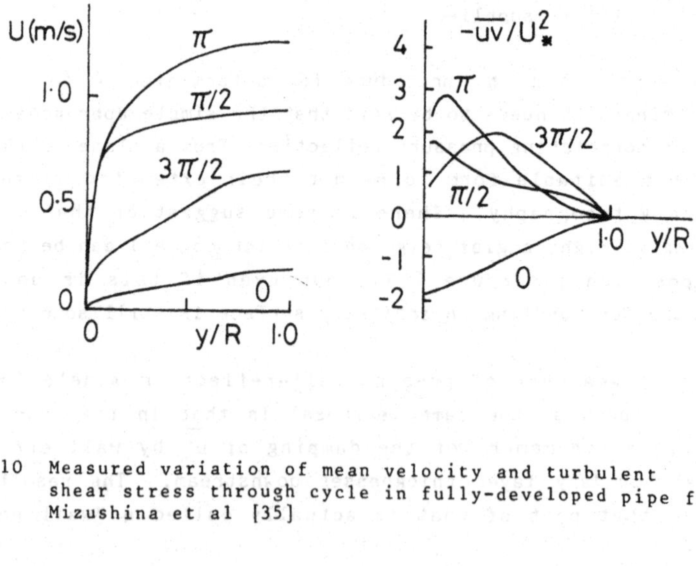

Figure 10 Measured variation of mean velocity and turbulent shear stress through cycle in fully-developed pipe flow, Mizushina et al [35]

agency is associated rather with the flattening of the near-wall eddies in the direction normal to the wall - a feature that, while clearly caused by the wall, will only gradually disappear as eddies flow past the downstream end of the plate into the wake region.

Pressure-Strain Modelling in Periodic Flows

At the moment we are still very much in the learning stage on how to model turbulence in flows with imposed periodicities of large amplitude. The effects of the unsteadiness will, of course, depend on the ratio of time scales of the imposed motion to that of the turbulence. It is usual to admit four frequency categories:

1. Quasi-steady : Changes so slow that time-dependent terms in mean-field and turbulent transport equations negligible.

2. Low : Time-dependent terms in mean momentum equations significant but unimportant in turbulent transport equations. (Thus, by definition, steady flow turbulence models will apply).

3. Intermediate : Effects of periodicities modify turbulence structure throughout shear flow. No agreement near wall with "universal" logarithmic law.

4. High : Turbulence energy "freezes" over major regions of shear flow (but shear stress does not). Mean flow becomes independent of turbulence.

The case of periodic, fully-developed turbulent flow through a pipe is the one for which experimental documentation is most complete. In laminar flow the magnitude of the (square of the) Stokes parameter governs the flow type: $\omega D^2/\nu$ where ω and D are respectively the angular frequency and the pipe diameter. In turbulent flow it is tempting to replace ν by a mean turbulent viscosity, taken proportional to $U_\tau D$, U_τ being the mean friction velocity. The appropriate frequency parameter is thus $\omega D/U_\tau$, Ramaprian and Tu [33]. These authors suggest a value of 5 to be typical of the 'intermediate' frequency and 50 of the high frequency behaviour. A more informative dimensionless scaling is achieved with the local turbulent turnover time scale, i.e. $(\omega k/\varepsilon)$. Since k/ε increases as one proceeds from the pipe wall to the centre it makes clear why the 'freezing' behaviour occurs first in the core region.

With the above remarks as background, current research at UMIST seems to indicate that existing second-moment closure methodology may allow the behaviour found at 'intermediate' frequencies to be satisfactorily simulated. The reported 'high' frequency data, however, seem incapable of prediction without major innovation in the modelling of ϕ_{ij}. The problem is indicated in Figure 10 which shows mean velocity and turbulent shear stress profiles across the pipe measured by Mizushina et al [35]. We notice that throughout the cycle (except possibly closer to the wall than resolved by the hot-wire probes) the mean velocity gradient does not change sign and hence neither does P_{12}. The sign of $-\overline{u_1 u_2}$ *does* change, however, being opposite from that of P_{12} during a portion of the cycle. Naturally, when faced with what seems such an unlikely occurrence, one questions whether the data are correct. While this matter is perhaps not entirely settled, the more recent data of Ramaprian and Tu [34] do show a similar type of behaviour - namely that the amplitude of the shear stress variation is substantially greater than that of the mean velocity.

The implication for ϕ_{ij} modelling seems to be that the 'return-to-isotropy' process, ϕ_{ij_1}, will need to be modified to allow 'overshoot'. While there are various ways of achieving this, it may not be worthwhile attempting to distinguish between alternatives until measurements of all non-zero components of the Reynolds stress tensor through a cycle have been obtained.

ASM versus DSM and the Implications for Closing the Stress Equations

Over the past five years the use of algebraic stress models has become widespread. They have been correctly seen as a major advance in concept from the $k\sim\varepsilon$ eddy viscosity model while retaining important elements of the EVM structure, in particular the solution of transport equations for k and ε. More recently at Imperial College London and at UMIST various separate strands of research have been probing the eligibility of ASMs to take over as the turbulence model "workhorse" of general flow-simulation codes.[†] Direct tests have been made on the most popular of the ASM transport approximations, eq(2). McGuirk (personal communication) has compared measured transport rates of $\overline{uv}$ with those given by eq(2) for the mixing layer of a disc held normal to the flow; broadly satisfactory agreement was found. Huang, Nemouchi and Younis (personal communication), however, have computed the four thin shear flows listed in Table 2 using both this ASM truncation and a full DSM (with eq(4) for D_{ij}). As the table indicates, they found serious differences in the rates of spread of these flows, the ASM producing generally higher levels than the DSM. When the magnitude of the transport terms was compared in the two sets of computations, it was found that in some regions of the wake and the round jet the ASM gave a different sign than indicated by the DSM.

From a computational point of view there do not in fact appear to be significant benefits from the use of an ASM rather than a DSM in recirculating or three-dimensional flows. *Both* pose considerably more problems in actually securing convergence than an EVM but when these are overcome, the approach to a fully converged solution seems practically as rapid as with an eddy viscosity model. Thus, in view of the fact that ASMs muddle the modelling issues, our efforts in software development (in common with those of a number of other European groups) are now mainly directed at incorporating *differential* stress models.

Before leaving the treatment of stress transport, two further observations are perhaps worth making. The first is that when written out in other than Cartesian co-ordinates the convective transport of $\overline{u_i u_j}$

$$C_{ij} \equiv \frac{\partial}{\partial x_k}\left(U_k\,\overline{u_i u_j}\right)$$

gives rise to certain source-like terms of the same form as elements of the production tensor P_{ij}. Some workers have, therefore, admitted the terms to full membership of the production tensor including their influence not just in P_{ij} but in the model of $\phi_{ij\ell}$ (eq(8)) where P_{ij} also appears. This practice has been found helpful in allowing, for example, the swirling jet to be predicted correctly with the Gibson-

† The work, which is at present largely unpublished, is being led by Dr. W.P. Jones and Dr. J.J. McGuirk at Imperial and Dr. Leschziner, Dr. Younis and the writer at UMIST.

Launder [4] values for c_1 and c_2 (Younis [27]). Unfortunately this practice destroys the invariance of one's model. One might, for example, decide to examine the swirling jet using x-y-z co-ordinates and this would return different results from those obtained with the r-θ-z system. These ill-advised tamperings with C_{ij} may, however, be an indication that we should, in fact, quite generally be including an effect of convective transport in ϕ_{ij2}, i.e.

$$\phi_{ij2} = -c_2\left(P_{ij} - \frac{\delta_{ij}}{3} P_{kk}\right) - c_2'\left(C_{ij} - \frac{\delta_{ij}}{3} C_{kk}\right) \tag{9}$$

If convective transport contributes to ϕ_{ij}, shouldn't diffusion also? Some years ago Lumley [36] recommended a re-arrangement of the traditional presentation of the pressure correlations to allow the traceless (i.e. re-distributive) part of the diffusion process to be identified. In consonance with this view, the velocity diffusion term may likewise be re-grouped:

$$\frac{\partial \overline{u_i u_j u_k}}{\partial x_k} = -\frac{\partial}{\partial x_k}\left(\overline{u_i u_j u_k} - \frac{1}{3}\delta_{ij}\,\overline{u_m u_m u_k}\right) - \frac{\delta_{ij}}{3}\frac{\partial}{\partial x_k}\left(\overline{u_m u_m u_k}\right) \tag{10}$$

One might thus regard the first term on the right of eq(10) as already a contributor to ϕ_{ij}. Lin and Wolfshtein's [37] analysis of the non-linear part of the pressure-strain term indicated that this term was *purely* associated with inhomogeneity. While this conclusion cannot be entirely correct (since, if it were, the anisotropy of non-isotropic grid-generated turbulence would increase as its energy decreased) it certainly supports the idea that the traceless pressure correlations will be modified by flow inhomogeneity. A further pointer to such effects may be found in a suggestion by the writer [38] for generalizing the representation of transport in an ASM. In place of eq(2) the paper proposes that the convection and diffusion processes be approximated by

$$C_{ij} = C_{kk}\left((1+\alpha)\frac{\overline{u_i u_j}}{\overline{u_k u_k}} - \frac{\alpha\delta_{ij}}{3}\right) \qquad D_{ij} = D_{kk}\left((1+\beta)\frac{\overline{u_i u_j}}{\overline{u_k u_k}} - \frac{\beta\delta_{ij}}{3}\right) \tag{11}$$

This modification has produced a modest but definite improvement in overall predictive accuracy for the half dozen or so flows against which it has been tested so far[†] even though no serious optimization of the constants α and β has been made (the values adopted α = 0.3

[†] Computations with this scheme are shown as 'ASM2' in Figures 1, 3 and 4.

β = —0.8, being those suggested in [38] without the benefit of numerical calculations). Yet, it seems unlikely that what we are seeing is really an improvement in modelling transport (indeed, direct comparisons suggest that it is not more successful than eq(2) as an approximation of that process). But the net change produced by adopting eq(11) rather than (2) is precisely the same as that which results from replacing the simplest model of ϕ_{ijl}, eq(6), by

$$\phi_{ijl} = -\left\{c_1 + \alpha(\tfrac{1}{2}P_{kk}/\varepsilon - 1) + \tfrac{1}{2}(\alpha - \beta)D_{kk}/\varepsilon\right\}\frac{\varepsilon}{k}\left(\overline{u_i u_j} - \frac{\delta_{ij}}{3}\overline{u_k u_k}\right) \tag{12}$$

In this form the effective return-to-isotropy coefficient depends on both the energy production:dissipation rate and the dimensionless net gain of turbulence energy by diffusion, a by no means implausible dependence.

The writer has no firm conclusions to draw from the above observations but he must admit to certain inclinations. Firstly, it seems to him that the posture - for that's what it now appears to be - of closing the stress transport equations by making term-by-term approximations for the different processes should be dropped - at least until there is available a substantial body of data in which all the terms have been measured accurately. Instead, we should consider the unknown turbulence products in the transport equation as a single entity to be tuned to give best overall agreement with as wide a range of shear flows as possible. Secondly, perhaps less weight should be given to modelling experiments of simply strained homogeneous flows; more to inhomogeneous flows with complex strain fields. The former approach too easily attracts attention to aspects of the flow and the turbulence model that are of little or no practical interest.

4. DETERMINING THE ENERGY DISSIPATION RATE

The scalar ε remains as an unknown in the models so far discussed; it is to be obtained by solving a transport equation. Before discussing proposed forms for that equation, however, it may be helpful to address the following frequently recurring question: how, in the absence of reliable measurements of ε, can one tell whether poor Reynolds stress and mean velocity predictions arise from errors in determining the level of the energy dissipation rate or from weaknesses in modelling the unknown processes in the $\overline{u_i u_j}$ equations, especially ϕ_{ij}? While difficulties in discrimination do sometimes occur, it is often possible to distinguish the source of the problem. Errors in ε will tend to give too high or too low energy levels;

errors in ϕ_{ij} will tend to give the wrong relative levels of individual Reynolds stresses. Thus, in ref[30], the failure to predict the swirling free jet was easily traced to the pressure-strain process ϕ_{ij_2} because the correlation between streamwise and azimuthal velocity fluctuations was of the wrong sign. For the (non-swirling) axisymmetric jet, however, the relative stress levels are reasonably correct but the predicted level of kinetic energy is too large - a result that points to the dissipation rate equation as the main source of error.

Of course, if one is interested only in the mean flow behaviour, one may sometimes *introduce* errors into the ε equation to compensate for weaknesses in other parts of the model. This is particularly common when an EVM is used. As we saw in Figure 3-5, the Newtonian stress-strain law used in EVMs does not give the correct response of the Reynolds stresses to streamline curvature. To compensate for this, users of such schemes add an empirical source/sink term to the ε rate equation proportional to the local curved flow Richardson number [39]. Thus, by raising or lowering ε the Reynolds stresses obtained from the EVM constitutive equation *are* rendered sensitive to streamline curvature. Figures 3-5 also showed that if closure is made at ASM or DSM level, the most important effects of curvature are accommodated automatically via the exact stress production tensor P_{ij}. Accordingly, the discussion below is mainly aimed at providing an equation suitable for use in such second-moment treatments.

Chou [20] proposed the first equation for a variable proportional to ε but current concepts in modelling the energy dissipation rate really spring from Davidov's work [40]. Although an exact transport equation for ε may be derived from the Navier Stokes equations [40,9] this is by no means as useful as the corresponding equation for $\overline{u_i u_j}$. The reason for this is that the major terms in the equation consist of fine-scale correlations describing the detailed mechanics of the dissipation process. However, the large-scale, stress-bearing eddies are usually held to be essentially independent of the fine-scale motion. There is ample evidence to support this view, for example the insensitivity of the rate of spread of turbulent jets to Reynolds number. Thus, the fine-scale motions, so far as the Reynolds stress field is concerned, are passive; they adjust in size as required to dissipate energy at the rate dictated by a system of substantially larger eddies whose structure is largely independent of viscosity. The conclusion to be drawn from all this is that, in devising a transport equation to mimic the spatial variation of ε, one is resorting mainly to dimensional analysis and intuition. All proposals can be written in the form

$$T_\varepsilon = c_{\varepsilon 1} \frac{P_{kk}\,\varepsilon}{2k} + \frac{\varepsilon^2}{k}\left[f(A_2) - c_{\varepsilon 2}\right] + \mathrm{EST} \tag{13}$$

where T_ε denotes the net transport of ε and EST stands for "extra strain terms". The influence of the second invariant A_2 was first introduced about ten years ago by Lumley [23]. If $f(A_2)$ is omitted the optimum constant values for $c_{\varepsilon 1}$ and $c_{\varepsilon 2}$ are approximately 1.32 and 1.8.

Although Lumley had proposed the introduction of $f(A_2)$ as a *replacement* for the first term on the right of (12) containing the turbulence energy generation rate, the subsequent studies on buoyant diffusion undertaken by his group [3] recommended the retention of both types of process. Specifically they proposed

$$f(A_2) = \frac{3.75A_2}{1+1.5A_2^{\frac{1}{2}}} \qquad \text{and} \qquad c_{\varepsilon 1} = 0.475$$

There are sound physical reasons for preferring to base the source term in the ε transport equation on the anisotropy of the stress field alone but Lumley's experiences and those of Morse [29] show that this is not possible. The latter found that in predicting free shear flows the use of a function linearly proportional to A_2 led to poor profile shapes. Worse, the loose coupling between the $\overline{u_i u_j}$ and ε fields that resulted produced strongly oscillatory rates of spread. Current work at UMIST, which at the moment has been limited to free flows and focused mainly on removing the anomalous prediction of the axisymmetric jet, suggests there is some overall benefit in reducing the value of $c_{\varepsilon 1}$ to 1.1 and adopting $f(A_2) = 0.6A_2$. The Zeman-Lumley proposal for $f(A_2)$ and $c_{\varepsilon 1}$ [3] is found to give poor predictions for the jet flows.

The presence of the terms marked EST is an indication that, like the Reynolds stresses, the energy dissipation rate (and thus effective time and length scales) appear to be rather sensitive to small secondary strains. Unfortunately while, in the $\overline{u_i u_j}$ equations, the production tensor P_{ij} shows clearly the origin and the approximate sensitivity to particular strains, no such help is available in the case of the ε equation. So, as remarked earlier, one proceeds by intuition. Pope [41], in a well argued proposal, suggests the inclusion of an additional source term in the ε equation proportional to

$$\left(\overline{u_i u_j} - \frac{\delta_{ij}}{3}\,\overline{u_k u_k}\right)\left(\frac{\partial U_i}{\partial x_m} - \frac{\partial U_m}{\partial x_i}\right)\left(\frac{\partial U_j}{\partial x_m} - \frac{\partial U_m}{\partial x_j}\right)$$

This term is zero in plane two-dimensional flows but not in an axisymmetric flow. The size of the coefficient was thus tuned to give the correct rate of spread of the round jet in stagnant surroundings (using the k~ε EVM). Recently P.G. Huang (personal communication) has made further tests on Pope's correction re-optimizing the coefficient to suit the DSM he was using. When, however, he shifted from the round jet in stagnant surroundings to the Forstall-Shapiro data [42] of coaxial jets, he found that these flows were predicted better with the extra term deleted.

More recently Hanjalic and Launder [43] recommended the addition of a term that is superficially similar in appearance to Pope's:

$$EST \propto -k\,\frac{\partial U_i}{\partial x_j}\,\frac{\partial U_\ell}{\partial x_m}\,\varepsilon_{ijk}\varepsilon_{\ell mk} \tag{14}$$

The action of the term is quite different, however. As indicated in (13) the coefficient of proportionality is negative and, in a straight two-dimensional shear flow, the main contributor to the term is $-k(\partial U_1/\partial x_2)^2$. The term thus acts to reduce the sensitivity of the ε source to shear strain relative to normal strain. The additional term has been found to be helpful in boundary layers in strongly adverse pressure gradients as well as in the round jet for it allows the term $-2(\overline{u_1^2} - \overline{u_2^2})\partial U_1/\partial x_1$ (in P_{kk}) to make a larger contribution (in comparison with $-2\overline{u_1 u_2}\,\partial U_1/\partial x_2$) to the generation rate of ε. Since $(\partial U_1/\partial x_1)$ is predominantly negative in these flows and $\overline{u_1^2} > \overline{u_2^2}$ the result is that ε becomes relatively larger (length scales smaller), reducing the Reynolds stress levels. The originators tested their proposal over several flows but unfortunately not in curved flows. When that comparison was made (Sindir [44]) the modification that the new term brought to the ε equation was substantial and its effect was to worsen agreement with experiment.

Leschziner and Rodi [45] adopted a modification of Hanjalic and Launder's [43] proposal in which at every node the orientation of the mean velocity was determined; this direction was designated as x_1 with

x_2 orthogonal to it. With that choice made, they could unambiguously discriminate between "normal" strains and "shear" strains. The contributions of the former to P_{kk} in eq(12) were simply multiplied by a coefficient approximately three times as large as $c_{\varepsilon 1}$ (which was retained as the coefficient of the shear strain). With this re-interpretation, Leschziner and Rodi [45] effectively changed the sign of the sensitivity of the equation to streamline curvature while retaining the benefits that ref[43] had demonstrated for straight flows. Leschziner and Rodi used this adaptation successfully in the context of an EVM calculation of curved recirculating free shear layers. If instead an ASM approach is used, however, this corrected form, while producing an effect of the correct sign, gives changes in ε that are too large in magnitude.

The above examples illustrate the difficulty with the ε equation. The standard version is naively simple in form and has well known failures. Yet, so far, all attempts at improving it have produced modifications which, when subjected to wider testing, have produced worse agreement than the original for some other flow.

The best way forward is probably to take a different route. One interesting possibility is to close an *exact* equation for the energy transfer rate across some appropriate (low) wave number of the spectrum [46] - logically one proportional to $\varepsilon/k^{3/2}$. This spectral transfer rate would, at least as a first step, be taken as equal to the rate of energy dissipation. The rationale of such an approach is that the energy transfer rate is associated with interactions that are predominantly *independent* of viscosity; the exact equation thus provides much more direct guidance of how the effects of mean strain and anisotropy should be modelled. In fact, the mean strain generation rate comprises products of mean velocity gradients and ε_{ij}, the spectral transfer rate of $\overline{u_i u_j}$. A major philosophical decision to be taken is thus whether to admit the ε_{ij} as dependent variables of a further set of transport equations or whether to relate the anisotropy of the spectral transfer rate to that of $\overline{u_i u_j}$ and other invariants of the turbulence field. In this connection we note that Lin and Wolfshtein [37] and Morse [29] have proposed a set of transport equations for ε_{ij} (strictly, in their articles, for the *dissipation* rate of $\overline{u_i u_j}$).

Other approaches to a scale-determining equation are possible. Sandri and Cerasoli [47] (see also Donaldson and Sandri [48]) have developed a transport equation for $k\Lambda_{ij}$ where Λ_{ij} is a vectorial macro-length scale defined in terms of the volume integral of the two-point correlation of u_i and u_j. The generation rate of $k\Lambda_{ij}$ by mean strain is equal to

$$-k\left\{\Lambda_{ik}\frac{\partial U_j}{\partial x_k} + \Lambda_{jk}\frac{\partial U_k}{\partial x_j}\right\}$$

Sandri and Cerasoli point out that if an equation for $\Lambda\ (\equiv \Lambda_{kk}/3)$ is formed by contracting that for Λ_{ij} and using the turbulent kinetic energy equation, the mean strain generation rate emerges as

$$-2\Lambda\left(\frac{\Lambda_{ij}}{\Lambda_{kk}} - \frac{\overline{u_i u_j}}{\overline{u_k u_k}}\right)\frac{\partial U_i}{\partial x_j} \tag{15}$$

This may be compared with the source due to mean strain in the equation for length scale formed from the ε and k equation (i.e. $\ell \equiv k^{3/2}/\varepsilon$):

$$-2\ell\,(1.5 - c_{\varepsilon 1})\,\frac{\overline{u_i u_j}}{\overline{u_k u_k}}\,\frac{\partial U_i}{\partial x_j} \tag{16}$$

If the dimensionless length scale tensor is taken proportional to the dimensionless Reynolds stress form, (14) and (15) are clearly equivalent. As an interesting halfway measure between retaining just the equation for Λ and the full Λ_{ij} as dependent variables, ref[48] suggests approximating Λ_{ij} by

$$\Lambda_{ij} = \Lambda\left[\delta_{ij} + \sigma\left(\overline{u_i u_j} - \frac{\delta_{ij}}{3}\,\overline{u_k u_k}\right)\Big/\overline{u_k u_k}\right]$$

where σ is a dimensionless scalar to be obtained from a transport equation (again formed by manipulating the equations for $k\Lambda_{ij}$ and k).

Finally, we mention briefly a group of more elaborate schemes that have been tested or are in the course of development. Hanjalic et al [49] proposed a "multi-scale" model in which the turbulence energy spectrum was divided into two parts with separate equations provided for the energy dissipation rate and the rate at which energy passed from the low-to-high-wave-number part of the spectrum. In practice the scheme had few advantages over a conventional single-scale approach because the division between the two regions was made at too high a wave number (in order to allow the assumption that the Reynolds stresses in the high wave number region were isotropic). The work has been extended by Cler [50] who dropped the assumption of isotropy in the finer scales but without, however, making the major overhaul of the model that ought really to have accompanied such a move. Schiestel [51] has made a number of interesting discoveries in connection with a multi-scale approach including providing a connection between various models for energy transfer in wave-number space and the coefficient $c_{\varepsilon 1}$ in eq(12). His multi-scale EVM scheme, most encouragingly, gives correctly the rate of spread of the axisymmetric and plane jet as well as that of the far turbulent wake. It will not, it seems safe to suppose, deal any better with the effects of streamline curvature than other EVMs. The model could, however, be extended in an obvious way to a closure of second-moment type which would allow this effect to be better accounted for.

5. CONCLUDING REMARKS

The preceding sections have tried to give an accurate flavour of what can be done and where more needs to be achieved in single-point closure. However, for the rest of the '80's, at least, the two biggest factors in advancing CFD in the field of turbulent flow are not connected with turbulence modelling directly: they are the continuing advance in computing power and multifarious improvements in what might be termed "numerical analysis". In this latter category are included more accurate discretization schemes, the use of faster solvers, boundary-fitted co-ordinates, flow adaptive meshes and various core-saving devices. This is perhaps not the place to elaborate on these advances. Their coming has meant, however, that *numerically* accurate solutions are now routinely possible for many complex three-dimensional flows - the class in which most flows of practical interest fall (even if academics like to pretend otherwise).

A conclusion this writer draws is that, increasingly, single-point turbulence-model development (as opposed to application) will need to include computations of such complex flows. It seems likely also that the greater scope that these advances in numerics offer will lead to a gradual blurring between single-point and two-point closures or between single-point and sub-grid-scale models. Already this trend is evident in the split-spectrum approaches discussed briefly above and in the efforts being made to develop more realistic sub-grid schemes.

ACKNOWLEDGEMENTS

Thanks are due to George Huang, Zoubir Nemouchi and Bassam Younis at UMIST, to Dr. J. McGuirk at Imperial College and Dr. F. Boysan of Sheffield University for making available their currently unpublished results.

The camera-ready manuscript was prepared by Mrs. L.J. Ball and the artwork by Mr. J. Batty.

REFERENCES

1. Launder, B.E. and Spalding, D.B. Comp. Meth. in Appl. Mech. & Engrg., 1974

2. Launder, B.E. J. Fluid Mech. 67, 569, 1975

3. Lumley, J.L., Zeman, O. and Siess, J. J. Fluid Mech. 84, 581, 1978

4. Gibson, M.M. and Launder, B.E. J. Fluid Mech. 86, 491, 1978

5. Rodi, W. ZAMM 56, 219, 1976

6. Lumley, J.L. Advances in Applied Mechanics 18, 123, 1978

7. Ettestad, D. and Lumley, J.L. Turbulent Shear Flows-4 (ed. F.J. Durst et al), 87-101, Springer Verlag, New York, 1984

8. Dekeyser, I. and Launder, B.E. Turbulent Shear Flows-4 (ed. F.J. Durst et al), 102, Springer Verlag, New York, 1984

9. Daly, B.J. and Harlow, F.H. Phys. Fluids, 1970

10. Irwin, H.P.A.H. PhD Thesis, Dept. Mech. Engrg., McGill University, Montreal, 1974

11. Leschziner, M.A. "An introduction and guide to the computer code 'PASSABLE'" UMIST Mech. Eng. Dept. Report, 1982

12. Enayet, M.M., Gibson, M.M., Taylor, A.M. and Yianneskis, M. Int. J. Heat and Fluid Flow 3, 213, 1982

13. Iacovides, H. and Launder, B.E. Proc. 1st UK National Heat Transfer Conference, I. Chem. Eng. Symp. Series No.86, 1097, 1984

14. Kline, S.J., Cantwell, B. and Lilley, G.M. (editors) Proceedings 1980-81 AFOSR-HTTM-Stanford Conference on Complex Turbulent Flows (in 3 volumes), Dept. Mech. Engrg., Stanford University, 1981

15. Rodi, W. and Scheuerer, G. Phys. Fluids 26, 1422-1436, 1983

16. Khezzar, L. "Computation of two-dimensional boundary layers and wall jets over convex surfaces" MSc Dissertation, Univ. of Manchester Faculty of Engineering (Mech. Eng. Dept.), 1983

17. Gillis, J.C. and Johnson, J.J. "Turbulent boundary layer on a convex curved surface" Rep. HMT 31, Mech. Eng. Dept., Stanford University

18. Castro, I. and Bradshaw, P. J. Fluid Mech. 73, 265, 1976

19. Gibson, M.M.. Jones, W.P. and Younis, B. Phys. Fluids 24, 386, 1981

20. Chou, P.Y. Quart. J. Appl. Math. 3, 38-54, 1945

21. Rotta, J.C. Z. Phys. 129, 547, 1951

22. Lumley, J.L. and Newman, G. J. Fluid Mech. 72, 161, 1977

23. Lumley, J.L. "Prediction Methods for Turbulent Flow - Introduction", Lecture Series 76, Von Karman Inst., Rhode-St-Genese, Belgium

24. Launder, B.E., Morse, A., Rodi, W. and Spalding, D.B. Proc. 1972 Langley Free Shear Flow Conference, NASA SP320, 1973

25. Launder, B.E., Reece, G.J. and Rodi, W. J. Fluid Mech. 68, 537, 1975

26. Naot, D., Shavit, A. and Wolfshtein, M. Phys. Fluids 16, 738, 1973

27. Younis, B. PhD Thesis, Faculty of Engineering, University of London, 1984

28. Gibson, M.M. and Younis, B. Manuscript submitted to J. Fluid Mech., 1984

29. Morse, A.P. PhD Thesis, Faculty of Engineering, University of London, 1980

30. Launder, B.E. and Morse, A.P. Turbulent Shear Flows-1, 279-294 (ed. F.J. Durst et al) Springer Verlag, New York, 1979

31. Schumann, U. and Patterson, G.S. J. Fluid Mech. 88, 711, 1978

32. Champagne, F.H., Harris, V.G. and Corrsin, S.C. J. Fluid Mech. 41, 81, 1970

33. Reece, G.J. PhD Thesis, Faculty of Engineering, University of London, 1977

34. Ramaprian, B.R. and Tu, S.W. "Study of periodic turbulent pipe flow" IIHR Report No.238, Institute of Hydraulic Research, Univ. of Iowa, Iowa City, 1982

35. Mizushina, T., Maruyama, T. and Hirasawa, H. "Structure of the turbulence in pulsating pipe flows" J. Chemical Engineering of Japan 8, 3, 210-216, 1975

36. Lumley, J.L. Phys. Fluids 18, 750, 1975

37. Lin, A. and Wolfshtein, M. Turbulent Shear Flows-2, (ed. L.J.S. Bradbury et al) Springer Verlag, New York, 1980

38. Launder, B.E. AIAAJ 20, 436, 1982

39. Launder, B.E., Priddin, C.H. and Sharma, B.I. J. Fluids Engrg. 99, 237, 1977

40. Davidov, B. Soviet Physics Doklady 4, 10, 1961

41. Pope, S.B. AIAAJ 16, 279, 1978

42. Forstall, W. and Shapiro, A.H. J. Appl. Mech. 17, 399, 1950

43. Hanjalic, K. and Launder, B.E. ASME J. Fluids Engrg. 102, 1980

44. Sindir, M. PhD Thesis, Dept. Mech. Engrg., Univ. California, Davis, 1981

45. Leschziner, M.A. and Rodi, W. ASME J. Fluids Engrg. 103, 352, 1981

46. Launder, B.E. Turbulent Shear Flows-2, 3-6 (ed. L.J.S. Bradbury et al) Springer Verlag, New York, 1980

47. Sandri, G. and Cerasoli, C. "Fundamental research in turbulent modelling" Aeronautical Research Associates of Princeton, Inc. Report No.438, 1981

48. Donaldson, C. du P. and Sandri, G. "On the inclusion of eddy structure in second-order-closure models of turbulent flows" AGARD Report, 1982

49. Hanjalic, K., Launder, B.E. and Schiestel, R. Turbulent Shear Flows-2, 36-49 (ed. L.J.S. Bradbury et al) Springer Verlag, New York, 1980

50. Cler, A. Thèse Docteur Ingénieur, Ecole Nat. Sup. Aero. Espace., Toulouse, 1982

51. Schiestel, R. Jo. de Méc. Th. et Appl. 2, 601, 1983

52. Donaldson, C. du.P. AIAA J. 10, 4, 1972

53. Lumley, J.L. and Khajeh Nouri, B. Adv. Geophysics 18A, 169, Academic, 1974

54. Rodi, W. "Prediction of free turbulent boundary layers using a 2-equation model of turbulence" PhD Thesis, Faculty of Engineering, University of London, 1972

CHAPTER VIII. Renormalisation Group Methods Applied to the Numerical Simulation of Fluid Turbulence

W.D. McComb

1. INTRODUCTION

Renormalisation group (or RG) is a powerful new method of tackling physical problems involving many length or energy scales. Originally a procedure for removing the divergences of quantum electrodynamics, it has recently come to prominence in the study of critical phenomena. In the context of fluid turbulence (which is what interests us here) RG may be seen (in principle, at least) as a systematic way of progressively eliminating the effect of the smallest eddies; then the next smallest eddies; and so on: and replacing their mean effect by an effective turbulent viscosity. In other words, the molecular kinematic viscosity of the fluid ν_0 becomes renormalised by the collective interaction of the turbulent eddies. Or, if we prefer to think in terms of 'modes' of wavenumber k(rather than 'eddies'), then for all $k < k_c$ we may replace ν_0 by $\nu(k)$, where $\nu(k)$ represents the mean effect of modes $k > k_c$. Here k_c is some cut-off.

This is a rather general definition of RG applied to turbulence but we shall be more specific later on. In the rest of this section we shall briefly discuss the background in critical phenomena, then we shall give a more specific definition of RG, discussing the various ways in which the method has been applied to turbulence. We shall conclude the section by outlining the contents of the succeeding sections.

The study of critical phenomena deals with matter in the neighbourhood of a phase transition. Familiar examples are a liquid-gas system near the critical point (the upper limit on temperature and pressure at which a liquid and its vapour can coexist)or a ferromagnet at the Curie point (the highest temperature for zero magnetic field at which there can be a finite overall magnetisation). We shall discuss the ferromagnet as an example of the application of RG.

Magnetism arises because spins at lattice sites become aligned with each other. In general such alignments occur as random fluctuations, with length scales ranging from the lattice spacing (L_0, say) up to some correlation length ξ. The correlation length is a function of temperature and as the temperature goes to the Curie point, $\xi \to \infty$. Thus at the critical point fluctuations occur on all wavelengths from L_0 (about 1 Å) to infinity.

The theoretical objective is to calculate the Hamiltonian (and hence the partition function and hence the thermodynamics properties) of the system. But the Hamiltonain H is dominated by a collective term which is the sum (over configurations) of all spin interactions. This leads to formidable problems and in practice the exact form has to be modelled; the classic approach being 'mean field theories', in which one spin is supposed to experience a mean field due to the collective effect of all the others. In passing we should note that mean field theories do not agree with experiment, but they are not wildly out either.

Successful RG approaches handle the interaction contained in H as a perturbation expansion about the mean field theory. The basic RG method is to start with the interaction Hamiltonian H_0 associated with two spins, separated by L_0. Then one calculates an effective Hamiltonian H_1, associated with a region of size $2L_0$ (the factor 2 is arbitrary), which means averaging out effects of scales L_0. Then one calculates H_2, associated with a region of size $4L_0$, with effects of scale less than $2L_0$ averaged out. An so on.

The process may be expressed as a transformation τ which is applied repeatedly

$$\tau(H_0) = H_1, \ \tau(H_1) = H_2, \ \tau(H_2) = H_3 \ \text{etc.} \tag{1.1}$$

At each stage the length scales are changed $L_0 \to 2L_0$, $2L_0 \to 4L_0$ etc. In order to compensate, spin variables are also scaled in an appropriate fashion, such that the Hamiltonian always looked the same in scaled variables. It is this re-scaling which leads to renormalisation and the transformations (1.1) define a simple group (strictly, a semi-group).

If iterating the transformation leads to the situation

$$H_{n+1} = H_n, \tag{1.2}$$

where $H_{n+1} = \tau(H_n)$, then $H_n \equiv H_N$ (say) is a fixed point and corresponds to the critical point. Intuitively this may be understood in terms of the fact that fluctuations of infinite wavelength (which occur at the critical point) will be invariant under scaling transformations.

If we think of the physical system as being repeated by a point in a multidimensional space, the "coordinates" being the various interaction forces. Then scaling moves the representative point. Thus the action of RG results in the system evolving along a trajectory, with the enlargement ratio playing the part of time. The resulting fixed point is determined by the solution of the equation

$$\tau(H_n) = H_n \tag{1.3}$$

and is a property of the transformation τ, rather than the initial interaction H_o. This is associated with the idea of universality of critical behaviour. And in the case of turbulence would imply that the renormalised effective viscosity $\nu(k)$ would not depend on ν_o.

If we repeat the above discussion in terms of wavenumber k, then the Fourier transform of the spin variables (strictly, spin field), is introduced and this brings with it the space dimension d as an explicit part of the analysis. We now talk about forming H_1 from H_0 by integrating out modes $k \gtrsim 2\pi/L_o$. Thereafter, modes are eliminated in bands $\pi/L_o \lesssim k \lesssim 2\pi/L_o$, $^{\pi}/2L_o \lesssim k \lesssim \pi/L_o$ etc. as we form H_2, H_3 etc. It is found that there are two fixed points. For d>4, the fixed point corresponds to the classical mean field theory; for d<4 there are non-trivial corrections to the classical results. These are obtained by expansion in $\varepsilon = 4-d$, where the perturbation expansions are justified for small ε. Then ε is set equal to unity and excellent agreement with experiment is obtained for d = 3!

For a proper account of RG and the ε-expansion in critical phenomena, reference should be made to the superb review by Wilson and Kogut (1974). The above discussion is only intended to provide a little background, but it is perhaps worth making the final point that the ε-expansion is not intrinsic to RG. Rather, it is a powerful method of implementing RG in certain problems.

At this stage it seems appropriate to give a more specific definition of RG, which will help us to approach the problem of turbulence. We shall adapt the form given by Ma and Mazenko (1975) for dynamical critical phenomena (i.e. where the 'conserved' variables of a system vary slowly with time), to the case of the Navier-Stokes equations for the Fourier components of the velocity field: $U_\alpha(\underline{k},t)$. (We shall deal with this equation in Section 2). The system will be taken to have random stirring forces $f_\alpha(\underline{k},t)$ with prescribed statistics and "noise" spectrum. The RG procedure involves two stages:

(1) The Fourier decomposition of the velocity field is taken to be cut off for $k < \Lambda$. Divide the velocity field into modes $U^<_\alpha(\underline{k},t)$ and $U^>_\alpha(\underline{k},t)$ where $U^>_\alpha(\underline{k},t)$ are the modes such that $a\,\Lambda < |\underline{k}| < \Lambda$. Eliminate

the high-k modes by solving the equation for $U_\alpha^>(\underline{k},t)$ and substituting the solution into the equation for $U_\alpha^<(\underline{k},t)$. [Note, because the nonlinear mixing induces a sum over modes, the solution for $U_\alpha^>$ will contain $U_\alpha^<$]. Average over $f_\alpha^>(\underline{k},t)$.
(2) Rescale $\underline{k}$, t, $U_\alpha^<$ and $f_\alpha^<$ so that the new equation looks like the original Navier-Stokes equation. This last step involves the introduction of renormalised transport coefficients.

It is somewhat easier to write down the above procedure than to implement it. Indeed attempts to apply RG methods to turbulence face formidable difficulties and, despite some encouraging results, all published theories, for one reason or another, may fairly be regarded as still rather tentative in character. On the other hand, if one wishes to take an optimistic view, Wilson (1975) points out that there is no cookbook for RG methods and that it is generally difficult to formulate renormalisation group methods for new problems. The result of this is that RG can seem as hopeless as any other method until someone succeeds in solving the problem by the renormalisation group approach!

For the purpose of this introduction we shall classify attempts to apply RG to turbulence into three broad groups, according to the nature of the problem studied. These are as follows: (a) the laminar-turbulent transition; (b) calculation of scaling laws for asymptotic turbulent energy spectra, with specified noise inputs; and (c) numerical simulation of well-developed turbulence.

The first of these topics involves a study of the transition from quasi-periodic behaviour to chaos, under the influence of external noise. RG has been used to establish the existence of universal behaviour near the transition, with good agreement being obtained with numerical (and other) experiments. This problem is, of course, closer to the area of critical phenomena where RG has had some striking successes. It is also a specialised discipline in itself and, as our interest here is restricted to well-developed turbulence, we shall not try to disucss it further. But some recent references are given for completeness (Feigenbaum 1979, Shraiman et al. 1981, Crutchfield et al. 1981, Shraiman 1984).

In topic (b), the dynamic RG scheme outlined above is implemented using low-order perturbation theory. The pioneers were Forster, Nelson and Stephan (1976, 1977: to be referred to as FNS) who chose the ultraviolet cut-off Λ low enough to exclude cascade effects. With this restriction, they pose a rather artificial problem which is susceptible to Wilson-type theory, complete with cross-over dimension and ε-expansion.

Their theory is an impressively rigorous approach to statistical hydrodynamics, with renormalisation of the viscosity, the stirring forces and the coupling constants. Under RG iteration, the Navier-Stokes equation is reduced to a (linear) Langevin equation, valid in the limit $k \to 0$.

FNS have been followed by various other asymptotic theories (e.g. Pauquet et al. 1978, De Dominicis and Martin 1979) for scaling behaviour in the infrared. A common feature is that the nonlinear terms are swamped (and hence made tractable) by arbitrarily chosen noise inputs which are unbounded in $\underline{k}$-space. For example, De Dominicis and Martin (1979) obtain the Kolmogorov "$-^5/3$" spectrum by assuming that the noise spectrum goes as k^{-3} in three dimensions. This may be compared with the ability of renormalised perturbation theory to yield the Kolmogorov spectrum with a noise spectrum given by a Dirac delta-function at the origin in k-space (Edwards and McComb 1969). The delta-function also has dimensionality $d = -3$ but is bounded and does not "force" the result. (And, of course, other RPTs also give the Kolmogorov result with stirring forces that are well-behaved, but otherwise arbitrary: Kraichnan 1965, McComb 1978).
Scaling behaviour has also been studied in the ultraviolet, where $k \to \infty$ (e.g. Grossman and Schnedler 1977, Levich 1980, Yakhot 1981, Levich and Tsinober 1984). A particular motivation has been to overcome specific problems with the FNS approach, viz: (1) the restriction to stirring forces which are unbounded in $\underline{k}$-space (which we have just discussed); and (2) the fact that the FNS theory connot be applied to the high-frequency spectrum because the effective coupling constant increases with increasing frequency. In these studies, the usual RG procedure is reversed, and it is the low wavenumbers which are progressively eliminated. In all cases, the theories yeild corrections to the Kolmogorov spectrum. An interesting side-issue is that Levich and Tsinober (1984) stress that their noninteger dimension is the fractal dimension D (Mandelbrot 1977) and not the noninteger dimension used for analytic continuation in Wilson-type theory (i.e. the ε-expansion).

It is difficult to know how to assess these theories. On the one hand, they bring a welcome (and novel!) rigour into turbulence theory. Yet, as Kraichnan has pointed out in a critical review (1982), they only lead to scaling exponents (at best) and do not fix constants of proportionality. Moreover, they are only asymptotically valid for situations remote from the local-in-wavenumber energy transfer which is the dominant feature of turbulence. Again, for the defence (so to speak), Sulem et al. (1979) point out that the FNS renormalised viscosity and energy spectrum (the solution of their Langevin equation) together satisfy the Heisenberg formula, which was obtained on phenomenological grounds (without the restriction to small ε) when the energy transfer is local. This could be taken as a hopeful sign that the FNS theory may someday be extended to the inertial range. Or it could just be a trivial consequence of dimensionality! In any case, we shall show later, that RG can yield the Heisenberg formula for the effective viscosity (complete with constant of proportionality) without the need for perturbation expansions. For, as Kraichnan (1982) notes, there is nothing intrinsic to tie the RG approach to perturbation methods. He suggests that the most apt model for turbulence may be Wilson's treatment of the Kondo problem (Wilson 1975) and instances Siggia's (1981) suggestion that numerical simulation of turbulence could be treated in a similar way. This brings us nicely to topic (c) - numerical simulation - which is indeed the field of interest of this lecture.

It is well known that turbulant flows of any practical significance lie far beyond the scope of full numerical simulation. For this reason, much attention has recently been given to the idea of large-eddy simulation (or LES). Roughly speaking this technique poses two problems. First, there is what one may call the "software aspect". That is, the numerical and computational methods needed to actually simulate the large eddies on a computer. Second, there is the problem of modelling the "sub-grid drain"; or transfer of energy from the explicit scales to the unresolved sub-grid scales.

When one thinks about it, the very idea of LES sounds like a crude realisation of the renormalisation group. And the subgrid modelling problem is clearly an ideal candidate for an RG approach.

This was first recognised by Rose (1977), who applied RG methods to the subgrid modelling of passive scalar convection. He was able to obtain a renormalised (eddy) diffusivity, which had a weak dependence on the explicit wavenumber, and represented the mean effect of the subgrid scales on the scalar-field explicit scales. He also found two additional terms, which (a) represented the noise injected from the subgrid scales and which (b) represented the coupling between the large eddies and the eddies just below the resolution of the grid. I shall refer again to these results presently.

My own approach to this problem (McComb 1982) has been from a somewhat different direction. Statistical equations (analogous to the Reynolds equation) were derived by progressively averaging the Navier-Stokes equation over a series of increasing time periods. Averaging over the shortest period (with a suitable weight function) smooths out that part of the field which corresponds to the highest-frequency fluctuations. The mean effect of these fluctuations was calculated from the time-averaged equation of motion and so eliminated from the equation for the rest of the velocity field (that is, the unaveraged part). An iteration process then led to equations for the mean and covariance of the fluctuating field, in which the Reynolds stresses do not appear explicitly. They are represented in each cycle of the iteration by the sum of the following (1) a constitutive relation expressing the mean-square fluctuation in terms of the mean rate of strain; and (2) the unaveraged portion of the nonlinear term. In the limit of long averaging times, the mean velocity equation would become equivalent to the Reynolds equation with a constitutive relationship for theReynolds stress as a functional of the mean rate of strain.

The motivation for this work was a feeling that Reynolds averaging was too inflexible to allow one to consider intermittent effects (e.g. "bursts") or even slow external time variations.

The above method of iterative averaging may be shown to be equivalent to the renormalisation group (McComb 1982). Connection with RG was made by: (1) Fourier-transforming into ω-space; and (2) invoking the Taylor hypothesis of frozen connection to take the analysis into $\underline{k}$-space. Formally, therefore, I regarded the calculation as giving the isotropic part of an (in general) anisotropic eddy viscosity and this was shown to take the Heisnberg functional form (McComb and Shanmugasundaram 1983 a).

However, the belated realisation that this RG analysis should be applied to LES (McComb and Shanmugasundaram 1983 b) had a liberating effect! It then became clear that the method could be formulated directly for isotropic turbulence, provided we were using filtered variables. The iterative averaging then became based on a combination filtering - (in $\underline{k}$-space) - and-ensemble average (McComb and Shanmugasundaram 1984 a), rather than a time-average, and the formal extension to the LES equation has also been made (McComb and Shanmugasundaram 1984 b,c).

With these changes, the method of iterative averaging has become very much closer to that of Rose (1977). We have previously shown (McComb and Shanmugasundaram 1983 a) that, if Rose's method is applied to the velocity-field problem, then it differs from ours in that a triple-moment involving only the explicit scales appears in the RG ineration. Unlike triple-moments involving only subgrid scales, there is no basis for neglecting such term. The underlying reason for this difference between the two methods lies in the order of the various operations, rather than in the approximations (which are, in fact, identical). Accordingly I feel that the terms (a) and (b) referred to above in the discussion of Rose's work are an artifact of his method. This does not mean that I am sceptical about eddy noise. Far from it! But I don't believe that it acts on explicit scales from subgrid scales. In short, I suspect that terms (a) and (b) probably should cancel, and that their apparent presence is only due to Rose's method failing to give a clean separation of high and low frequencies.

In section 2 we introduce the basic equations, in $\underline{k}$-space and discuss the formulation of large-eddy simulation for isotropic turbulence. Section 3 is in two parts. First we show that our averaging procedure can, in effect, simplify the LES equations when the cut-off wavenumber is very large (i.e. in the viscous region). Second, we show that iterative averaging extends this simplification to ever-decreasing cut-off wavenumbers, with an effective viscosity that depends on the sub-grid energy spectrum. In section 4 we introduce the RG transformation and present the calculations of the effective viscosity. Finally, section 6 summarises the results, and states our plans for future work.

2. Formulation Of LES Equations For Isotropic Turbulence.

Full discussions of the practical problems of LES may be found in the many excellent reviews (e.g. Ferziger 1981, Voke and Collins 1983; Rogallo amd Moin 1984). In the present work we shall concentrate on the statistical treatment of the small eddies and, in particular, on the fundamental aspects of the subgrid-scale modelling problem. This means that we shall consider only incompressible fluid turbulence which is both homogeneous and isotropic. We begin by introducing the basic equations in wavenumber (k) space.

If we let the velocity field be $U_\alpha(\underline{x},t)$ then the Fourier components in a cubical box of side L are defined by

$$U_\alpha(\underline{x},t) = \sum_k U_\alpha(\underline{k},t)\, e^{i\underline{k}\cdot\underline{x}} . \tag{2.1}$$

The equation of motion for a fluid of kinematic viscosity ν_o may be written as

$$\left(\frac{\partial}{\partial t} + \nu_o k^2\right) U_\alpha(\underline{k},t) = \sum_j M_{\alpha\beta\gamma}(\underline{k})\, U_\beta(\underline{j}),t)\, U_\gamma(\underline{k}-\underline{j},t) \tag{2.2}$$

and the continuity equation becomes

$$k_\alpha \, U_\alpha(\underline{k},t) = 0, \tag{2.3}$$

where the inertial-transfer operator $M_{\alpha\beta\gamma}(\underline{k})$ is defined by

$$M_{\alpha\beta\gamma}(\underline{k}) = (2i)^{-1}\{k_\beta \, D_{\alpha\gamma}(\underline{k}) + k_\gamma \, D_{\alpha\beta}(\underline{k})\} \tag{2.4}$$

and

$$D_{\alpha\beta}(\underline{k}) = \delta_{\alpha\beta} - k_\alpha \, k_\beta \, |\underline{k}|^{-2}. \tag{2.5}$$

The pair-correlation of velocities may be defined, thus:

$$\begin{aligned}(\tfrac{L}{2\pi})^3 \langle U_\alpha(\underline{k},t) \; U_\beta(-\underline{k},t')\rangle &= Q_{\alpha\beta}(\underline{k};t,t') \\ &= D_{\alpha\beta}(\underline{k}) \; Q(k;t,t'),\end{aligned} \tag{2.6}$$

the last step being appropriate to isotropic turbulence, and the relationship between the energy spectrum E(k,t) and the correlation function follows in the usual way:

$$E(k,t) = 4\pi \, k^2 \, Q(k;t,t) \, . \tag{2.7}$$

Finally, we complete our set of basic equations by multiplying each side of equation (1.2) by $U_\alpha(-\underline{k},t')$ and averaging to obtain an equation for the correlation function, thus:

$$(\frac{\partial}{\partial t} + \nu_0 k^2) \; Q(k;t,t') = P(k;t,t') \, . \tag{2.8}$$

The inertial transfer term P(k;t,t') contains the unknown triple moment $\langle U_\beta(\underline{j},t) \; U_\gamma(\underline{k}-\underline{j},t) \; U_\alpha(-\underline{k},t')\rangle$; and so calculation of the correlation function depends on a suitable closure approximation (e.g. McComb 1978; McComb and Shanmugasundaram 1984 d). However, for our purposes here, an explicity form for P(k;t,t') will not be needed; and it is only necessary to bear in mind that this term involves a sum over modes as indicated on the right-hand side of equation (2.2).

The LES equations are conventionally formulated by dividing the velocity field up at $k = k_c$ into explicit and sub-grid scales, thus:

$$\begin{aligned}U_\alpha(\underline{k},t) &= U^<_\alpha(\underline{k},t), \; k \le k_c \; ; \\ &= U^>_\alpha(\underline{k},t), \; k \ge k_c \, .\end{aligned} \tag{2.9}$$

Substitution of (2.9) into (2.2) yields an immediate separation into low and high frequencies on the LHS. However, on the RHS, all the modes are coupled and we have to apply a filter (cut-off at $k = k_c$) to the whole nonlinear term. Thus, with an obvious extension of the notation, equation (2.2) is resolved into:

$$(\frac{\partial}{\partial t} + \nu_0 k^2)\ U^<_\alpha(\underline{k},t) = \sum_{\underline{j}} M^<_{\alpha\beta\gamma}(\underline{k})\ U^<_\beta(\underline{j},t)\ U^<_\gamma(\underline{k}-\underline{j},t)$$

$$+ \sum_{\underline{j}} M^<_{\alpha\beta\gamma}(\underline{k})\ U_\beta(\underline{j},t)\ U_\gamma(\underline{k}-\underline{j},t); \qquad (2.10)$$

$$(j \text{ and/or } |\underline{k}-\underline{j}| > k_c)$$

and

$$(\frac{\partial}{\partial t} + \nu_0 k^2)\ U^>_\alpha(\underline{k},t) = \sum_{\underline{j}} M^>_{\alpha\beta\gamma}(\underline{k})\ U^>_\beta(\underline{j},t)\ U^>_\gamma(\underline{k}-\underline{j},t)$$

$$+ \sum_{\underline{j}} M^>_{\alpha\beta\gamma}(\underline{k})\ U_\beta(\underline{j},t)\ U_\gamma(\underline{k}-\underline{j},t). \qquad (2.11)$$

$$(j \text{ and/or } |k-j| < k_c)$$

Corresponding to these last two equation, we can derive equations (as in going from (2.1) to (2.8)) for the correlation functions of the explicit and subgrid scales. Again, with an obvious extension of the notation, we have:

$$(\frac{\partial}{\partial t} + \nu_0 k^2)Q^<(k;t,t') = P^<(k;t,t') + P^{<>}(k;t,t') \qquad (2.12)$$

and

$$(\frac{\partial}{\partial t} + \nu_0 k^2)Q^>(k;t,t') = P^>(k;t,t') + P^{><}(k;t,t'), \qquad (2.13)$$

where $P^{<>}$ gives the inertial coupling to $k \leq k_c$ from sums over j with j and/or $|\underline{k}-\underline{j}| > k_c$ and $P^{><}$ gives the inertial coupling to $k \geq k_c$ from sums over j with j and/or $|\underline{k}-\underline{j}| < k_c$.

The standard method of subgrid modelling is to assume that the subgrid scale transfer function on the RHS of (2.10) may be represented in terms of an eddy viscosity $\nu(k|k_c)$. Thus equation (2.10) can be written in the form:

$$\{\frac{\partial}{\partial t} + \nu(k|k_c)k^2\}\ U^<(\underline{k},t) = \sum_{\underline{j}} M^<_{\alpha\beta\gamma}(\underline{k})\ U^<_\beta(\underline{j},t)\ U^<_\gamma(\underline{k}-\underline{j},t), \qquad (2.14)$$

which governs the evolution of the explicit scales, once $\nu(k|k_c)$ is prescribed.

Kraichnan (1976) proposed that the effective eddy viscosity should be defined statistically by the relation

$$P^{<>}(k;t,t) = -2\nu(k|k_c)k^2\ Q^<(k;t,t) \qquad (2.15)$$

and obtained an expression for $\nu(k|k_c)$ in terms of the energy spectrum, on the basis of a two-point closure (the Test Field Model). In principle, given a satisfactory closure approximation, equation (2.14) can be simulated numerically, with equations (2.15) and (2.13) being solved for $\nu(k|k_c)$. We note that the coupling terms $P^{<>}$ and $P^{><}$ depend on the spectrum at all wavenumbers. In a practical calculation this will be made up of discrete contributions for $k \leq k_c$ but will take a continuous form for $k \geq k_c$.

Kraichnan (1976) considered the more restricted problem of quasi-stationary turbulence, with the spectrum at large Reynolds numbers taking the Kolmogorov form. Thus in terms of the correlation (or spectral density) functions, we have

$$E(k) = 4\pi k^2 Q(k) = \alpha \varepsilon^{2/3} k^{-5/3} , \tag{2.16}$$

where α is a constant and ε is the dissipation rate. For this case, Kraichnan's effective eddy viscosity took the asymptotic form

$$\nu(k|k_c) = 0.267 \alpha^{1/2} \varepsilon^{1/3} k_c^{-4/3} , \quad k \ll k_c . \tag{2.17}$$

A more extensive analysis using several spectra has been given by Leslie and Quarini (1979) and various recent LES studies have been based on this approach (e.g. Chollet and Lesieur 1981). We shall return to these results later when we discuss our own calculations.

3. Iterative Averaging And RG.

Large-eddy simulation could well be regarded as a rather crude realisation of the RG approach. The effect of large wavenumber ($k \geq k_c$) is handled as an effective viscosity, acting on the smaller wavenumber modes ($k \leq k_c$). Equally we can regard iterative averaging and RG as a formal (and perhaps even rigorous) way of formulating LES equations by eliminating the high-k modes progressively and systematically in bands. Then the subgrid effective viscosity comes about in a very natural way. In any case, a consideration of the LES equations, with the cut-off wavenumber k_c taken, to be very large, makes quite a good lead-in to the RG approach.

Let us consider the covariance equations of the explicit and subgrid scales. Clearly if we take $k_c \to \infty$, then $P^{<>}$ in equation (2.12) becomes negligible, as do all the terms of equation (2.13). This is, therefore, the case of full simulation. The interesting case to consider is when k_c is of the order of the dissipation wavenumber $k_d = (\varepsilon/\nu_0{}^3)^{1/4}$: then $P^{<>}$ represents a small, but finite subgrid drain. However, the effect on (2.13) for the subgrid scales is more significant. We can neglect the inertial transfer to higher modes when $k \geq k_c (= k_d)$. That is, the term $P^>$ in equation (2.13). Further, for stationary turbulence we may put the local time-derivative equal to zero. Then $Q^>$ is determined by the local (in wavenumber) viscous damping and the energy injected from the explicit scales. Thus it is possible to solve for $Q^>$ and, if the problem is properly set up, to eliminate $Q^>$ (or, strictly, from equation (2.10) for $U^<$).

We shall see how these ideas may be implemented in the following sub-sections. However, first they need to be supplemented by one specific approximation. We shall assume tht explicit and subgrid modes are statistically independent, and that this remains so throughout the iterative calculations to be presented.

Our procedure will be to divide the velocity field up at $k = k_o$, where $k_o \sim k_d$, and eliminate the mean effect of the high-frequency modes ($k \geq k_o$). Then we shall repeat this operation for $k_1 < k_o$, $k_2 < k_1$, and so on, until a recursion relation for the effective viscosity reaches a fixed point.

3.1 The Filtered Equation at Large Wavenumbers

Let us divide-up the velocity field at $k = k_o$, but from now on we shall use a different notation to help distinguish our own approach from the RG analysis discussed proviously. We put

$$U_\alpha(\underline{k},t) = U_\alpha^-(\underline{k},t), \; k \leq k_o \;;$$

$$= U_\alpha^+(\underline{k},t), \; k \geq k_o \;. \tag{3.1}$$

The averaging process is taken to be an ensemble average over the small scales such that

$$<U_\alpha(\underline{k},t)> = U_\alpha^-(\underline{k},t);$$

$$<U_\alpha^-(\underline{k},t)> = U_\alpha^-(\underline{k},t) \;; \quad <U_\alpha^+(\underline{k},t)> = 0, \tag{3.2a}$$

and, with the assumption of statistical independence,

$$<U_\alpha^-(\underline{k},t)\, U_\alpha^+(\underline{k},t)> = 0 \tag{3.2b}$$

Substituting (3.1) into (2.2) and averaging according to (3.2) we obtain the low-frequency equation

$$(\frac{\partial}{\partial t} + \nu_o k^2)\, U_\alpha^-(\underline{k},t) - \sum_{\underline{j}} M_{\alpha\beta\gamma}(\underline{k}) \langle U_\beta^+(\underline{j},t)\, U_\gamma^+(\underline{k}-\underline{j},t)\rangle$$

$$k_o \le \underline{j},\ |\underline{k}-\underline{j}| \le \infty$$

$$= \sum_{\underline{j}} M_{\alpha\beta\gamma}(\underline{k})\, U_\beta^-(\underline{j},t)\, U_\gamma^-(\underline{k}-\underline{j},t) \qquad (3.3)$$

This is now our basic equation and our main objective is to eliminate the terms with wavenumbers greater than k_o, which already only appear in a statistical sense through $\langle U^+ U^+\rangle$. We form the high-frequency equation by substituting (3.3) back into equation (2.2), with the result:

$$(\frac{\partial}{\partial t} + \nu_o k^2)\, U_\alpha^+(\underline{k},t) = \sum_{\underline{j}} M_{\alpha\beta\gamma}(\underline{k}) \{2U_\beta^-(\underline{j},t)\, U_\gamma^+(\underline{k}-\underline{j},t)$$

$$+ U_\beta^+(\underline{j},t)\, U_\gamma^+(\underline{k}-\underline{j},t) - \langle U_\beta^+(\underline{j},t)\, U_\gamma^+(\underline{k}-\underline{j},t)\rangle\} \qquad (3.4)$$

Our objective now is to solve (3.4) in order to obtain an expression for the covariance $\langle U^+ U^+\rangle$ which occurs on the left-hand side of (3.3) To do this, we multiply each term in (3.4) by $U_{\alpha'}^+(\underline{k}',t)$ and average. As the covariance is single-time and we take the field to be stationary, the time-derivative is eliminated. The last term on the right-hand side vanishes under the averaging, and the term involving $\langle U^+ U^+ U^+\rangle$ (i.e. cf $P^>$ in the preceding discussion) may be neglected for $k_o \sim k_d$. Thus (3.4) yields for the high-wavenumber covariance

$$\langle U_\alpha^+(\underline{k},t) U_{\alpha'}^+(\underline{k}',t)\rangle = 2(\nu_o k^2)^{-1} \sum_{\underline{j}} M_{\alpha\beta\gamma}(\underline{k}) \langle U_\beta^-(\underline{j},t) U_\gamma^+(k-j,t) U_{\alpha'}(\underline{k}',t)\rangle \qquad (3.5)$$

We now wish to obtain an expression specifically for $\langle U_\beta^+(\underline{j},t) U_\gamma^+(\underline{k}-\underline{j},t)\rangle$ which occurs in the RHS of equation (3.3). We therefore re-label as appropriate and name dummy variables to avoid confusion, to obtain

$$\langle U_\beta^+(\underline{j},t)\, U_\gamma^+(\underline{k}-\underline{j},t)\rangle = 2(\nu_o\, j^2)^{-1} \sum_{\underline{p}} M_{\beta\rho\delta}(\underline{j})\, U_\rho^-(\underline{p},t)$$

$$\times\ U_\delta^+(\underline{j}-\underline{p},t)\, U_\gamma^+(k-j,t)\rangle \qquad (3.6)$$

Now, from (2.6), with an obvious extension of the notation, we have

$$\langle U_\delta^+(\underline{j}-\underline{p},t)\, U_\gamma^+(\underline{k}-\underline{j},t)\rangle = (\frac{2\pi}{L})^3\, D_{\delta\gamma}(\underline{k}-\underline{j})\, Q_o^+(|\underline{k}-\underline{j}|)\, \delta_{\underline{k}\,\underline{p}}\ , \qquad (3.7)$$

where homogeneity implies:

$$\underline{j} - \underline{p} + \underline{k} - \underline{j} = 0 \quad \therefore\ \underline{k} = \underline{p}\ .$$

Substituting (3.7) into (3.6) and summing over $\underline{p}$ to eliminate $\delta_{\underline{k}\,\underline{p}}$ we obtain

$$<U^+_\beta(\underline{j},t)\; U^+_\gamma(\underline{k}-\underline{j},t)>_o = 2(\nu_o\, j^2)^{-1}\; (\frac{2\pi}{L})^3\; M_{\beta\rho\delta}(\underline{j})\; D_{\delta\gamma}(\underline{k}-\underline{j})$$

$$\times\; Q^+_o(|\underline{k}-\underline{j}|)\quad U^-_\rho(\underline{k},t). \qquad (3.8)$$

Finally we substitute (3.8) into the RHS of equation (3.3). In the process we take two steps, thus:

(1) $M_{\alpha\beta\gamma}(\underline{k})\; U^-_\rho(\underline{k},t) = M_{\rho\beta\gamma}(\underline{k})\; U^-_\alpha(\underline{k},t)$;

(2) $({}^{2\pi}/_L)^3 \sum_{\underline{j}} \rightarrow \int d^3\underline{j}$ as $L \rightarrow \infty$;

with the result

$$\{\frac{\partial}{\partial t} + \nu_o\, k^2 + \delta\nu_o(k)k^2\}U^-_\alpha(\underline{k},t) = \sum_{\underline{j}} M_{\alpha\beta\gamma}(\underline{k})\; U^-_\beta(\underline{j},t)\; U^-_\gamma(k-j,t), \qquad (3.9)$$

$$0 \leq k \leq k_o$$

where

$$\delta\nu_o(k) = \int d^3\underline{j}\; \frac{L_{kj}\; Q^+_o\;(|\underline{k}-\underline{j}|)}{\nu_o\; j^2\; k^2} \qquad (3.10)$$

for $k_o \leq j,\ |\underline{k}-\underline{j}| \leq \infty$ and $0 \leq k \leq k_o$,

and

$$L_{kj} = -2\; M_{\rho\beta\gamma}(\underline{k})\; M_{\beta\rho\delta}(j)\; D_{\delta\gamma}(\underline{k}-\underline{j})\ . \qquad (3.11)$$

The main point to be noted here is that the "subgrid" term in equation (3.3) had no restriction on the labelling wavevector $\underline{k}$. In going from (3.3) to (3.4), we have identified that part of the "subgrid" term which receives energy only from modes $k \leq k_o$. As this new term is seen to be linear in $U^-_\alpha(\underline{k},t)$, it is correctly interpreted as an increase in the coefficient of viscosity. It should also be noted that the transition from discrete to continuous wavenumber has only been made for wavenumbers greater than k_o.

3.2 Iteractive averaging at decreasing wavenumber.

Equation (3.9) now takes the place of the original Navier-Stokes equation. If we replace the U^- by U, then it is just equation (2.2), with ν_o replaced by $\nu_1 = \nu_o + \delta\nu_o$ and the wavenumber variables restricted $0 \leq k,\ j,\ |\underline{k}-\underline{j}| \leq k_o$. The effect of the wavenumbers greater than k_o is only felt inside the expression for $\delta\nu_o$, where

$k_o \leq j, |\underline{k}-\underline{j}| \leq \infty$.

If we now repeat the above procedure for a wavenumber cut-off $k_1 < k_o$, then we may hope to eliminate the meam effect of fluctuations in the band $k_1 \leq k \leq k_o$. It should be clear that we now divide the new velocity field, described by equation (3.9), into $U_\alpha(\underline{k},t) = U_\alpha^-(\underline{k},t) + U_\alpha^+(\underline{k},t)$ at $k = k_1$. Then it seems reasonable to expect the procedure described in Section 3.1 to go through again, for a coordinate system in which k_d is replaced (in effect) by its renormalised form $k_\alpha^{(1)} = (\varepsilon/\nu_1^3)^{1/4}$. In particular, this implies that in re-calculating $\langle U^+ U^+\rangle$ we may suppose that it is mainly governed by a balance between inertial transfer <u>from</u> modes $0 \leq k \leq k_1$ and "direct dissipation", with viscosity coefficient ν_1 in the band $k_1 \leq k \leq k_o$.

The actual re-scaling process will be carried out in the next section. Here we shall just note that if we carry on the iterative averaging process for successive wavenumbers $k_1, k_2 \ldots\ldots k_n \ldots\ldots$, such that $k_o > k_1 > k_2 > \ldots\ldots > k_n > \ldots\ldots$, then by induction, the iteration for k_n yields

$$\left\{\frac{\partial}{\partial t} + \nu_n(k)k^2 + \delta\nu_n(k)k^2\right\}U_\alpha^-(\underline{k},t) = \sum_{\underline{j}} M_{\alpha\beta\gamma}(\underline{k})\, U_\beta^-(\underline{j},t)\, U_\gamma^-(\underline{k}-\underline{j},t)$$

$$0 \leq k \leq k_n \tag{3.12}$$

where the effective viscosity satisfies:

$$\nu_{n+1}(k) = \nu_n(k) + \delta\nu_n(k) \tag{3.13}$$

and

$$\delta\nu_n(k) = \int \frac{d^3\underline{j}\; L_{kj}\; Q_n^+(|\underline{k}-\underline{j}|,t)}{\nu_n(j)\; j^2\, k^2} \tag{3.14}$$

for

$$k_n \leq j,\ |\underline{k}-\underline{j}| \leq k_{n-1} \quad \text{and} \quad 0 \leq k \leq k_n$$

4. RG Calculation of the Effective Viscosity

Calculations have been carried out for the stationary case where the spectrum may be taken as the Kolmogorov form (2.16). The wavenumber bands are arbitrarily chosen as

$$k_n = h^n k_o\ ; \quad 0 \leq h \leq 1 \tag{4.1}$$

and we make the scaling transformation

$$k \to k_n \tilde{k}\ . \tag{4.2}$$

The scaled effective viscosity ν_n^* is defined by

$$\nu_n(k_n \tilde{k}) = \alpha^{1/2}\, \varepsilon^{1/3}\, k_n^{-4/3}\, \nu_n^*(\tilde{k}) \tag{4.3}$$

and equations (3.13) and (3.14) become

$$\nu_{n+1}^*(k) = h^{4/3}\, \{\nu_n^*(h\tilde{k}) + \delta\nu_n^*(h\tilde{k})\} \tag{4.4}$$

and

$$\delta\nu_n^*(\tilde{k}) = \frac{1}{4\pi}\int \frac{d^3\tilde{j}\; L_{\tilde{k}\tilde{j}}\; |\tilde{k}-\tilde{j}|^{-11/3}}{\nu_n^*(\tilde{j})\; \tilde{k}^2\; \tilde{j}^2} \tag{4.5}$$

for

$$1 \leq \tilde{j},\; |\tilde{k}-\tilde{j}| \leq h^{-1}$$

(N.B. Previously (McComb and Shanmugasundaram 1983 a) we have given equation (4.5) with the incorrect wavenumber limits $1 < \tilde{j} < \infty$, although the limits on $|\tilde{k}-\tilde{j}|$ were correct. This resulted in a numerical error. For instance, we calculated the Kolmogorov constant to be $\alpha \sim 5.57$. With the correct wavenumber limits this result becomes $\alpha = 1.8$).

Our earlier calculations have been for $\tilde{k} \geq 1$, in order to establish the high-k asymptotic form $\nu_N(k) \sim k^{-4/3}$. Here we are concerned with the sub-grid-scale eddy viscosity and the calculation of equations (4.3) - (4.5) for $0 \leq \tilde{k} \leq 1$.

Some problems were encountered with the extension of the calculation to small $\tilde{k}$. Interaction between the RG analysis and the basic numerics led to instabilities. These were cured by using analytic approximations to low-k asymptotic expressions. Details are given elsewhere (McComb and Shanmugasundaram 1984 c).

The recursion relation was found to reach a fixed point $\nu_{n+1}^*(\tilde{k}) = \nu_n^*(\tilde{k}) \equiv \nu_N^*(\tilde{k})$ for various initial values of ν_0^*. This is shown in Fig. 1 for $\tilde{k} = 0.01$ and $h = 0.6$. Note that N lies between 4 and 6, according to the value of ν_0^*.

In Fig. 2, we show the evolution of the actual eddy viscosity (calculated from ν_n^* using (4.3)), as it would appear in equation (3.12) for the purposes of LES. Here $\nu_N(k|k_n)$ is plotted, as a multiple of the molecular viscosity ν_0, against wavenumber k scaled by the Kolmogorov dissipation wavenumber, k_0. The smooth development of $\nu(k|k_n)$ towards a constant value as $k \to 0$ is present at all stages of the iteration.

The calculations presented in Fig. 2 were for $h = 0.6$. The scaling parameter h was chosen arbitrarily. Clearly the whole procedure must break down as $h \to 1$ or $h \to 0$. Somewhere between these limits one would hope for a region where the sub-grid-scale eddy viscosity was reasonably insensitive to arbitrary choices on h. In practice, this was found to be so for $0.45 \leq h \leq 0.95$.

Finally, in Fig. 3 we plot the (scaled) evolved eddy viscosity $\nu_N^*(\tilde{k})$ against $\tilde{k}$ for $h = 0.6$. This may be compared with result from Chollet and Lesieur (1981: calculated using the work of Kraichnan (1976)), which is plotted on the same figure. Two points are noteworthy. First, both theories give $\nu_N^*(\tilde{k}) \rightarrow$ constant as $k \rightarrow 0$, and the constant values are of the same order of magnitude. (In making a quantitative comparison one should bear in mind that Chollet and Lesieur (1981) have fitted an adjustable constant in order to make their underlying closure theory yield a Kolmogorov constant $\alpha = 1.4$. Our own results are for $h = 0.6$ which would predict a value $\alpha = 2.06$). Second, both theories show the effect of local interactions in the neighbourhood of the cut ($\tilde{k} = 1$). However, one should also bear in mind that different definitions of eddy viscosity are involved, when comparing their behaviour near $\tilde{k} = 1$.

Finally, for $h = 0.6$, we may write our asymptotic form of sub-grid-scales effective viscosity as

$$\nu(k|k)_N = 0.367\ \alpha^{1/2}\ \varepsilon^{1/3} k_N^{-4/3} \tag{4.6}$$

where we would take $k_c = k_N$ for purposes of large-eddy simulation. This may be compared with the RPT form given earlier in equation (2.16)

5. CONCLUSION

In view of the results presented, it seems that Iterative Averaging is a promising method of applying RG theory to the statistical description of well-developed turbulence. There seem to be three key points in the method. Thus:

(1) In formulating the low-frequency equation, the high frequencies only appear in an average sense, through the variance : see equation (3.3).

(2) When the variance of the high-frequency modes is calculated, the labelling k-vector is found to be restricted to $0 \leq k \leq k_n$) : see equations (3.8) and (3.9). Thus we have correctly identified the contribution to the eddy viscosity in the low-frequency equation.

(3) No spurious low-frequency terms are introduced into the low-frequency equation in the process of eliminating the high-frequency modes : again, see equations (3.8) and (3.9).

The next stage is to carry out a large-eddy simulation based on the present formulation. This would allow us to check the underlying approximations (outlined in Section 1) in some detail. For instance, neglect of terms $O(U^+)^2$ and higher may accumulate a finite error during the course of the iteration, even though such an approximation will be good initially. It is, however, worth noting that a higher order of approximation - if needed - would not lead to an intolerable level of complication. This may be seen as a major strength of the RG approach.

ACKNOWLEDGEMENTS

We are pleased to acknowledge that this work was carried out with the financial support of the United Kingdom Atomic Energy Authority.

REFERENCES

Chollet, J.P. and Lesieur, M. (1981) J. Atm. Sci 38, 2747.
Crutchfield, J., Navenberg, M. and Rudnich, J. (1981) Phys. Rev. Lett. 46, 933.
De Dominicis, C. and Martin, P.C. (1979) Phys. Rev. A. 19, 419.
Edwards, S.F. and McComb, W.D. (1969) J. Phys. A: Gen. Phys. 2, 157.
Feigenbaum, M.J. (1979) Phys. Lett. A. 74, 375.
Ferziger, J.H. (1981), "Higher-level simulations of turbulent flows" V.K. Inst. for Fluid Dynamics, Lecture series 1981-5.
Forster, D., Nelson. D.R. and Stephen, M.J. (1976) Phys. Rev. Lett. 36, 867.
Forster, D., Nelson, D.R. and Stephen, M.J. (1977) Phys. Rev. A 16, 732.
Grossmann, S. and Schnedler, E. (1977) A. Phys. B 26, 307.
Kraichnan, R.H. (1965) Phys. Fluids 8, 1385.
Kraichnan, R.H. (1976) J. Atm. Sci. 33, 1521.
Kraichnan, R.H. (1982) Phys. Rev. A. 25, 3281.
Leslie, D.C. and Quarini, G.L. (1979) J. Fluid Mech. 79, 65.
Levich, E. (1980) Phys. Lett. A 79, 171.
Levich, E. and Tsinober, A. (1984) Phys. Lett. A 101, 265.
Ma, S.K. and Mazenko, G.F. (1975) Phys. Rev. B 11, 4077.
Mandelbrot, B. (1977) "The fractal geometry of nature" (Freeman, San Francisco, 1977).
McComb, W.D. (1978) J. Phys. A. : Math. Gen. 11, 613.
McComb, W.D. (1982) Phys. Rev. A. 26, 1078.
McComb, W.D. and Shanmugasundaram, V. (1983a) Phys. Rev. A. 28, 2588.
McComb, W.D. and Shanmugasundaram, V. (1983b) "Some developments in the application of renormalisation methods to turbulence theory". Paper presented to the Fourth Symp. on Turb. Shear Flows, Karlsruhe, September 12-14, 1983.
McComb, W.D. and Shanmugasundaram, V. (1984a) J. Phys. A.: submitted for publication.
McComb, W.D. and Shanmugasundaram, V. (1984b) "Renormalisation methods applied to the calculation of the subgrid-scale eddy viscosity for isotropic turbulence." Paper presented to the XVIth Int. Congress on Theor. and Appl. Mechanics (IUTAM), Lyngby, Denmark, August 20-25, 1984.
McComb, W.D. and Shanmugasundaram, V. (1984c) J. Fluid Mech.: submitted for publication.
McComb, W.D. and Shanmugasundaram, V. (1984d) J. Fluid Mech. 143, 95.
Pouquet, A.,Fournier, J.D. and Sulem, P.L. (1978) J. Phys. (Paris) Lett. 39, 199.
Rogallo, R.S. and Moin, P. (1984) Ann. Rev. Fluid Mech. 16, 99.
Rose, H.A. (1977) J. Fluid Mech. 81, 719.
Shraiman, B., Wayne, E.C. and Martin, P.C. (1981) Phys. Rev. Lett. 46, 935.
Shraiman, B.I. (1984). Phys. Rev. A. 29, 3464.
Siggia, E.D. (1981) J. Fluid Mech. 107, 375.
Sulem, P.L., Fournier, J.D. and Pouquet, A. (1979) "Fully developed turbulence and renormalisation froup" in Dynamical Critical Phenomena and Related Topics, Proc. Int. Conf. Geneva, Switzerland, April 2-6, 1979.

Voke, P.R. and Collins, M.W. (1983) PCH Physiochem. Hydrodym. 4, 119.
Wilson, K.G. and Kogut, J. (1974) Phys. Rep. 12C, 75.
Wilson, K.G. (1975) Adv. Math. 16, 170.
Yakhot, V. (1981) Phys. Rev. A 23, 1486.

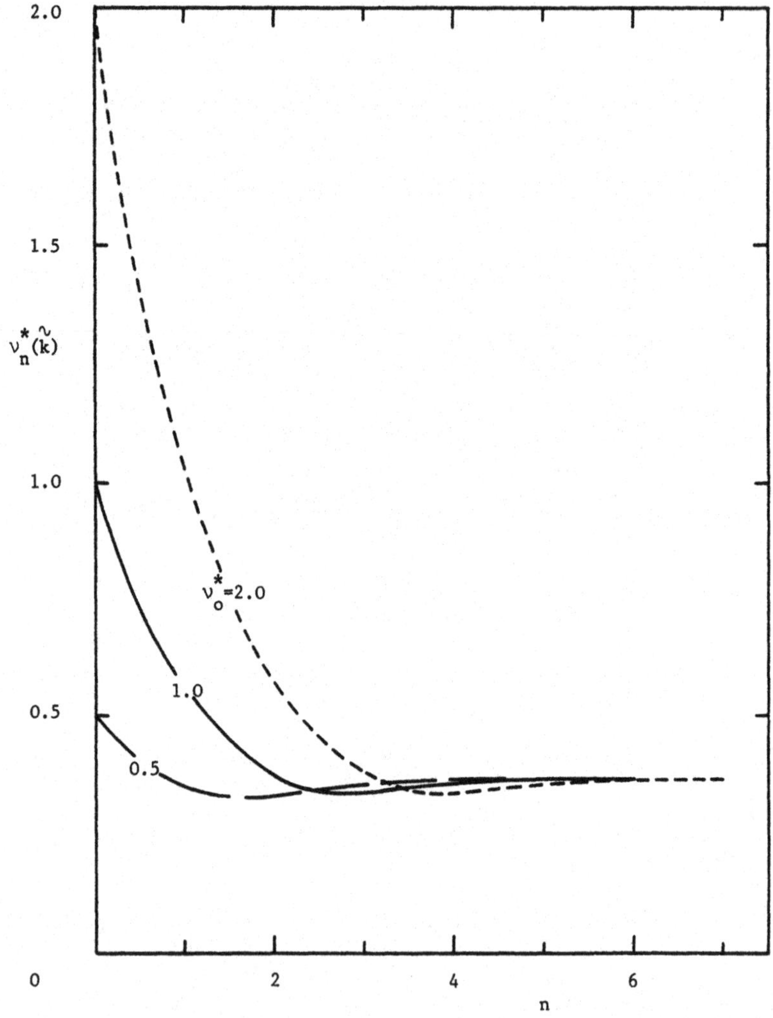

Figure 1. Demonstration of the scaled eddy viscosity reaching a fixed point, independent of its initial value: $h = 0.6$ and $\tilde{k} = 0.01$.

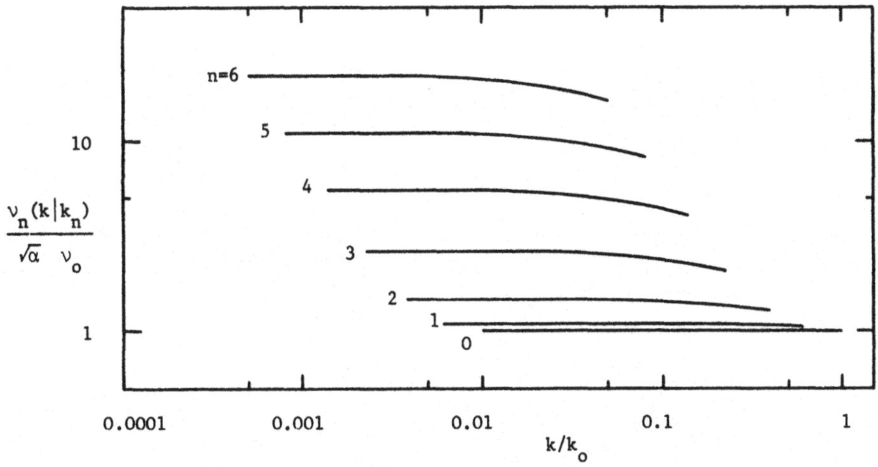

Figure 2. Evolution of the eddy viscosity for h = 0.6.

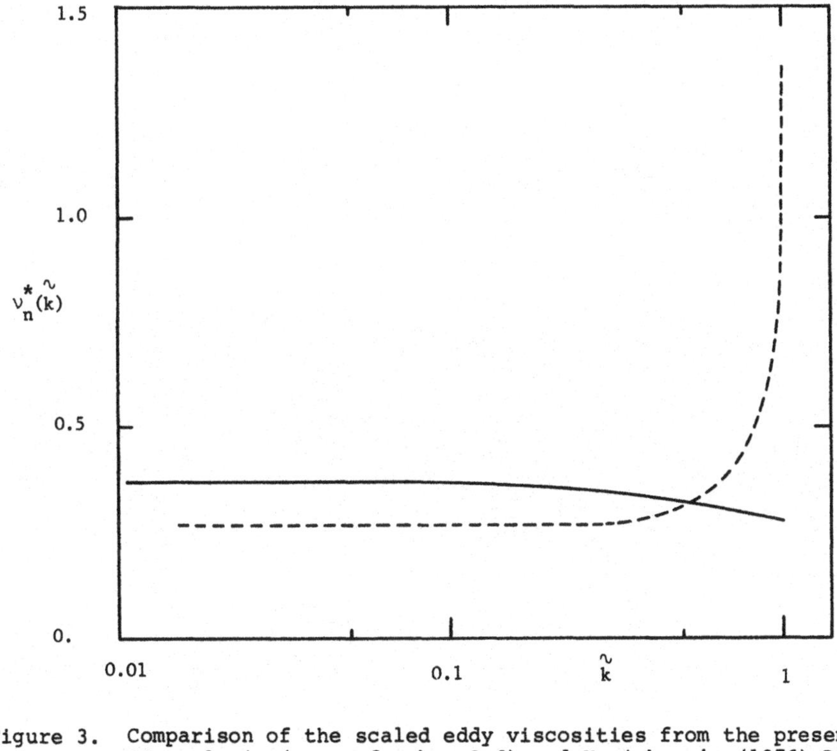

Figure 3. Comparison of the scaled eddy viscosities from the present RG analysis (—— for h = 0.6) and Kraichnan's (1976) DI approximation (----).

CHAPTER IX. Statistical Methods in Turbulence

A. Pouquet

I INTRODUCTION.

Statistical methods in turbulence can be traced back to the ideas of the beginning of this century in statistical mechanics. Can they nevertheless cope with 1984 modern physics, as is sometimes asked? We shall try to show that indeed they can.
One of the main dilemma of turbulence research today is to try to encompass in a unique formalism both the chaotic aspect of turbulence, conspicuous in a temporal display of any physical variable (temperature in the atmosphere, velocity measured in a wind tunnel, magnetic field fluctuations observed in the solar wind are but a few examples) and the persistence of large scale coherent structures. Is it possible that an appropriate interplay of exact solutions of the inviscid hydrodynamical equations, such as point vortices, can lead the way to a deep understanding of turbulence in the presence of small but non-zero viscosity?

Several methods for dealing with turbulent flows will be described in this conference. It is the purpose of this paper to show what are the possibilities and limitations of the statistical methods. I shall not give a thorough derivation of closures, since it can be found in several review articles /1 - 4/. I shall rather attempt to give an idea of what are the strong points of statistical theories and where lie their possibilities.

One aspect of statistical mechanics with which I shall not deal in this paper is that of phase transition and its connexion to singularities. It can be loosely said that the reason why water boils is linked to the fact that the zeros of the partition function lie, in the complex plane, on a circle of radius one that intersects the the real axis at two points corresponding to the liquid and vapor phases. For systems more complicated than the liquid/vapor transition (for example for the Ising model on a hierarchical lattice), the structure of zeros can become intricate. There exists for fluid mechanics an analogous approach. Bessis and Fournier /5/ have studied on the Burgers equation (one-dimensional model of Navier-Stokes) the temporal evolution of the complex singularities of a particular initial condition. They obtained coalescence of the singularities on the real axis at a finite time for zero viscosity. The statistical case, in which an average over an ensemble of initial conditions is taken, is now being studied. Another field in which singularities play a central role is linked to the extension of the Painlevé conjecture to partial differential equations (P.D.E.). What is at stake here is the possibility of predicting by examining the properties of the movable singularities of a PDE whether the system is fully integrable (with an infinite number of conserved quantities) and therefore possesses soliton-like solutions, or not (see /6/ for a recent overview, and references therein). This problem will not be dealt with here.

The paper proceeds as follows. In the next Section, the properties of a non-dissipative flow are mentionned. Statistical closures of turbulence are described in Section III and the last Section is devoted to statistical methods and chaotic behaviour.

II STATISTICAL EQUILIBRIA.

The simplest analogy one can make with classical statistical mechanics is when no dissipation is taken into account. In this case the representative point of the system moves in phase space on the surface of constant energy, with sometimes more restrictions due to the existence of more than one quadratic integral constraint (see below). Let us write first, for reference, the Navier-Stokes equation :

$$\frac{\partial \vec{u}}{\partial t} + u.\nabla\vec{u} = -\vec{\nabla}p + \nu\nabla^2\vec{u}$$
$$\nabla.u = 0 \ ; \qquad (2.1)$$

u is the velocity field (incompressible, here), ∂p the pressure and ν the kinematic viscosity. For zero viscosity, the Euler equation has several invariants among which energy

$$E^V = 1/2\langle u^2\rangle$$

and helicity

$$H^V = \langle\vec{u}.\vec{\omega}\rangle$$

(where ω = curl u is the vorticity, and brackets indicate integration over the whole box). For purely mirror symmetric turbulence, helicity is exactly zero and we shall suppose we are in that case. Helicity has not yet been measured in the laboratory, because one needs to determine at small scales the velocity field in three directions and the vorticity. Observations in the Solar Wind however give indications of the spatial spectrum of magnetic helicity /7/

$$H^M = \langle\vec{a}.\vec{b}\rangle$$

(where b = curl a is the magnetic field), an important dynamical variable for the dynamo problem (see Moffatt /8/ for review). For a given energy E, the canonical distribution will be $f(E) \sim \exp(-\beta E)$ where β is analogous to an inverse temperature. This yields, for a truncated system, to prediction of equipartion among the various modes. Since the number of modes is, in three dimensions, proportional to $4\pi k^2$, the energy spectrum $E(k)$ has a k^2 dependence, yielding to the well known ultraviolet catastrophe in the absence of high wavenumber cut-off. In this example, it is seen that the effect of the nonlinear interactions alone is to transfer energy to the small scales. When viscosity is turned on, its main effects will be to prevent the energy accumulation at high wavenumber and allow a quasi-equilibrium between transfer through nonlinear interactions and drain through viscosity. In the absence of energy forcing term, energy decays with time in a self-similar way.

Statistical equilibria may be more rich in structure when more than one quadratic invariant (those that survive truncation) exists. This is the case for the two-dimensional Euler equations, in which the squared vorticity (or enstrophy)

$$\Omega = \langle\omega^2\rangle$$

is also preserved by nonlinear interactions. It also occurs in magnetohydrodynamics, for which the equations read :

$$\frac{\partial \vec{u}}{\partial t} + u.\nabla\vec{u} = -\vec{\nabla}p^* + b.\nabla\vec{b} + \nu\nabla^2\vec{u},$$

$$\nabla.u = 0,$$

$$\frac{\partial \vec{b}}{\partial t} + u.\nabla\vec{b} = b.\nabla\vec{u} + \lambda\nabla^2\vec{b}, \qquad (2.2)$$

$$\nabla.b = 0,$$

where b is the magnetic field (in Alfven velocity units), λ the magnetic diffusivity and p^* the total pressure. In three dimensions, total energy

$$E = 1/2(\langle u^2\rangle + \langle b^2\rangle)$$

and magnetic helicity

$$H^M = \langle\vec{a}.\vec{b}\rangle$$

are preserved. In two dimensions, the helicity is identically zero, and it is the squared magnetic potential

$$A = 1/2\langle a^2\rangle.$$

which is also preserved.

When two invariants are to be taken into account, the canonical distribution is defined with two temperatures associated with the two invariants and other possibilities but that of equipartition may occur. The energy spectral distribution varies according to the values of the invariants ; in some cases, the energy will condense, for the two-dimensional Euler problem, on the lowest mode. For example, Seyler et al. /9/ compared the results of a direct numerical simulation on a 32x32 points grid to the predicted spectra, performing an average over angle (isotropic spectrum) and over time to dampen the strong fluctuations. When viscosity is restored, it will alter the lowest mode only for very long times and thus should not prevent the nonlinear transfer to large scales. One may thus predict the existence of a double-cascade system in which enstrophy cascades towards small scales (the power-law of which will be discussed further in Section III) and energy cascades towards the large scales. This has now been verified by several authors /10-12/ (with resolution increasing with date of publication). One possible consequence of the phenomenon of the energy piling up in the lowest mode has been the prediction /13/ of what is the most likely state in which the system may be found for a very small viscosity. The argument was extended to MHD (see /14/ for a review in two dimensions, and /15/).

These predictions about the probable state of a turbulent flow make no use of the nonlinear terms in the equations governing the evolution of the flow, except for their global conservation properties. To go beyond these heuristic arguments, two possibilities can be seen : experiments and theories. Experimentation can be done in the laboratory, or on the computer. The latter is of particular interest in two dimensions both for obvious experimental difficulties and because high resolutions can then be achieved. Observations of natural flows (tidal flow in an open channel, atmosphere, Solar Wind) complement the laboratory experiments and yield information at high Reynolds numbers. In the three-dimensional case, however, the experimental situation is less clear : numerically, flows only slightly above critical can be simulated ; and in the laboratory, data is still lacking on the vorticity field (and hence the helicity) which may play an important role in the development of hurricanes, and in MHD, in the dynamo problem.

A global theory of turbulence still being in demand, paths have been followed some of which will be reviewed at this conference. The next section deals with infinite expansions in a parameter that is not necessarily small.

III STATISTICAL CLOSURES OF TURBULENCE

3.1. Introduction.

It is not the purpose of this paper to give a derivation of closures in turbulence, developed mainly by R.H. Kraichnan (/11/,/16-17/). Closures have been reviewed several times (/2/, /3/, /18/). and their connection to various techniques of theoretical physics have already been emphasized (|19|, |20|), in particular with the Renormalization Group in its dynamical version (|21| - |24|). I will rather attempt to show where they have been most useful, and to give a few examples. Closures can be viewed as a model of turbulence, in which high-order correlations are expressed, in a more or less ad hoc fashion, in terms of low-order moments of the velocity field. The simplicity of this approach is counter-balanced by the possible pitfalls : the quasinormal approximation, in which fourth-order cumulants are set equal to zero, yield negative energy spectra, a physically unacceptable result. The most widely used closure today introduces a linear relaxation of triple correlations through an operator that may be taken constant (Markovian Random Coupling Model), or determined from phenomenological considerations (Eddy Damped Quasi Normal Markovian-EDQNM, in brief- approximation) or from an auxiliary problem (Test Field Model, see /17/ and /2/ for review). However this is not really a fair treatment of closures, and one may be more sophisticated than that. For example, the Direct Interaction Approximation (DIA ; /16/) can be seen as a Dyson-Schwinger type of theory in which the vertex (nonlinear term) is not renormalized. Note that in principle, the renormalization group produces at the same order of the expansion (second order) a modification of the vertex. It was shown that to all orders /21/, /25/ and /23/ for a discussion of that point) that the Navier-Stokes vertex remains unrenormalized due to the Galilean invariance. In MHD however, a large scale magnetic field acts on small scale turbulence (Alfven waves), and the renormalization of the Lorentz force vertex /26/ is non zero. An important property of the DIA is that it is the exact solution to a model problem, which insures its realisability (in particular, the positivity of the energy spectrum). Higher order closures, such as vertex renormalized expansion, become horrendous in the involved algebra. Use of a computer language dealing with symbolic manipulation would be necessary if one wants to extend to P.D.E. the work of Gauthier, Brachet and Fournier done on an algebraic cubic equation /27/. These authors studied the following nonlinear Langevin equation in the limit of very slow varying force

$$\mu v(t) + \lambda v^3(t) = f(t) \tag{3.1}$$

where $f(t)$ is a stationary Gaussian function with zero mean; μ is a positive damping and λ a coupling constant.

This can be taken as a model of several problems, for example a Van der Pol oscillator. The cubic non linearity was a source of difficulty because the DIA approximation used could not distinguish between positive and negative λ, i.e. between damped and self-excited solutions. This caused a loss of realizability of the solution above a threshold in Reynolds number. The problem was solved by using an auxiliary variable $x(t) = v^2(t) + h(t)$ (where $h(t)$ is a forcing term) which rendered the problem quadratic in (x, v).

Closure schemes yield to a set of nonlinearly coupled integro-differential equations. In the simplest case, that of a Markovian second-order closure, for an isotropic and homogeneous flow, the set of variables are the modal energy densities depending only on the modulus of the wavenumber. To derive them may require considerable algebra in problems involving more than one field without particular symmetry. But again a computer may be useful in that matter. The equations depend on geometrical coefficients which weigh the various nonlinear interactions.

Relationships between such coefficients which transcribe the global conservation properties that are preserved by the closure are helpful in reducing their number, which is necessary for storage purposes on the computer when simulating high Reynolds number flows. Since the geometrical coefficients depend on space dimension d, it is convenient to compute them first with variable d and set d equal to its value afterwards.

In the appendix are given the closure equations for d-dimensional MHD turbulence for zero correlation between velocity and magnetic field and zero helicity.

The closure equations possess the same quadratic invariants as the primitive equations (since detailed conservation per triad of interacting modes holds and thus also have the statistical equilibria mentionned in Section II. In the presence of viscosity, the inertial range that develops follows in the EDQNM for three-dimensional Navier-Stokes turbulence, a $-5/3$ range corresponding to the Kolmogorov (1941) prediction. No intermittency effects (due to fluctuations of the energy transfer rate along the cascade) is taken into account in the closure and no departure from the Kolmogorov prediction is thus observed. A study of intermittency in the dissipation range on a simple nonlinear Langevin model can be found in /28/. However, a modification of the closure scheme that retains only some of the interactions between modes has shown departures from the pure Kolmogorov spectrum (see Section IV). Spectral laws that are compatible with the Kolmogorov phenomenology but which take into account other characteristic times than the eddy turn-over time can also obtain with closures, either in two-dimensional Navier-Stokes turbulence in the enstrophy range /29-31/ or in MHD /32/.

Comparison between closures and direct numerical simulations have been performed in three dimensions at low resolution on a 32^3-points grid /33/ and in two dimensions /34-35/. On the computers appearing today, an order of magnitude in Reynolds number can be gained and this type of work should be redone, paying particular attention to the high order cumulants which can also be compared against laboratory experiments (such as in /36/) and to intermittency /37/.

The closure equations are well-behaved ; this is likely to be connected to the underlying model which ensures realizability. An exponential discretisation in wavenumber $k_n = 2^{n/F}$ is thus possible but it has the defect of suppressing the non-local interactions (i.e. interactions between greatly differing scales). The cut-off depends on the resolution F. For the commonly used value $F = 4$, triad interactions down to a ratio r of roughly .20 between medium and large wavenumber are included. To circumvent this drawback of the numerical scheme, it is convenient to extract the information from the closure equations through an analytic expansion (in r) and include by hand in the numerical integration these expansions in a conservative way /38/. One may thus obtain, in a self-consistent way, an analytic expression for transport coefficients.

3.2 Transport coefficients.

The effect of small scale turbulence on large scales (and vice versa) can be computed from the closure equations and yield useful information. It is generally thought that small scales play the role of a drain on the larger scale containing eddies ; hence, the time-dependence of the large scale energy spectrum E(K) can be modelled by :

$$\frac{\partial E(K)}{\partial t} = -(\nu + \nu_{turb})K^2 E(K) \quad ; \tag{3.2}$$

ν is the viscosity and ν_{turb} the eddy viscosity the expression of which for the EDQNM closure in dimension three is :

$$\nu_{turb} = \frac{1}{15}\int_{>>} \theta_{(o,q,q,)} (5E(q) + q\frac{\partial E}{\partial q}) dq \tag{3.3}$$

where θ_{oqq} is the decorrelation time of third-order cumulants $\theta(K,p,q)$ revaluated at $K = 0$ and $p = q$ and where >> indicates integration over the large wavenumber range (above a given cut-off). For situations involving more than one field, other transport coefficients can be computed. For example in the problem of the advection of a passive scalar, it was found /39/ that the small scales produce both an eddy viscosity and an eddy noise. For a conducting fluid (MHD case), transport coefficients can be found in /40/ and /41/. These coefficients provide a simple explanation for the phenomenon of inverse cascades mentionned in the previous Section. The destabilization of large-scale turbulence by small scale turbulence is linked to a negative eddy viscosity /40/ or, for three-dimensional helical MHD, to the combined effect of kinetic /41/ and magnetic helicity /42/, leading to the appearance of magnetic fields at the size of the system (dynamo problem).

Transport coefficients can also be used to modelize the small scale turbulence in a large eddy simulation (see lectures of Chollet, McComb, Launder and Reynolds, this conference, for related topics). In particular, the expression given in (3.3) has been used in conjonction with a 32^3 simulation using periodic boundary conditions /43/ (see also /44/.

3.3. Closures : a few results.

It would be both an awesome and fastidious task to review the various two-point closures that have been derived (DIA, EDQNM, LET, TFM to name a few) or to review the many problems to which they have been applied, ranging from purely theoretical (turbulence in variable dimension) to applied (advection of a chemical pollutant by winds). I will concentrate on three examples of application of a closure scheme : the persistence of a spectral gap, problem that cannot be treated on the computer given the range of wave numbers needed ; the interplay of two-dimensional coherent structures and three-dimensional small - scale noise ; and finally the role of interacting vortices in two-dimensional Navier Stokes turbulence.

In the case of a spectral gap /45/, suppose that at a given time two inertial ranges with typical wavenumber k_1 and $k_2 (k_2 \gg k_1)$ and typical energy injection rate ε_1 and ε_2 are well separated (see fig. 1).

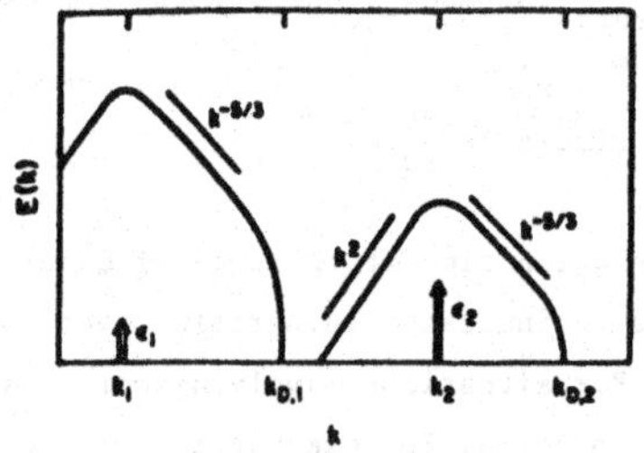

Fig. 1.- Schematic representation of steady-state Kolmogorov inertial ranges separated by a spectral gap.

The condition for them to remain so can be found by stating that the dissipation wavenumber k_{D1} computed with the turbulent viscosity due to the small scale range -viz $\nu_{turb} = E_2 T_2$ where E_2 and T_2 are typical energy and time of the second range- be smaller than the energy-containing wavenumber in the small range, i.e.:

$$k_{D1} < k_2 \quad .$$

Assuming a Kolmogorov law for the energy range of both spectra, for which:

$$k_D \sim (\varepsilon/\nu^3)^{1/4} \quad ,$$

one finds that the large scale range will not be swamped by small scales provided :

$$\varepsilon_1 \ll \varepsilon_2 \tag{3.4}$$

This phenomenological result is based on an analysis of non local effects between widely separated scales. To confirm the plausibility of the condition (3.4), resort to a numerical integration of the closure EDQNM equation was necessary.

In figure 2 is shown the steady states that develop for different ratios of the

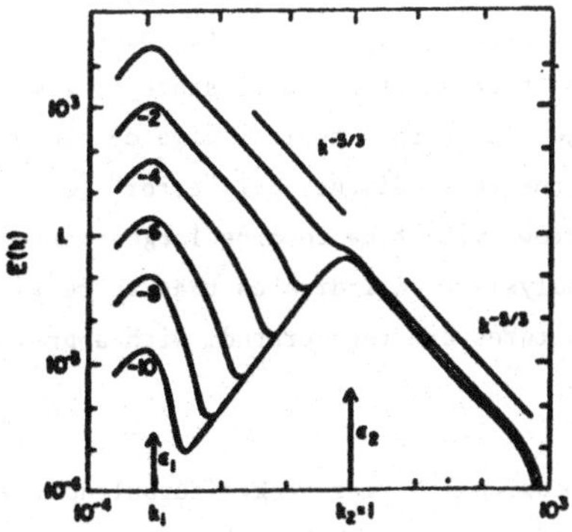

Fig. 2.- Steady-state energy spectra E(k) when the ratio of the energy injection rates $\varepsilon_1/\varepsilon_2$ varies from 1 to 10^{-10}.

two injection rates. Figures 1 and 2 are taken from reference /45/. When this ratio is very small, a large gap is obtained. When it is only relatively small, the gap is not so clear-cut and an identation of the large scale spectrum forms before the small scale Kolmogorov range develops. This type of multi-instability spectrum has sometimes been invoked in oceanography, in atmospheric turbulence and in astrophysics for solar granulation and supergranulation (for which condition (3.4) is fulfilled when we take for velocities in granules and supergranules respectively 1 km s^{-1} and 0.5 km s^{-1} and for typical length scale 1 Mm and 30 Mm respectively).

One of the most interesting aspects of a turbulent flow is the cohabitation of well-defined coherent structures (generally in the large scales, but not necessarily) and of a temporal and spatial chaos (strongly fluctuating signals). Few studies deal with the interplay of structures and chaos. In /46/, the authors look at the temporal chaos in a Sine-Gordon breather soliton forced by a periodic function ; although the breather does not visually appear to be affected by the forcing, a temporal analysis of the derivative field shows a broad spectrum corresponding to chaotic motions in the small scales of the soliton. Another example is that of reconnection at a neutral X-point of the magnetic field. The large scale hyperbolic configuration does not preclude the existence of small scale turbulent motions, visible for example in the chaotic temporal variation of the magnetic current as shown by recent numerical simulations in dimension two.

The interplay of turbulence and coherent structures has been studied with the EDQNM closure /47/. The large scales are supposed two-dimensional (for example in a free shear - layer) ; the small scales are three-dimensional and are considered as a decorrelation between two subsequent cross-sections of the flow, described by velocities u_1 and u_2 with the same energy spectrum ($E_1(k) = E_2(k)$). Stated as such, the problem is related to that of predictability in meteorology in which small scale errors (between the observed state and the actual state of the atmosphere) progressively invade larger and larger scales in a typical time of the order of a few eddy turnover times. Here, it is the three dimensional "error" (or decorrelation of the flow at small scales) that grows with time towards larger scales and destroy the coherent structures. This analysis then indicates that there must be a mechanism by which these large scale structures are regenerated, with approximatively the same characteristic time.

Inverse cascades (in two-dimensional Navier-Stokes turbulence, and in two-dimensional and three-dimensional helical MHD turbulence) have been observed in closure calculations. These cascades are also seen in direct numerical simulations with drastically less spatial and temporal resolution (needed because typical times for build-up of large scale excitation varies as the inverse of the wavenumber to some power). Inverse transfer to scales larger than the typical scale of the instability are indicative of the formation of large scale coherent structures (through for example vortex pairing). Closures give information only on the spectral indices and on the characteristic time of growth of the structures. Resort to direct numerical simulations is needed to obtain information about the geometry of these structures and their interactions. For example, it was found in /67/ that in 2D-MHD it is the occurence of quasi one-dimensional current sheets that prevent the formation of a finite-time singularity in the inviscid ($\nu \equiv 0$, $\lambda \equiv 0$ in (2.2)) limit contrary to the prediction of closures.

Finally, let us now consider the case of two-dimensional Navier-Stokes turbulence. The enstrophy inertial range may be following a log-corrected K^{-1} law /17/, or equivalently a log-corrected K^{-3} for the energy spectrum. This corresponds to a filamentation of the vorticity field which yields on average a K^{-1} spectrum, spectrum with which closures agree, except that the log-correction is not resolved. On the other hand, Saffman /48/ depicts 2D Navier Stokes turbulence as a series of jumps of vorticity, which yields a K^{-4} inertial range. It is not clear whether a turbulent flow will evolve towards a configuration in which order is preserved (latter case) or whether mixing among eddies will yield the shallower spectrum. To help resolve this problem of vortex turbulence versus statistical turbulence, use of numerical simulations is necessary but difficult. Indeed, a large resolution is needed because of the non locality of transfer among modes. The highest resolution achieved today (grid of 1024^2 points on a Cray 1 computer), utilizes a pseudo-spectral method /49/. Computations at lower resolutions (/50/, /35/) may yield somewhat steeper spectra than those of /49/, but using a higher power of the Laplacian. The computed energy spectrum is expressed as a power law with an exponential tail

$$E(k, t) = C(t)\, k^{-n(t)} \exp(-\beta(t)\, k) \qquad (3.5)$$

with the three parameters C, n and β fitted through a least square algorithm as a function of time. The numerical error is evaluated by computing the energy dissipation

$$D(t) = \int_0^\infty k^2\, E(k, t)\, dk$$

using the expression (3.5) and comparing it to the actual dissipation of the run (i.e. (3.5) with a truncation at the maximum wavenumber k_{max}). The error is worse at the time at which the palinstrophy

$$P(t) = \langle \mathrm{curl}(\omega)^2 \rangle$$

is maximum (t = 6) and, for a Reynolds number of 30 000 is 1 % at t = 6. The spectral index n(t) varies strongly in time (see figure 3) but on average there is at first a (- 4) law followed by a (- 3) spectrum with a rather sharp transition at a time close to the maximum of P(t). The fluctuations in n(t) are damped for a calculation at a higher Reynolds number (250,000), but the errors are more important because the exponential decrease in the small scales is only barely sufficient to drain the enstrophy.

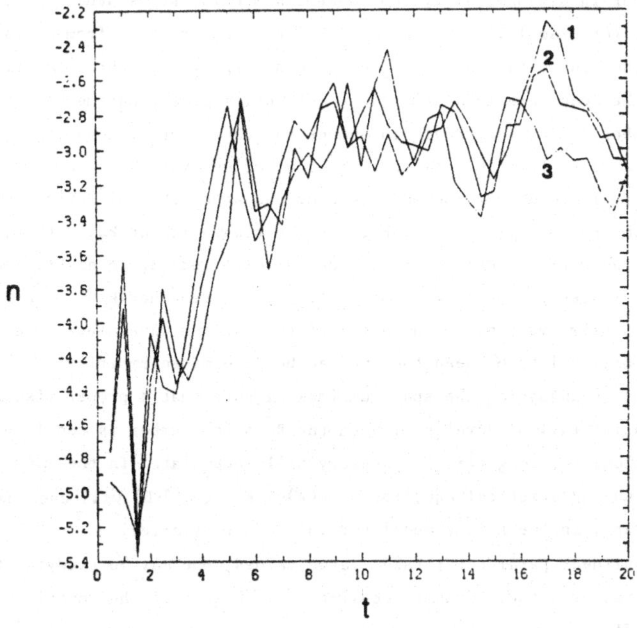

Fig. 3.- Spectral index n of energy spectrum $E(k) \sim k^{-n} \exp(-\beta k)$ as a function of time for a 2D-Navier Stokes pseudo-spectral simulation at a Reynolds number of 3×10^4. Curves labelled 1, 2 and 3 correspond to different intervals of fit in wavenumber. Note the sharp transition in n around t = 5. (Taken from ref. /49/).

Closures have proven versatile enough to accomodate new situations: nonlinear Rossby waves (see /51/), for which one includes the characteristic frequency of the wave in the operator of the decorrelation of triple moments; anisotropic case (see for example /52/); inhomogeneous flows (in the Boussinesq approximation to convection, see /53/ for the DIA and /54/ for the EDQNM). However, closures lack one essential feature of a turbulent flow, that of a chaotic behavior, merely by construction (through an averaging procedure). A modification of the closure equations through severe truncations of the nonlinear interactions, permits one to recover a chaotic behavior and is described in the next Section.

IV STATISTICAL METHODS AND CHAOTIC BEHAVIOUR.

For a Kolmogorov spectrum, the number of modes at a given Reynolds number R varies as $R^{9/4}$ in three dimensions. Are all those modes necessary to describe the flow, or is a subset sufficient? This question of the number of dynamically relevant modes has been revived recently because of the progress made in the comprehension of chaotic behaviour in a dynamical system. Most dynamical systems studied today have very few modes: for example, the celebrated Lorenz model results from a drastic truncation of the convective Boussinesq equations keeping only three modes. The properties of the model are then studied as a function of the parameter r (ratio of Rayleigh number to critical Rayleigh number for the onset of convection); chaotic behaviour (as seen from the temporal spectrum of one variable) obtains above a threshold in r. In so doing, one must be cautious about the following fact. Take a set of wavenumbers k $|k_{min}, k_{max}|$. For a given truncation at k_{max}, there is a maximum Reynolds number that can be treated accurately. Indeed, to ensure the existence of a proper dissipation range, the dissipation wavenumber $k_D \sim (\varepsilon/\nu^3)^{1/4}$ (for a Kolmogorov law) must be of the order of the maximum wavenumber. The constant of proportionality between the two can be determined empirically in a numerical simulation by adjusting the spectrum (making sure that a proper viscous range with exponential fall-of develops): when the Reynolds number is too large for a given truncation, for long times the energy will accumulate in the large wavenumber domain and a statistical equilibrium $E(k) \sim k^2$ (in 3-D) will eventually obtain. In letting, in the Lorenz model (or in another dynamical system), the nonlinear parameter r take large values, one may be studying the chaotic properties of a thermodynamic equilibrium, instead of the chaotic properties of a turbulent flow.

Instead of truncating the number of modes, one can truncate the range of interaction between the modes by assuming the existence of a screening effect. A trivial case is that of an initially symmetrical flow for which it is assumed that the evolution equation will preserve the symmetries. Another possibility stems from the fact that, for three-dimensional turbulence, nonlinear interactions are known (from closure studies for example and from phenomenological considerations) to be local in wave-number space. One may thus devise a scheme to truncate the interactions in Fourier space on the closure equations to obtain a simpler model. It is also possible to write from first principles a set of coupled O.D.E. which have in common with the Navier-Stokes equations structural properties (quadratically non linearities, energy conservation, gradient-type time scale, linear laplacian dissipation) and which include only nearest-neighbours interactions (mode n interacts only with itself and modes n ± 1). This type of scalar model has been studied by several authors /55 - 58/. The model has been extended to MHD /59 - 60/. The equations in the MHD case read:

$$\frac{d\, u_n}{dt} = \alpha(k_n\, u_{n-1}^2 - k_{n+1}\, u_n\, u_{n+1} - k_n\, b_{n-1}^2 + k_{n+1}\, b_n\, b_{n+1})$$
$$+ \beta(k_n\, u_{n-1}\, u_n - k_{n+1}\, u_{n+1}^2 - k_n\, b_n\, b_{n-1} + k_{n+1}\, b_{n+1}^2) \qquad (4.1)$$
$$- \nu k_n^2\, u_n + C_n$$

$$\frac{d\, b_n}{dt} = \alpha\, k_{n+1}(u_{n+1}\, b_n - u_n\, b_{n+1}) + \beta\, k_n(u_n\, b_{n-1} - u_{n-1}\, b_n) - \lambda\, k_n^2\, b_n \qquad (4.2)$$

where α and β are parameters of the model, $C_n = \delta_{n1}$ is the forcing term, u_n and b_n represent the velocity and magnetic field modes (b_n can also be interpreted as a second velocity variable for mode n) and $k_n = 2^n$ is the wavenumber for mode $n(n\ |1,\ N|)$. Such equations, when the viscosity ν, the magnetic diffusivity λ and the forcing term C_n are set equal to zero, conserve both the total energy

$$E = \tfrac{1}{2} \sum (u_n^2 + b_n^2)$$

and the correlation

$$E_C = \sum (u_n\, b_n).$$

Although nonlinear interactions are restricted to neighbouring shells, the terms in (4.1) and (4.2) mimic the phenomenology of turbulence. For example, in the r.h.s. of (4.1), the first term represents energy production in mode n due to mode (n-1) (large scales feed small scales) and the second term represents an energy drain (when the variables are positive) from mode n due to mode (n+1) (small scales enhance the dissipation of energy). The case $b_n \equiv 0$ was studied in /55 - 56/ with the restriction that the field variables remain positive, and no chaos obtains in that case. In MHD, however, this restriction on the positivity of the u_n and b_n variables was not maintained. The Kolmogorov fixed point $u_n \sim k_n^{-1/3}$ destabilizes above a threshold in Reynolds number and chaotic behaviour appears, as can be seen in figures 4 and 5.

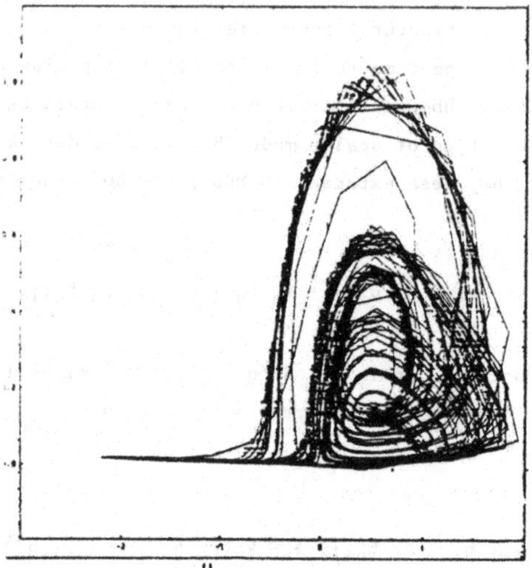

Fig. 4.- Phase portrait $u_3 = f(u_2)$ for the scalar model (4.1 and 4.2); N = 3, $\nu = \lambda = 0.13$; sampling interval $\Delta T = 0.2$.

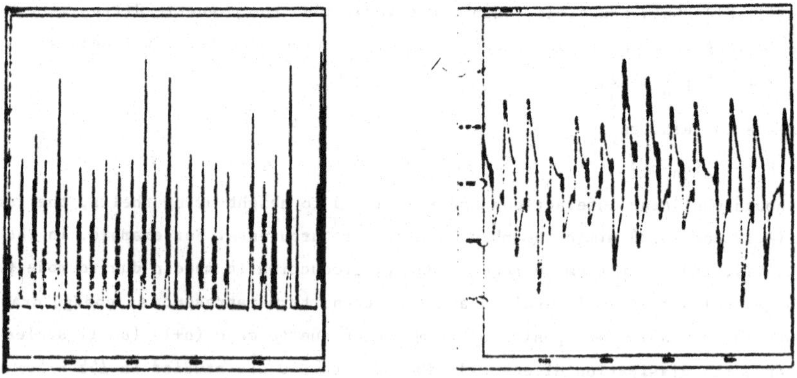

Fig. 5.- Temporal variation of two modes in the scalar model, same conditions as in fig. 4. (Both figures taken from ref. /59/).

To go beyond this visual appearance, one can try to measure the degree of chaos by tacking one or two numbers to the strange attractor that is presumably there. The most commonly used today is its dimension (see /61/ for another parameter, the lacunarity). Several definitions of dimensions have been proposed. The purpose is to measure in a simple way how much of phase space (all of which is a priori accessible) is occupied by the attractor. It is easy to see from eqs. (4.1) and (4.2) that, for the dissipative case, the system is shrinking in volume at a rate given by

$$\frac{d\,\dot{x}_n}{d\,x_n} = -\,(\nu + \lambda) \quad \sum_n k_n \quad . \tag{4.3}$$

(Note that when $\nu \equiv 0$, $\lambda \equiv 0$, a Liouville theorem holds for the scalar model, as well as for the primitive equations). However, the representative set of points in phase space under the action of the dynamical equations, is also stretched in one (or more) direction corresponding to the largest Lyapounov exponent (the sum of which is given by (4.3))

The several practical ways to compute a dimension which are reviewed in /62/ fall in two main categories: geometry and probability. The capacity dimension (or fractal dimension) d_g pertains to the first category. It is defined as:

$$d_g = \lim_{\varepsilon \to 0} \frac{\log N(\varepsilon)}{\log (1/\varepsilon)} \tag{4.4}$$

where $N(\varepsilon)$ is the number of boxes of size ε needed to cover entirely the attractor (box-counting algorithm). The problem with this definition is that some points on the attractor are visited only extremely rarely (say with a probability 10^{-6}) and yet their weight is the same as all other points. The information dimension d_p takes into account the fraction of time $\mu(C_i)$ that is spent in a cube C_i on the attractor. It is defined as:

$$d_p = \lim_{\varepsilon \to 0} \frac{I(\varepsilon)}{\log (1/\varepsilon)} \tag{4.5}$$

where

$$I(\varepsilon) = \sum_{i=1}^{N(\varepsilon)} P_i \log \frac{1}{P_i} \tag{4.6}$$

with $P_i = \mu(Ci)$. When all cubes have the same probability to be visited ($P_i = a$, for all i), the two dimensions are equal: $d_g = d_p$. If the frequency of visit of cubes on the attractor varies with the position of the cube (nonlocality situation which is more likely), the two dimensions differ and it can be shown (see /62/) that $d_g > d_p$. Numerical determination of dimensions are probabilistic by nature (the rarer the occurences of some points, the longer the time of integration of the dynamical system needed to reach them). However, the inequality between the two dimensions becomes probably less significant when one deals with a system with a large number of modes. In the atmosphere for example what is at stake is whether the number of dynamically relevant modes is of the order of a few millions or a few tens, and the problem of computing a dimension is that of finding a workable algorithm.

To compute a dimension associated with the attractor, one may locally (at point P_o) observe how the number of points N on the attractor varies with distance ℓ to P_o. A power law fit $N(\ell) \sim \ell^d$ will give the "mass" dimension d of the attractor. This dimension indicates, in a loose sense, the number of dynamically relevant variables minimally needed to describe the evolution of the flow with a proper set, yet to be found, of variables. In particular, we would like to know how the dimension scales with Reynolds number in the turbulent regime. This work is now in progress on the scalar model.

The computation of a dimension for a system with many modes is quite costly in computer time if one uses direct methods. Efficient algorithms using embedding dimensions and random sampling /63, 64/ have been proposed but their application to partial differential equations still remain difficult. Another method to evaluate the number of relevant modes involves a conjecture /65/ which relates the properties of the Lyapounov spectrum to the (information) dimension.

One of the first investigations of this kind has been done /66/ on the Kuramoto equation:

$$\frac{\partial u}{\partial t} + u\frac{\partial u}{\partial x} = -\nu\frac{\partial^2 u}{\partial x^2} - \mu\frac{\partial^4 u}{\partial x^4} \qquad (4.7)$$

which models the formation of patterns in flame combustion, or the evolution of the phase of convective rolls in Boussinesq convection. When $\mu \equiv 0$, the equation (4.7) is that of Burgers, with an anti-laplacian: the second derivative term feeds energy in the intermediate range of scales of the problem. The higher derivative, on the contrary, drains energy at the smallest scales, and a steady-state can be reached. This equation is known to possess intrinsic chaos, the degree of which may be measured by the number of positive Lyapounov exponents N^+. The parameter of the problem is the length of the box L. The results are obtained using a numerical simulation with periodic boundary conditions; 80 Lyapounov exponents are computed, and N^+ is found to grow linearly with L. No saturation effect thus occurs. Another interesting result concerns the spatial spectrum of the Lyapounov vectors, obtained by sampling over the box. The spectra are shown in figure 6 for the 1st, 6th, 11th and 16th vectors. In the large scales, the spectra differ: the lowest power is associated to the first exponent. This indicates that the large scales are more stable. In the small scales of the flow, all Lyapounov exponents are on the same (approximately exponentially decreasing) spectrum which shows that small scales may undergo all kinds of motions (separation of trajectories as well as convergence) and in that sense are more chaotic than the large scales.

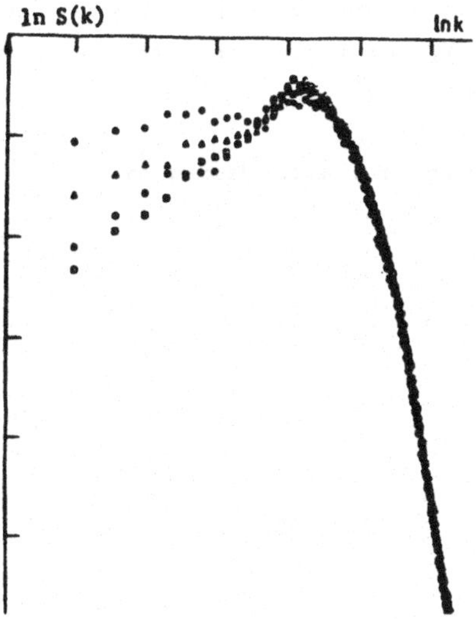

Fig. 6.- Spectra of various Lyapounov exponents (□ : 1st; o : 6th; △ : 11th and • : 16th) for the Kuramoto equation (4.7). (Taken from ref. /66/).

In conclusion, it is apparent that no matter what tool we use in studying turbulence, more often than not, we find ourselves using the computer, from algebraic expansions to modelling and simulations. The numerical experiments that are performed today on large computers pose several problems: one must evaluate the accuracy of the computation (through checks of integral constraints for example). Another problem is that of manipulation of data, reminiscent of the avalanche of data one is faced with in satellite experiments. For example a two-dimensional simulation on a 1024^2 grid for a compressible conducting fluid using a second-order accurate temporal scheme will use on the order of 10^8 bits of storage at each time-step saved on tape. This type of computation is envisageable today on a CRAY XMP using SSD memories (or an equivalent machine). The data created in the simulation must however, be reduced using powerful graphical displays (in particular in the case of a remote site).

However, it is an herculean as well as a frustrating task to deal with so many numbers and degrees of freedom when possibly much less will be required. Statistical methods are devised just for that reason , trying to discard unnecessary details and concentrating on averaged quantities because, for example, of the intermittency of the flow: large scales may be space-filling but small scales, the vision of which is enhanced by looking at derivatives of the velocity field, may be very spotty. Closures are one such example of reduction in the number of degrees of freedom through averaging. Their main drawback is that they are limited to spectral information and do not seem to be able to handle flows in which geometry plays an important role /67,35 /.

Their main advantage is that they permit to compute in a consistent way global parameters characterizing the flow, such as transport coefficients or Lyapounov exponents and dimensions of attractors.

I am thankful to J.-D. Fournier for a useful discussion.

APPENDIX

We shall give here for reference the EDQNM closure equations for mirror-symmetric MHD turbulence in variable dimension d and assuming that v - b correlations are zero. The case d= 3 can be found in /42, 68/ and the two-dimensional case can be found in /40/. Omitting the linear terms, we have:

$$\frac{\partial}{\partial t} U^V(k, t) = \int d^d \vec{p}\{a_1(k, p, q) \, |U^V(p, t) \, U^V(q, t) + U^M(p, t) \, U^M(q, t)|$$
$$+ b_1(k, p, q) \, U^V(k, t) \, U^V(q, t) + b_3(k, p, q) \, U^V(k, t) \, U^M(q, t)\}$$
$$\Theta(k, p, q) \tag{A1}$$
$$\frac{\partial U^M(k, t)}{\partial t} = \int d^d \vec{p}\{a_2(k, p, q) \, U^V(p, t) \, U^M(q, t)$$
$$+ b_2(k, p, q) \, U^M(k, t) \, U^M(q, t)$$
$$+ b_4(k, p, q) \, U^M(k, t) \, U^V(q, t)\} \quad \Theta(k, p, q) \quad ,$$

where the a_i and b_i geometrical coefficients are given by:

$$a_1(k, p, q) = \frac{k^2}{4} \, |d(2 - y^2 - z^2) - 4yz(2x + yz) - 3x^2 - 1|,$$
$$b_1(k, p, q) = \frac{+ kp}{4} \, |- z(2z^2 - 3) + xy - d(z + xy)|,$$
$$b_3(k, p, q) = \frac{kp}{4} \, |z(1 + 2y^2) + 3xy - d(z + xy)|,$$
$$a_2(k, p, q) = \frac{+ k^2}{4} \, |d(2 - y^2 - z^2) - 3 - x^2 + 2y^2 + 2z^2|, \tag{A2}$$
$$b_2(k, p, q) = \frac{kp}{4} \, |z(1 + 2x^2) + 3xy - d(z + xy)|,$$
$$b_4(k, p, q) = \frac{+ kp}{4} \, (1 - d) \, (z + xy);$$

k, p and q form a triangle with angles denoted by (α, β, γ) and their cosinus (x, y, z). $U^V(k, t)$ is the trace of the spatial Fourier transform of the kinetic energy covariance (single-time) $U^V_{ij}(x, t) = \langle v_i(x, t) \, v_j(x', t)\rangle$ and similarly $U^M(k, t)$ relates to the single-time magnetic energy covariance $\langle b_i(x, t) \, b_j(x', t)\rangle$; spatial homogeneity has been assumed together with isotropy so that only the modulus of wavevectors appears. The decorrelation time of third-order cumulants is denoted by $\Theta(k, p, q)$. Various choices are possible (see /68, 42, 40/ with $\Theta(k, p, q) \equiv C_0$ corresponding to the Markovian Random Coupling Model /69/.

To obtain the spectral closure equations in the usual form, several transformations must be performed. One needs to relate the Fourier covariance to the energy spectrum $E^V(k, t)$ (and $E^M(k, t)$) by using (see /70/ for details):

$$\tfrac{1}{2} S_d \, k^{d-1} \, U_{ij}(k, t) = \frac{1}{d-1} P_{ij}(k) \, E(k, t)$$

where

$$S_d = 2\,\Pi^{d/2}/\,\Gamma(d/2)$$

is the surface of the d-dimensional unit sphere (and $\Gamma(d/2)$ the Γ-function) and where

$$P_{ij}(k) = \delta_{ij} - \frac{k_i\,k_j}{k^2}$$

is the incompressibility operator: ($P_{ij}(k)$ projects a vector onto the plane perpendicular to the vector k, ensuring the field to be divergence free). One also needs an expression for the d-dimensional volume element, namely:

$$d^d\,\vec{p} = S_{d-1}\,\left(\frac{pq}{k}\right)^{d-2}\,(\sin\alpha)^{d-3}\,dp\,dq \quad .$$

In the form given here, six geometrical coefficients appear (a_1 and a_2; b_1 to b_4). To compute at high resolution, this poses a problem of storage since these coefficients are two-dimensional arrays. To reduce their number, one makes use of the conservation laws of total energy that relates $a_1(k, p, q)$ and $b_1(k, p, q)$ (kinetic energy conservation), $a_1(k, p, q)$ and $b_2(k, p, q)$ (squared terms in magnetic energy) and $b_3(k, p, q)$, $a_2(k, p, q)$ and $b_4(k, p, q)$ (through cross kinetic-magnetic terms). For example, one can show from general symmetry considerations that the following relationship holds:

$$a_1(k, p, q) = -\,(b_1(q, k, p) + b_1(p, k, q)) \quad .$$

The number of coefficients using conservation properties can be thus reduced to three in the non-helical case. Furthermore, nonlocal expansions can be performed which will give information for example on transport coefficients.

REFERENCES.

1) Kraichnan, R.H.: 1972, in "Statistical Mechanics: New Concepts, New Problems, New Applications", ed. J.A. Rice, K.F. Freed and J.L. Light, p. 20 (U. of Chicago Press).
2) Rose, H.A. and Sulem, P.-L.: 1978, J. de Physique (Paris) 39, 441.
3) Orszag, S.A.: 1974, in Les Houches Summer School, ed. R. Balian and J.-L. Peube, p. 235 (Gordon and Breach).
4) Frisch, U.: 1981, in Les Houches, ed. G. Iooss, R.H.G. Helleman and R. Stora, p. 667 (North-Holland).
5) Bessis, M. and Fournier, J.-D.: 1984, preprint Observatoire de Nice.
6) Tabor, M.: 1984, Nature 310, 277.
7) Matthaeus, W.H. and Goldstein, M.: 1982, J. Geophys. Res. 87, 6011.
8) Moffatt, H.K.: 1979, Magnetic Field Generation in Electrically Conducting Fluids, Cambridge University Press.
9) Seyler, C.E., Salu, Y., Montgomery, D. and Knorr, G.: 1975, Phys. Fluids 18, 803.
10) Herring, J., Orszag, S.A., Kraichnan, R.H. and Fox, J.: 1974, J. Fluid Mech. 66, 417.
11) Matthaeus, W.H. and Montgomery, D.: 1977, J. Plasma Phys. 17, 369.
12) Frisch, U. and Sulem, P.-L.: 1984, Phys. Fluids 27, 1921.
13) Bretherton, F. and Haidvogel, D.: 1976, J. Fluid Mech. 78, 129; Leith, C.E.: 1984, Phys. Fluids 27, 1388.
14) Kraichnan, R.H. and Montgomery, D.: 1979, Rep. Prog. Phys. 43, 547.
15) Matthaeus, W.H. and Montgomery, D.: 1980, Proc. N. Y. Acad. of Sci. 357, 203.
16) Kraichnan, R.H.: 1959, J. Fluid Mech. 5, 497.
17) Kraichnan, R.H.: 1971, J. Fluid Mech. 47, 513.
18) Leslie, D.C.: 1973, Development in the Theory of Turbulence (Oxford, Clarendon).
19) Martin, P., Siggia, E.D. and Rose, H.A.: 1973, Phys. Rev. A 8, 423.
20) Phythian, R.: 1977, J. Phys. A 10, 777.
21) Forster, D., Nelson, D.R. and Stephen, M.J.: 1977, Phys. Rev. A 16, 732.
22) Pouquet, A., Fournier, J.-D. and Sulem, P.-L.: 1978, J. Phys. Letters 39, 199.
23) Sulem, P.-L., Fournier, J.-D. and Pouquet, A.: 1979, Lecture Notes in Physics 104, 321 (Springer).
24) McComb, W.D. and Shanmugasundaram, V.: 1983, Phys. Rev. A 28, 2588.
25) de Dominicis, C. and Martin, P.C.: 1979, Phys. Rev. A 19, 419.
26) Fournier, J.-D., Sulem, P.-L. and Pouquet, A.: 1982, J. Phys. A 15, 1393.
27) Gauthier, S., Brachet, M.-E. and Fournier, J.-D.: 1981, J. Phys. A 14, 2969.
28) Batchelor, G.K.: 1969, Phys. Fluids 12, 233.
29) Kraichnan, R.H.: 1971, J. Fluid Mech. 47, 525.
30) Leith, C. and Kraichnan, R.H.: 1972, J. Atmos. Sci. 29, 1041.
31) Kraichnan, R.H.: 1965, Phys. Fluids 8, 1385.
32) Grappin, R., Pouquet, A. and Léorat, J.: 1983, Astron. Astrophys.
33) Herring, J. and Kraichnan, R.H.: 1972, in Statistical Models and Turbulence, p. 148, eds. M. Rosenblatt and C. von Atta, Springer.
34) Herring, J., Orszag, S.A., Kraichnan, R.H. and Fox, D.J.: 1974, J. Fluid Mech. 66, 417.
35) Herring, J. and McWilliams, J.C.: 1983, "Comparison of Direct Numerical Simulation of Randomly Stirred Two-Dimensional Turbulence with Two-point Closure (TFM): Preliminary Report", NCAR preprint (Boulder, Co., USA).
36) Anselmet, F., Gagne, Y., Hopfinger, E.J. and Antonia, R.A.: 1984, J. Fluid Mech. 140, 63.
37) Frisch, U. and Morf, R.: 1981, Phys. Rev. A 23, 2673.
38) Lesieur, M. and Schertzer, D.: 1978, J. Mécanique 17, 609; Grappin, R., 1985, Thèse d'Etat, Université Paris VII, Observatoire de Meudon.
39) Rose, H.A.: 1977, J. Fluid Mech. 81, 719.
40) Pouquet, A.: 1978, J. Fluid Mech. 88, 1.
41) Steenbeck, M., Krause, F. and Rädler, K.H.: 1966, Z. Naturforsch. 21a, 364.
42) Pouquet, A., Frisch, U. and Léorat, J.: 1976, J. Fluid Mech. 77, 321.

43) Chollet, J.-P.: 1984, Thèse d'Etat, Université de Grenoble.
44) Dang, K.: 1983, ONERA TP 1983-69.
45) Pouquet, A., Frisch, U. and Chollet, J.-P.: 1983, Phys. Fluids 26, 877.
46) Bishop, A.R., Fesser, K., Landahl, P.S., Kerr, W.C., Williams, M.B. and Trullinger, S.E.: 1983, Phys. Rev. Lett. 50, 1095.
47) Lesieur, M.: 1984, in Les Houches "Combustion and Nonlinear Phenomena", eds. P. Clavin, B. Larrouturou and P. Pelce, Les Editions de Physique.
48) Saffman, P.G.: 1971, Stud. Applied Math. 50, 377.
49) Brachet, M.-E. and Sulem, P.-L.: 1984, 9th I.C.N.F.D. Saclay (France); also 4th Beer Sheva Seminar on Magnetohydrodynamic Flows and Turbulence (Israël).
50) Basdevant, C., Legras, B., Sadourny, R. and Beland, M.: 1981, J. Atmos. Sci. 38, 2305.
51) Holloway, G. and Hendershott, M.: 1977, J. Fluid Mech. 82, 747.
52) Cambon, C., Jeandel, D. and Mathieu, J.: 1981, J. Fluid Mech. 104, 247.
53) Herring, J.: 1984, this conference.
54) Dolez, N., Legait, A. and Poyet, J.-P.: 1984, Mem. S. A. It. 55, 293.
55) Denianski, V.N. and Novikov, E.A.: 1974, Prikl. Mat. Mekh. 38, 507.
56) Bell, T.L. and Nelkin, M.: 1977, Phys. Fluids 20, 345.
57) Kerr, W. and Siggia, E.: 1978, J. Stat. Phys. 19, 543.
58) Lee, J.: 1980, J. Fluid Mech. 101, 349.
59) Gloaguen, C.: 1983, Thèse de 3ème Cycle, Université Paris VII.
60) Gloaguen, C., Léorat, J., Pouquet, A. and Grappin, R.: 1984, preprint, Observatoire de Meudon, "A Scalar Model for MHD Turbulence".
61) Mandelbrot, B.: 1984, this conference.
62) Farmer, J.D., Ott, E. and Yorke, J.A.: 1983, Physica 7D, 153.
63) Grassberger, P. and Procaccia, I.: 1983, Phys. Rev. Lett. 50, 345.
64) Termonia, Y. and Alexandrowicz, Z.: 1983, Phys. Rev. Lett. 51, 1265.
65) Kaplan, J. and Yorke, J.: 1979, Lecture Notes in Mathematics 730, 228, eds. H.O. Peitgen and H.O. Walther (Springer).
66) Pomeau, Y., Pumir, A. and Pelce, P.: 1984, J. Stat. Phys. in press.
67) Frisch, U., Pouquet, A., Sulem, P.-L. and Meneguzzi, M.: 1983, J. Mécanique Theor. Appli. Numéro Spécial sur la Turbulence Bi-dimensionnelle, p. 191 (Gauthier-Villars).
68) Kraichnan, R.H. and Nagarajan, S.: 1967, Phys. Fluids 10, 859.
69) Frisch, U., Lesieur, M. and Brissaud, A.: 1974, J. Fluid Mech. 65, 145.
70) Fournier, J.-D. and Frisch, U.: 1978, Phys. Rev. A 17, 747.

CHAPTER X. The Structure of Homogeneous Turbulence

William C. Reynolds and Moon J. Lee

Abstract

Full turbulence simulation has been conducted of homogeneous turbulence subject to irrotational strains and under relaxation from these. The behavior of the anisotropy in the Reynolds stress, dissipation and vorticity fields has been analyzed to reveal new physics in distorting and relaxing turbulence. A model for the relaxation process is proposed for the vorticity field from the study of the structure of homogeneous turbulence.

1. Introduction

Most of the present models for turbulent flows make reference to homogeneous turbulent flows, e.g., Launder *et al.* (1975) and Lumley (1978). However, many of the terms that must be modeled have never been measured. Specifically, the pressure strain rate term appearing in the Reynolds stress transport equations and the higher-order statistics such as the double vorticity correlation tensor have been unavailable. With the advent of computer technology, recent full turbulence simulations (FTS in short) have begun to fill this gap. The full simulation of homogeneous turbulence conducted by Rogallo (1981) on a 128 × 128 × 128 mesh is an excellent database which provides some informations on these terms. He computed a number of distorting flows such as axi-symmetric contraction flows and pure shear flows and also conducted the relaxation† runs from axisymmetrically contracted turbulence. However, informations such as the vorticity correlations, which we

† The term *relaxation* is used (instead of the conventional *return-to-isotropy*) to indicate that the turbulence field is 'relaxed from' straining (or shearing) by the mean field. The 'returning to' isotropy does *not* necessarily take place as a consequence.

believe are crucial both for understanding of the turbulence structure and for constructing a computational model, were not provided. And he did not run a very wide range of cases.

In order to provide the information mentioned above and to study the structural behavior of turbulence, a new FTS has been performed of homogeneous turbulence subject to irrotational strains (plane strain, axisymmetric contraction and axisymmetric expansion) and under relaxation from these strains. This paper reports outlines some of the key results of this work which sheds new lights on the fine-scale structure of turbulence. For more details, see Lee & Reynolds (1984) (hereafter referred to as R).

2. Results

The FTS reported here have been carried out using the CRAY version of Rogallo's (1981) computer code for homogeneous turbulence with a 128 × 128 × 128 mesh system on the supercomputer CRAY X-MP at NASA Ames Research Center. Summarized run specifications are given in table 1 for the irrotationally strained flows and in table 2 for the relaxation runs from the distorted turbulence fields included in table 1.

The initial fields were carefully developed and the parameters of the flows were examined to eliminate any flow fields that had eddies too large for the computational domain or too small to be detected by the mesh. For more detail see R.

In order to understand the structural behavior of turbulence, the evolution of the anisotropy tensors and their invariants has been investigated. The anisotropy tensor b_{ij} and the invariants II_b, III_b of the Reynolds stress tensor $R_{ij} = \overline{u_i u_j}$ can be defined as

$$b_{ij} = \frac{R_{ij}}{R_{kk}} - \frac{1}{3}\delta_{ij}, \quad II_b = -\frac{1}{2}b_{ij}b_{ji} \quad \text{and} \quad III_b = \frac{1}{3}b_{ij}b_{jk}b_{ki}. \tag{1}$$

By the exactly same procedure, we can specify the anisotropy tensors d_{ij} and v_{ij} (hence, their invariants II_d, III_d, II_v and III_v) of the dissipation rate tensor $D_{ij} = 2\nu\,\overline{u_{i,k}u_{j,k}}$ and the double vorticity correlation tensor $V_{ij} = \overline{\omega_i \omega_j}$ respectively. Notice that anisotropy tensors are trace-free by definition and that the three anisotropy invariant maps (the phase planes for the invariants) occupy the identical domain (see Lumley 1970 and Lumley 1978 for detailed discussion).

2.1. Irrotational distortion runs

All the irrotational distortion runs start with isotropic initial fields as listed in table 1. Runs PXF, AXM and EXQ were obtained by applying very large values of mean strain rates; for these runs, the behavior of the invariants, as shown in figures 3, 6 and 9 respectively, is in excellent agreement with the rapid distortion theory (Batchelor & Proudman 1954).

Rather diverse trajectories of b_{ij}- and d_{ij}- invariants during the plane strain runs PXB, PXD and PXF are presented in figures 1–3. From these figures, it is clearly seen that the stronger the mean strain rate imposed during the distortion, the more anisotropic D_{ij}- and V_{ij}-fields become.

Surprisingly enough, comparison of all the trajectories of R_{ij} and V_{ij} anisotropy invariants during distortion show that vorticity field is always ***more*** anisotropic than velocity field, contrary to the widely held belief. This might be attributed to the absence of a transfer term in the transport equation for V_{ij} analogous to the pressue-strain rate term in the Reynolds stress equation.

In the axisymmetric contraction runs AXK, AXL and AXM, most of the vorticity is concentrated only in the direction of the positive mean strain rate, so that 'one-dimensional vorticity' develops with a very big anisotropy $II_v \to -1/3$ as shown in figures 4–6.

In figures 10–12, the isocorrelation contour plots of the two-point velocity and vorticity correlation coefficients,

$$Q_{ii}(\mathbf{r}) \doteq \overline{u_i(\mathbf{x})u_i(\mathbf{x}+\mathbf{r})}/\overline{u_k(\mathbf{x})u_k(\mathbf{x})} \quad \text{and} \quad V_{ii}(\mathbf{r}) = \overline{\omega_i(\mathbf{x})\omega_i(\mathbf{x}+\mathbf{r})}/\overline{\omega_k(\mathbf{x})\omega_k(\mathbf{x})}, \tag{2}$$

are presented for the fields PXD4, AXL6 and EXP6 respectively. When the plane strain or the axisymmetric contraction is applied, the principal correlation ellipsoid (PCE) has a ***rod-like*** shape (figs. 10 & 11), while it has a ***disk-like*** shape for the axisymmetric expansion case (fig. 12). Rod-like turbulence consists of line vortices which become more consolidated (organized) as the mean strain rate increases. In the axisymmetric expansion flow, the disk-like structure of turbulence is developed mainly from suppression of the growth of line vortices in the axial direction which experiences the negative strain rate. In the other two directions of positive strain, the vorticity is randomly oriented due to axisymmetry and hence does not develop any organized structure in the form of line vortices.

In axisymmetric expansion flow, the growth of II_v toward the limit value of $-1/12$ as the mean strain rate gets large is a direct result of the suppression of the vorticity component in the negative strain direction (see fig. 9). The behavior of the Reynolds stress field is quite different, however. If the mean strain rate imposed is large, the vortices with initially random direction have less of an opportunity to orient themselves in the disk so that the induced velocity component perpendicular to the disk will be statistically smaller. Hence, larger mean strain rate of axisymmetric expansion results in smaller Reynolds stress anisotropy as is evident from the comparison of figures 7, 8 and 9 (see table 1 for values of S^* in the expansion runs).

Many other runs with complex combinations of strain have been run. For details see R.

2.2. Relaxation runs

Relaxation runs were performed by eliminating the mean strain rates from the fields having been submitted to the above strain modes. In figures 13–21, the faster return-to-isotropy of d_{ij}-invariants than b_{ij}- and v_{ij}-invariants is readily discernible, which confirms the traditional notion that the fine scale turbulence becomes isotropic faster than the large scale eddies.

The most general expression for the vorticity anisotropy transport model during relaxation can be written as

$$\frac{dv_{ij}}{dt} = -\alpha v_{ij} + \beta(v_{ik}v_{kj} + \frac{2}{3}II_v\delta_{ij}) \tag{3}$$

$$\alpha = \alpha(II_v, III_v) \quad \text{and} \quad \beta = \beta(II_v, III_v), \tag{4}$$

provided that the evolution of v_{ij} during relaxation is influenced by v_{ij} itself only. From analysis of all the simulations, of which figures 13 through 21 are a small sample, it is observed that regardless of prior history of strain, the relaxation trajectories of vorticity anisotropy invariants can be described as a family of hyperbolas,

$$\frac{d\mathrm{III}_v}{d\mathrm{II}_v} = \frac{3}{2}\frac{\mathrm{III}_v}{\mathrm{II}_v}. \tag{5}$$

See R for details. This observation leads to

$$\alpha = -\frac{1}{v_{ij}}\frac{dv_{ij}}{dt} = -\frac{1}{2}\frac{1}{\mathrm{II}_v}\frac{d\mathrm{II}_v}{dt} = -\frac{1}{3}\frac{1}{\mathrm{III}_v}\frac{d\mathrm{III}_v}{dt}, \qquad \beta = 0 \tag{6}$$

where α is the inverse of the relaxation time scale τ_v of vorticity anisotropy.

It is found that as turbulence relaxes the ratio of the anisotropy invariant of the vorticity field to that of R_{ij}-field aymptotically approaches to a limit which has strong dependence on the axisymmetry parameter, A.

$$\lim_{\mathrm{II}_v \to 0} \frac{\mathrm{II}_v}{\mathrm{II}_b} = 1 + \frac{(A+1)^2}{4}, \qquad A = \frac{\mathrm{III}_v/2}{(-\mathrm{II}_v/3)^{3/2}}. \tag{7}$$

Incorporating this asymptotic characteristics of the vorticity anosotropy relaxation time scale $\tau_v = 1/\alpha$, we found that the data points for the relaxation runs collapse onto a line expressed by

$$\alpha = 10^{-2}\frac{\omega}{(-\mathrm{II}_v)}\left[\frac{\mathrm{II}_v/\mathrm{II}_b}{1+(A+1)^2/4} - 1\right], \qquad \omega^2 = V_{ii} \tag{8}$$

In figure 22, a comparison is made between the inverse relaxation time scale α from (6) and the model expression from (8). Consequently, the model for vorticity anisotropy during relaxation is proposed as

$$\frac{dv_{ij}}{dt} = 10^{-2}\frac{\omega}{\mathrm{II}_v}\left[\frac{\mathrm{II}_v/\mathrm{II}_b}{1+(A+1)^2/4} - 1\right]v_{ij}. \tag{9}$$

This result has a very important implication that sheds new lights on the small scale turbulence. As the turbulence relaxes, the vorticity anisotropy adjusts 'rapidly' at first, but then becomes *locked on* to the anisotropy of the Reynolds stress (with coefficient A, see eq. (7)). After this time, the vorticity anisotropy (small-scale) relaxes 'slowly', locked on to the Reynolds stress (large-scale) anisotropy. This observation is quite different than the conventional concept of small-scale turbulence evolution.

Contrasted with the uniform relaxation trajectories of the voticity invariants, the R_{ij}-field shows quite complex behavior on the anisotropy invariant map. Especially in the case of runs D4R, F4R, P6R and Q6R (figures 14,15, 20 and 21 respectively), the values of II_b either remain unchanged or tend to increase, which no present turbulence models (Launder *et al.* 1975; Lumley 1978) have capabilities to predict. During relaxation, the large eddies become less organized and their role on the energy balance becomes less important. This leaves strong vorticity anisotropy to play a more dominant role and suggests that one should incorporate vorticity anisotropy in the R_{ij} anisotropy model for relaxation,

$$\frac{db_{ij}}{dt} = C_1 b_{ij} + C_2 v_{ij} + \text{higher-order-terms} \tag{10}$$

where C's are invariant functionals of b_{ij} and v_{ij}. We are now evaluating this suggestion.

3. Conclusions

The results of the present study revealed the effects of the strength and the mode of the mean strain rate on the development of the anisotropy of the Reynolds stress, the dissipation rate and the vorticity fields subject to irrotational distortion.

Regardless of the modes and the strength of the mean strain rate imposed, the vorticity field is more anisotropic than both b_{ij}- and d_{ij}-fields due to the absence of the transfer term in the V_{ij} balance equation.

During relaxation process, it was found that d_{ij}-field returns to isotropy faster than b_{ij}- and v_{ij}-fields. However, the vorticity anisotropy becomes *locked on* to the Reynolds stress anisotropy, after which the vorticity and Reynolds stress relax at the same rate. A simple model for the evolution of the vorticity anisotropy during relaxation fits the numerical data quite well.

This work has been supported by NASA-Ames Research Center under Grant NASA-NCC 2-15. The assistance of Dr. Robert S. Rogallo is gratefully acknowledged. Prof. Joel H. Ferziger and Dr. Anthony Leonard made several useful suggestions.

References

BATCHELOR, G. K. & PROUDMAN, I. 1954 The effect of rapid distortion on a fluid in turbulent motion. *Quart. Journ. Mech. Appl. Math.* **7**, 83–103.

LAUNDER, B. E., REECE, G. J. & RODI, W. 1975 Progress in the development of a Reynolds stress turbulence closure. *J. Fluid Mech.* **68**, 537–566.

LEE, M. J. & REYNOLDS, W. C. 1984 Numerical experiments on the structure of homogeneous turbulence. *Report* **TF**, Dept. Mech. Eng., Stanford University, Stanford, Calif. (in preparation).

LUMLEY, J. L. 1970 *Stochastic tools in turbulence.* Academic Press, New York.

LUMLEY, J. L. 1978 Computational modeling of turbulent flows. *Adv. in Appl. Mech.* **18**, 123–176.

ROGALLO, R. S. 1981 Numerical experiments in homogeneous turbulence. *NASA TM 81315.*

Run	Strain mode¶	Initial field†	Final field	Range of variations of $S^{*\ddagger}$	$Re_\ell^{\parallel}$	$-II_b \times 10^2$	$-II_d \times 10^2$	$-II_v \times 10^2$
PXB	PS	HIA9	PXB4	2.0 ~ 3.7	41.8 ~ 70.1	6.7	4.7	21.4
PXD		HIA9	PXD4	6.6 ~ 8.0	43.8 ~ 57.9	6.2	7.7	25.7
PXF		HIA9	PXF4	63.6 ~ 154.0	39.0 ~ 46.7	5.8	12.6	27.1
AXK	AC	HIC6	AXK6	1.0 ~ 2.1	48.1 ~ 61.2	5.3	4.6	20.7
AXL		HIC6	AXL6	5.7 ~ 9.7	54.4 ~ 61.2	6.6	6.8	28.9
AXM		EIC6	AXM5	38.2 ~ 96.5	49.2 ~ 61.2	6.8	7.3	30.3
EXO	AE	HID6	EXO6	0.7 ~ 3.0	44.7 ~ 69.7	5.6	0.5	5.1
EXP		HID6	EXP6	7.0 ~ 8.5	55.0 ~ 84.3	3.0	0.8	7.3
EXQ		HID6	EXQ6	47.6 ~ 70.7	55.0 ~ 58.2	1.9	2.5	7.8

¶ PS, plane strain; AC, axisymmetric contraction; AE, axisymmetric expansion.

† Isotropic fields, $II_b = II_d = II_v = 0$. ‡$S^* = Sq^2/\epsilon$ where $S = \sqrt{S_{ij}S_{ij}/2}$. ∥ $Re_\ell = q^4/(\nu\epsilon)$.

Mean velocity gradient tensors

Plane strain : $$U_{i,j} = S \begin{pmatrix} 0 & 0 & 0 \\ 0 & -1 & 0 \\ 0 & 0 & 1 \end{pmatrix}$$

Axisymmetric contraction : $$U_{i,j} = \frac{2}{\sqrt{3}} S \begin{pmatrix} 1 & 0 & 0 \\ 0 & -\frac{1}{2} & 0 \\ 0 & 0 & -\frac{1}{2} \end{pmatrix}$$

Axisymmetric expansion : $$U_{i,j} = \frac{2}{\sqrt{3}} S \begin{pmatrix} -1 & 0 & 0 \\ 0 & \frac{1}{2} & 0 \\ 0 & 0 & \frac{1}{2} \end{pmatrix}$$

TABLE 1. Specifications for irrotational strain runs

Run	Previous strain¶	Initial field	Final field	Re_ℓ ‖	Range of variations of $-II_b \times 10^2$	$-II_d \times 10^2$	$-II_v \times 10^2$
B4R	PS	PXB4	B4R6	65.7 ~ 70.1	6.6 ~ 6.7	2.9 ~ 4.7	17.5 ~ 21.4
D4R		PXD4	D4R6	57.9 ~ 68.9	6.1 ~ 6.3	3.3 ~ 7.7	18.6 ~ 25.7
F4R		PXF4	F4R6	39.0 ~ 64.1	5.8 ~ 6.0	3.6 ~ 12.6	19.3 ~ 27.1
K6R	AC	AXK6	K6R6	39.5 ~ 52.3	4.8 ~ 5.3	2.7 ~ 4.6	17.7 ~ 20.7
L6R		AXL6	L6R6	55.5 ~ 58.4	5.1 ~ 6.6	3.0 ~ 6.8	19.1 ~ 28.9
M5R		AXM5	M5R6	49.2 ~ 61.0	5.3 ~ 6.8	3.2 ~ 7.3	19.8 ~ 30.3
O6R	AE	EXO6	O6R6	74.5 ~ 88.5	5.0 ~ 5.6	0.0 ~ 0.5	4.5 ~ 5.1
P6R		EXP6	P6R6	84.3 ~ 93.7	3.0 ~ 3.7	0.2 ~ 0.8	4.3 ~ 7.3
Q6R		EXQ6	Q6R6	56.7 ~ 83.6	1.9 ~ 3.3	0.2 ~ 2.5	4.0 ~ 7.8

¶ PS, plane strain; AC, axisymm. contraction; AE, axisymm. expansion. ‖ $Re_\ell = q^4/(\nu\epsilon)$.

TABLE 2. Specifications for relaxation runs

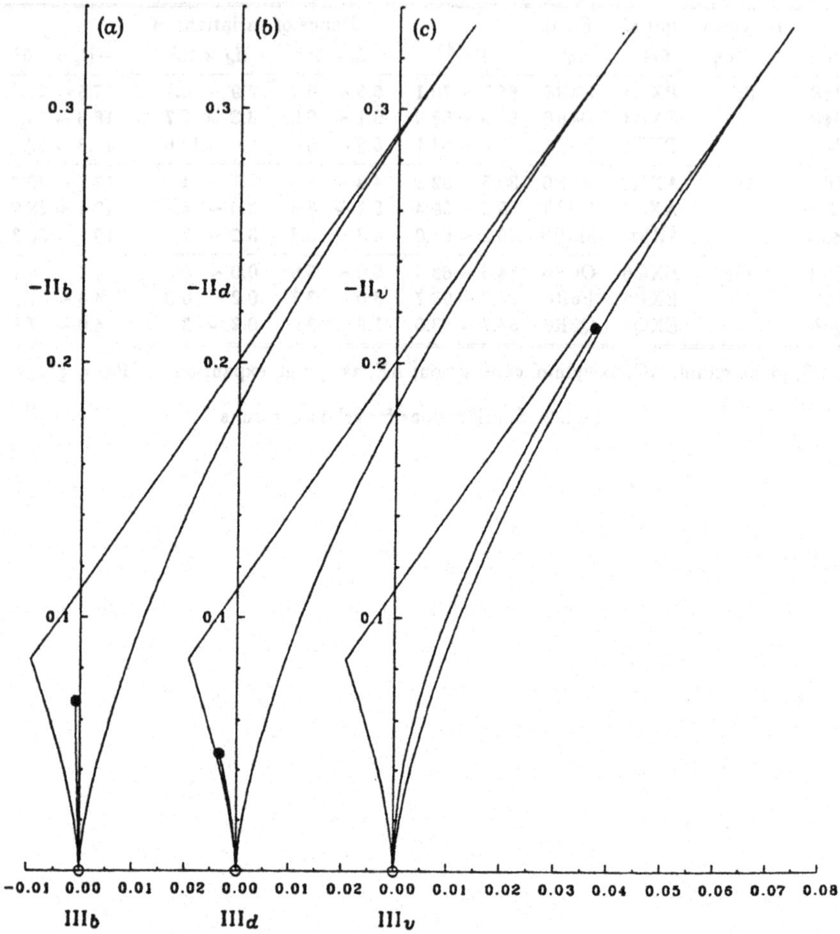

FIGURE 1. Anisotropy invariant maps of run PXB : ○ , initial field (HIA9); ● , final field (PXB4).

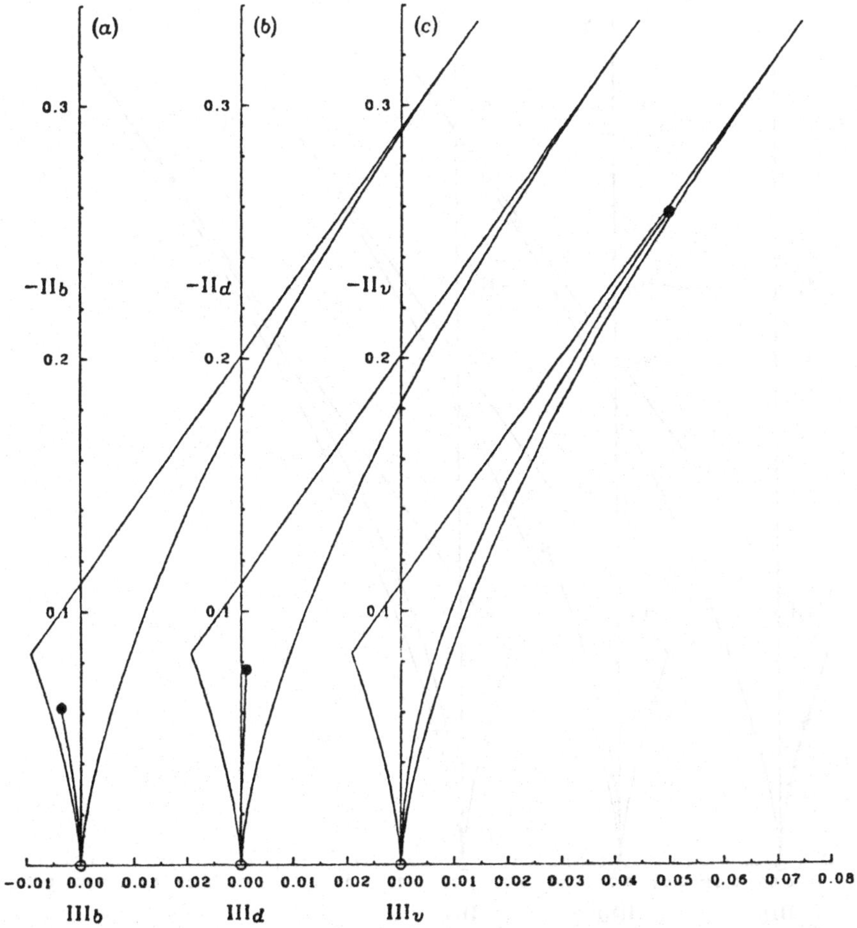

FIGURE 2. Anisotropy invariant maps of run PXD : ○ , initial field (HIA9); ● , final field (PXD4).

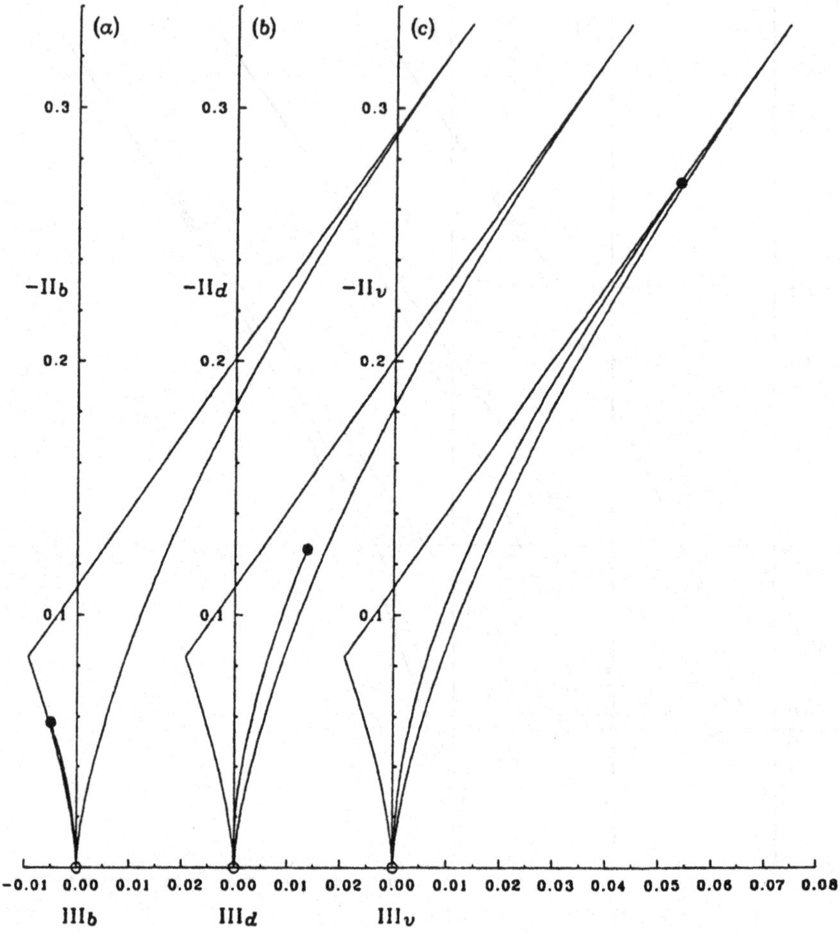

FIGURE 3. Anisotropy invariant maps of run PXF : ○ , initial field (HIA9); ● , final field (PXF4).

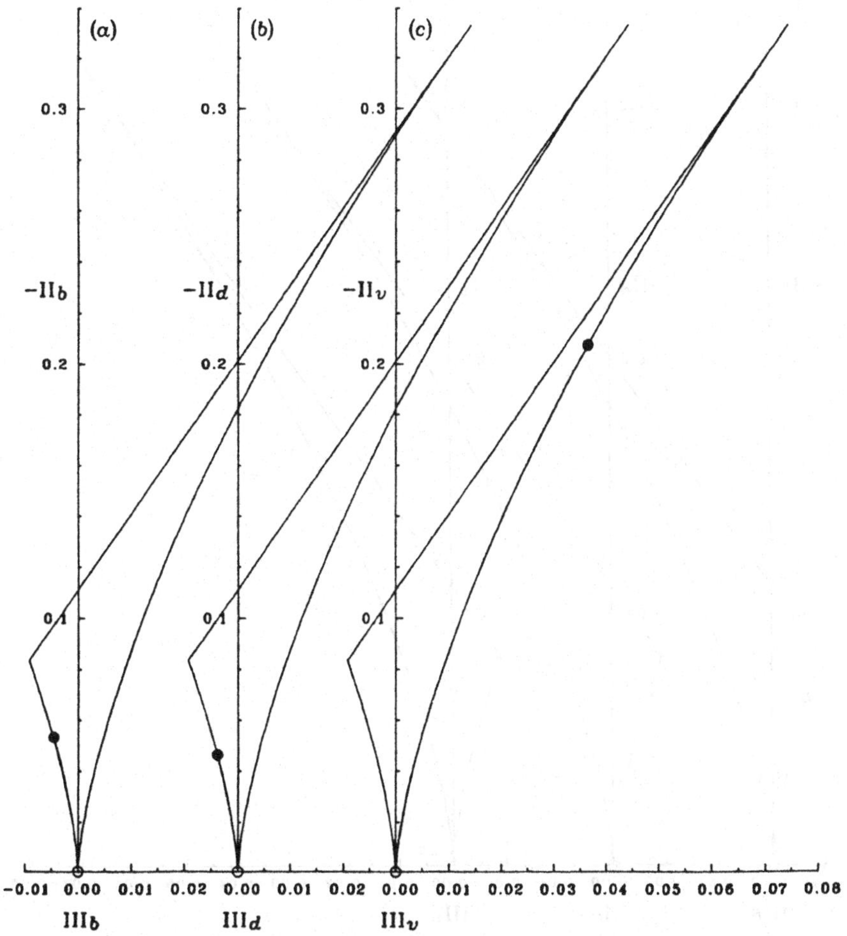

FIGURE 4. Anisotropy invariant maps of run AXK : ○ , initial field (HIC6); ● , final field (AXK6).

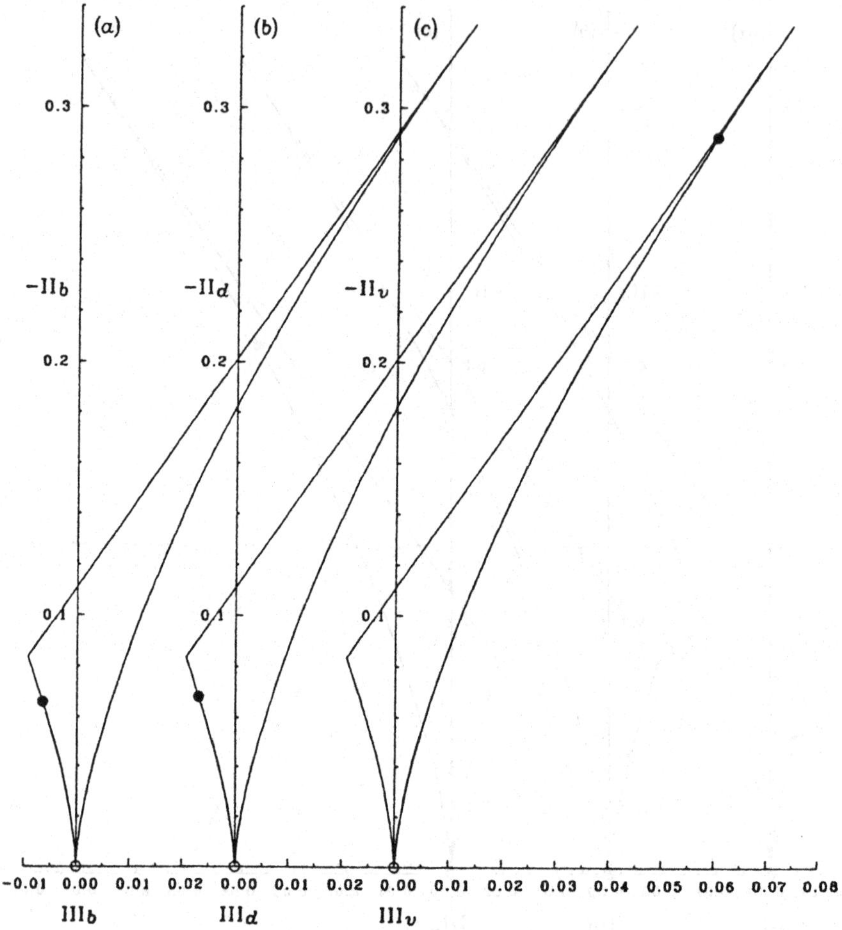

FIGURE 5. Anisotropy invariant maps of run AXL : ○ , initial field (HIC6); ● , final field (AXL6).

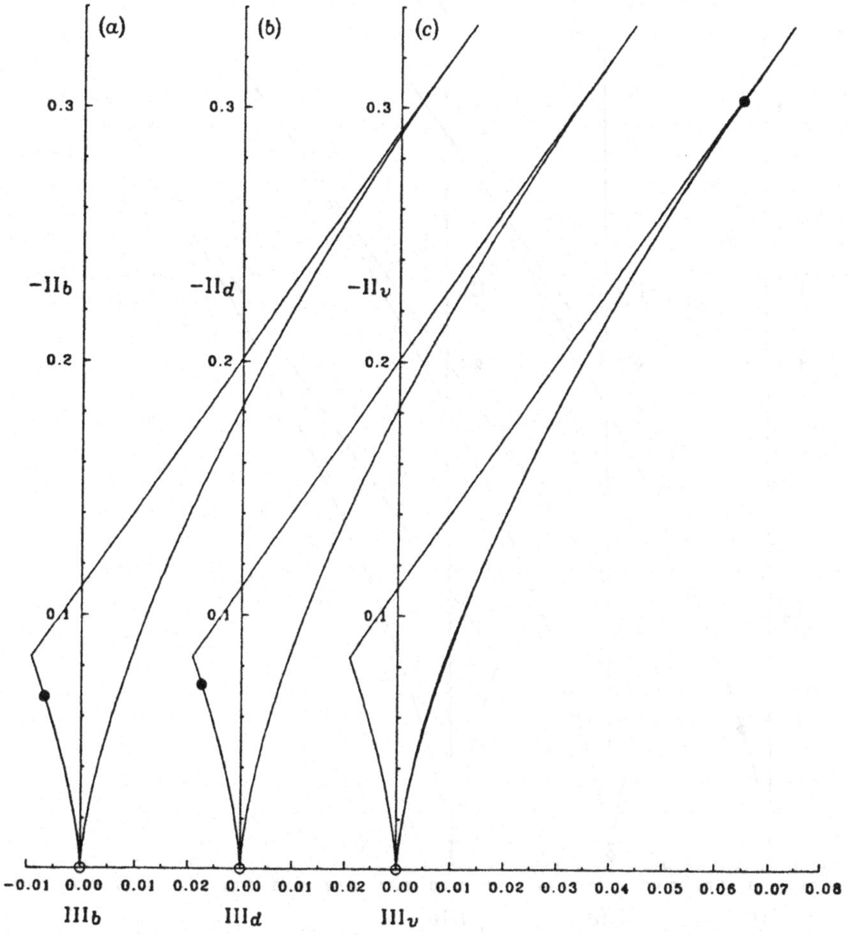

FIGURE 6. Anisotropy invariant maps of run AXM : ○ , initial field (HIC6); ● , final field (AXM5).

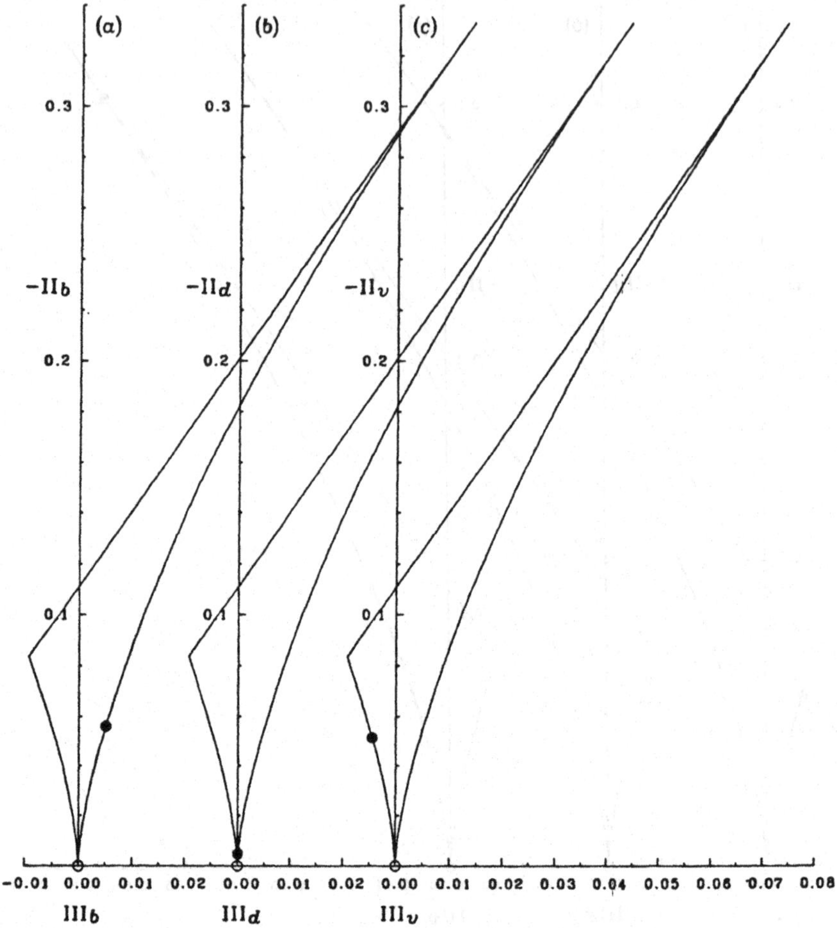

FIGURE 7. Anisotropy invariant maps of run EXO : ○ , initial field (HID6); ● , final field (EXO6).

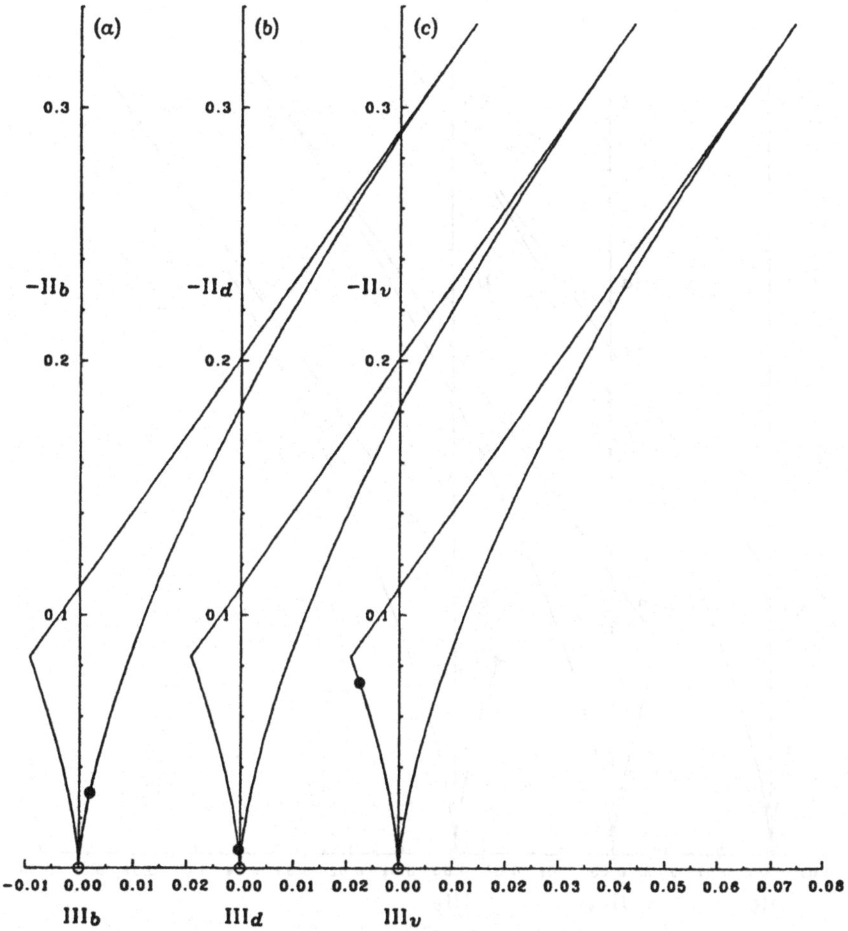

FIGURE 8. Anisotropy invariant maps of run EXP : ○ , initial field (HID6); ● , final field (EXP6).

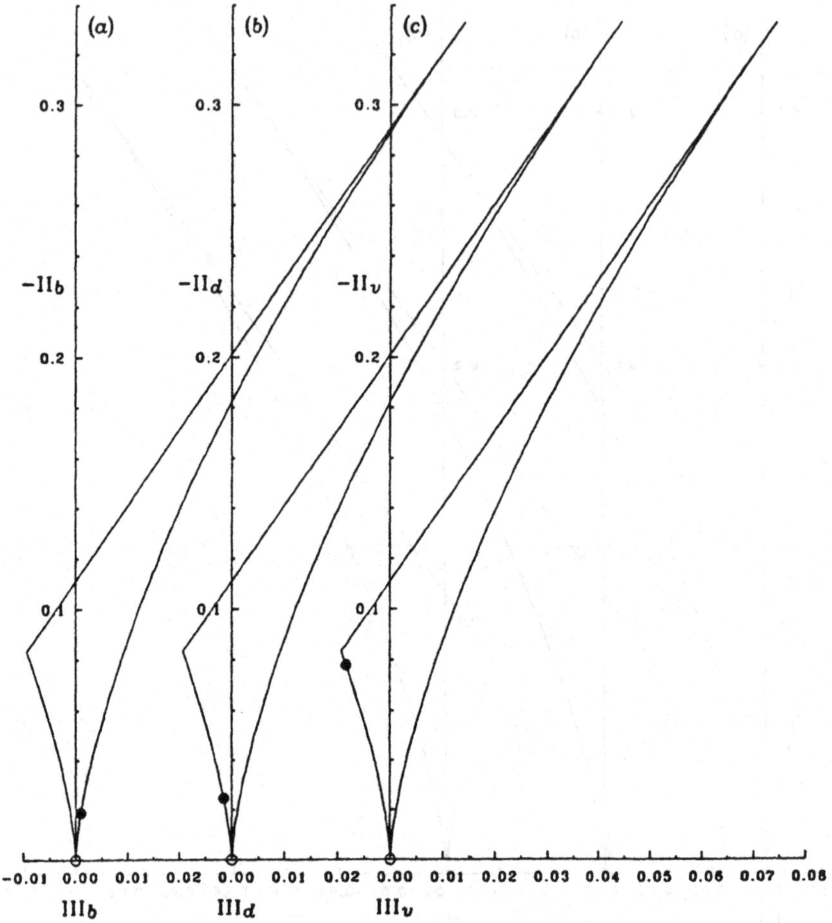

FIGURE 9. Anisotropy invariant maps of run EXQ : ○ , initial field (HID6); ● , final field (EXQ6).

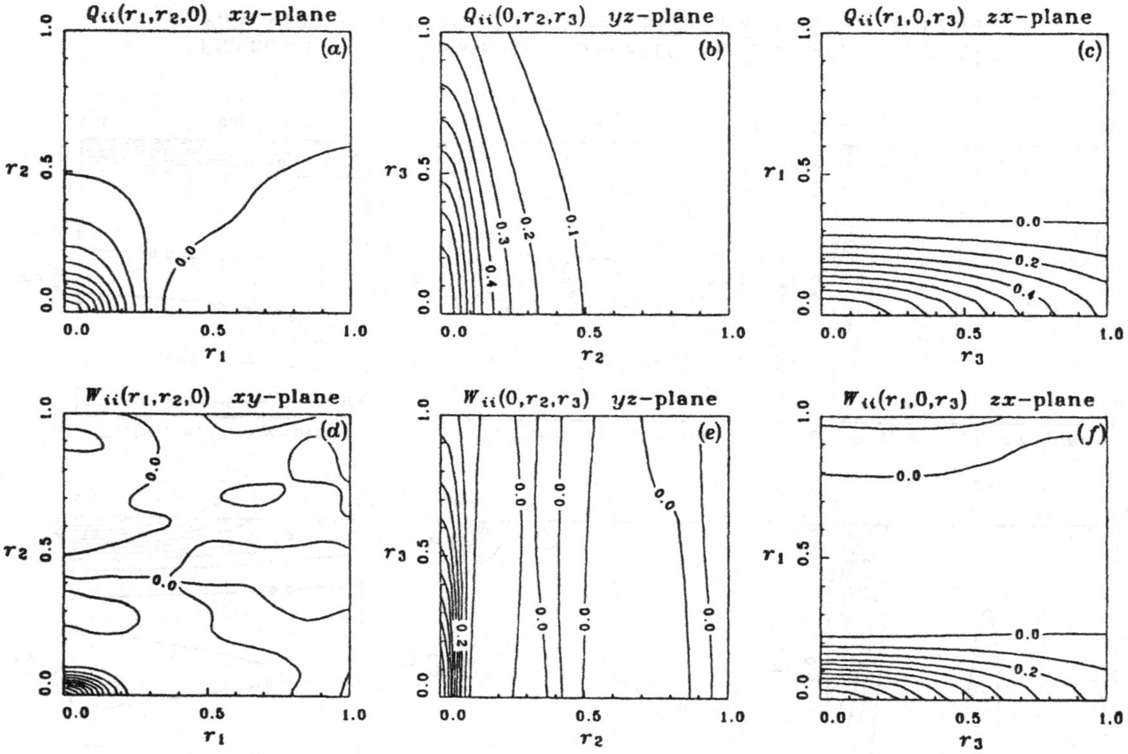

FIGURE 10. Isocorrelations of field PXD4 : Nstep = 522, t = 1.204E+00. R_{ii} = 1.970E−01, V_{ii} = 3.624E+01, half box size = (3.142, 1.568, 6.296).

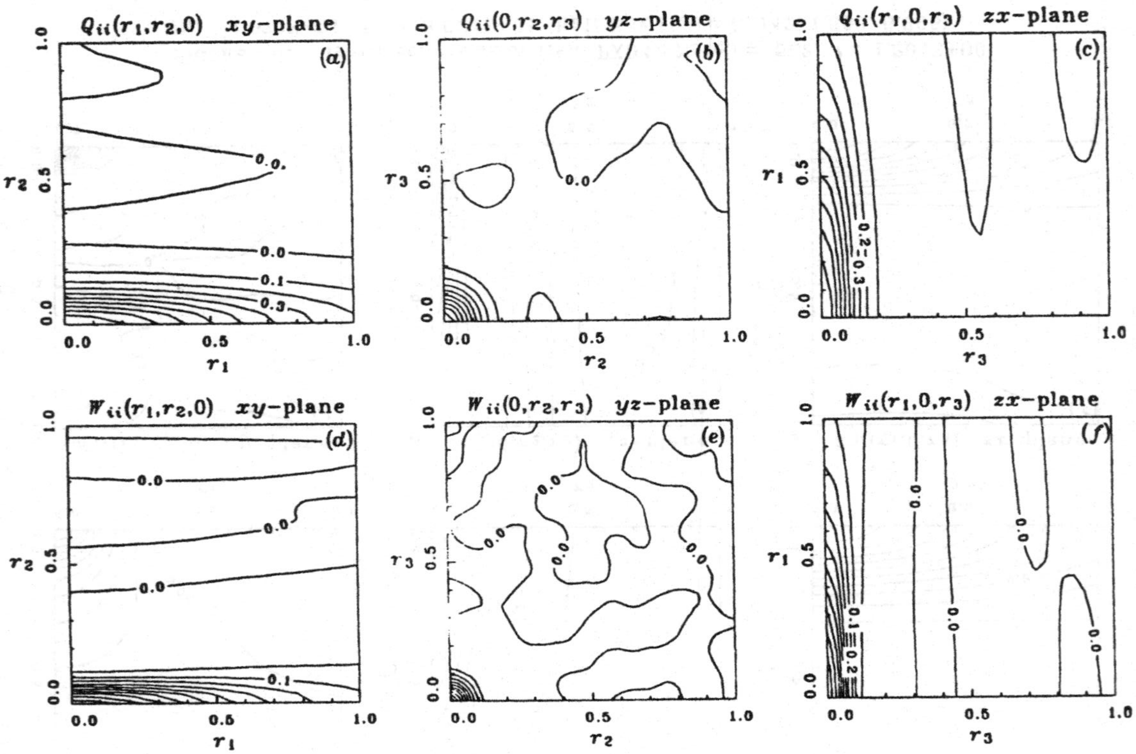

FIGURE 11. Isocorrelations of field AXL6 : Nstep = 348, t = 6.870E−01. R_{ii} = 3.606E−01, V_{ii} = 1.267E+02, half box size = (6.373, 2.206, 2.206).

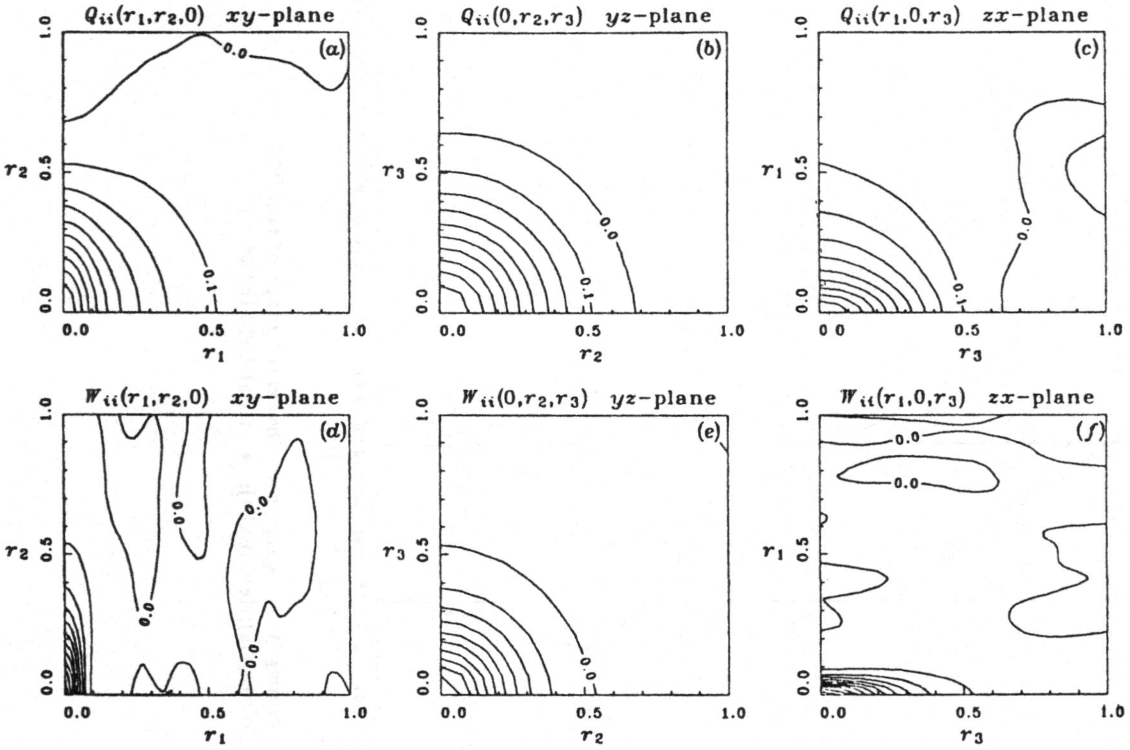

FIGURE 12. Isocorrelations of field EXP6 : Nstep = 364, $t = 7.821E-01$. $R_{ii} = 2.646E-01$, $V_{ii} = 4.493E+01$, half box size = (1.559, 4.459, 4.459).

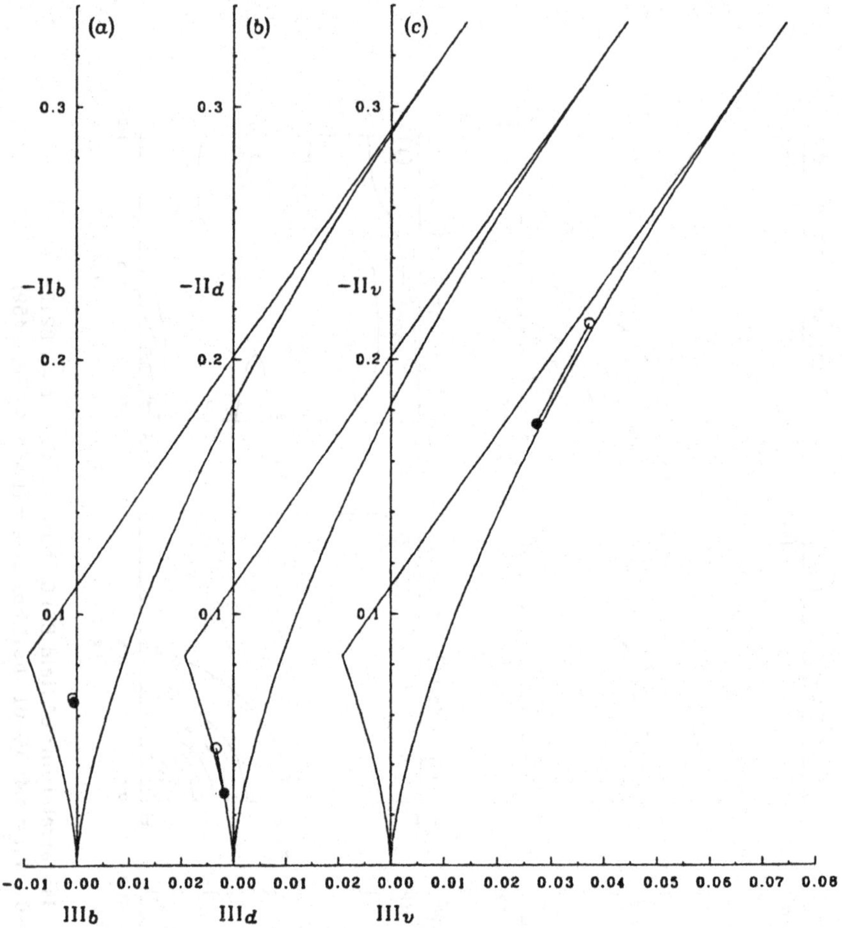

FIGURE 13. Anisotropy invariant maps of run B4R : ○ , initial field (PXB4); ● , final field (B4R6).

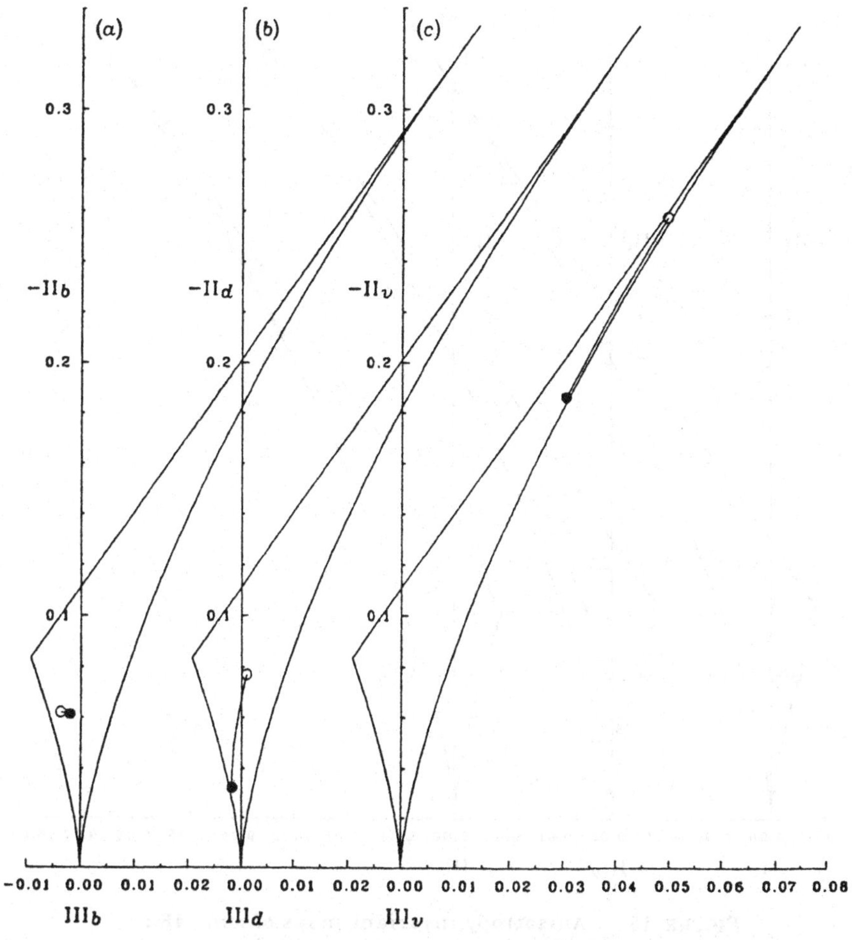

FIGURE 14. Anisotropy invariant maps of run D4R : ○ , initial field (PXD4); ● , final field (D4R6).

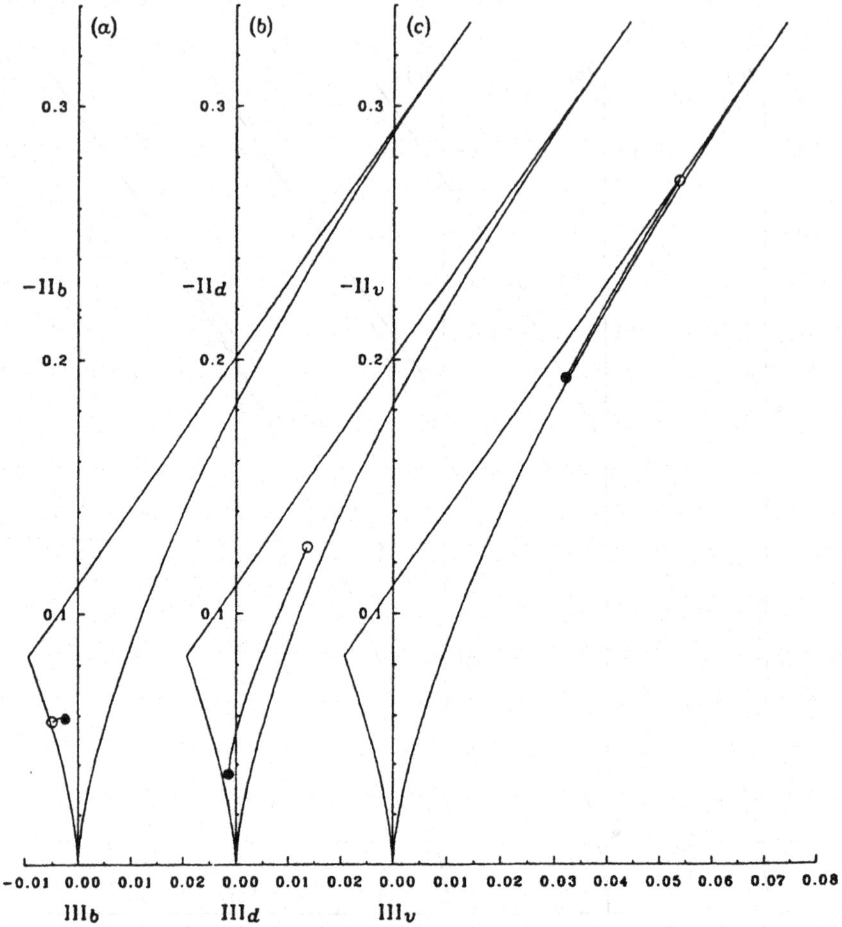

FIGURE 15. Anisotropy invariant maps of run F4R : ○ , initial field (PXF4); ● , final field (F4R6).

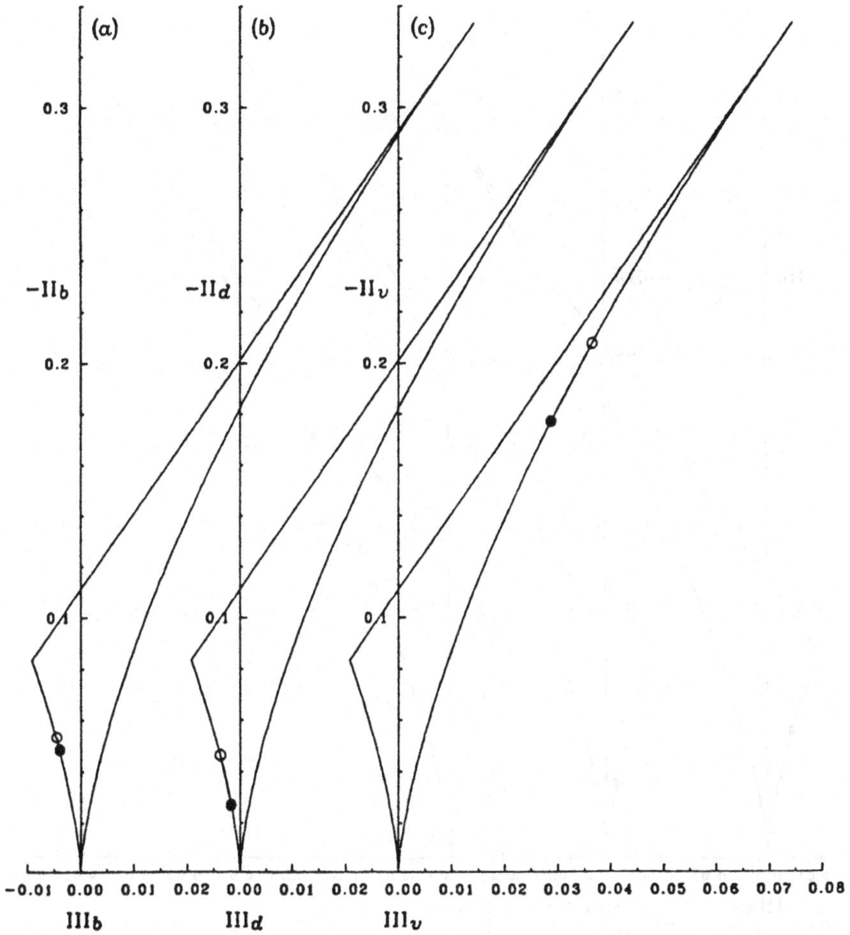

FIGURE 16. Anisotropy invariant maps of run K6R : ○ , initial field (AXK6); ● , final field (K6R6).

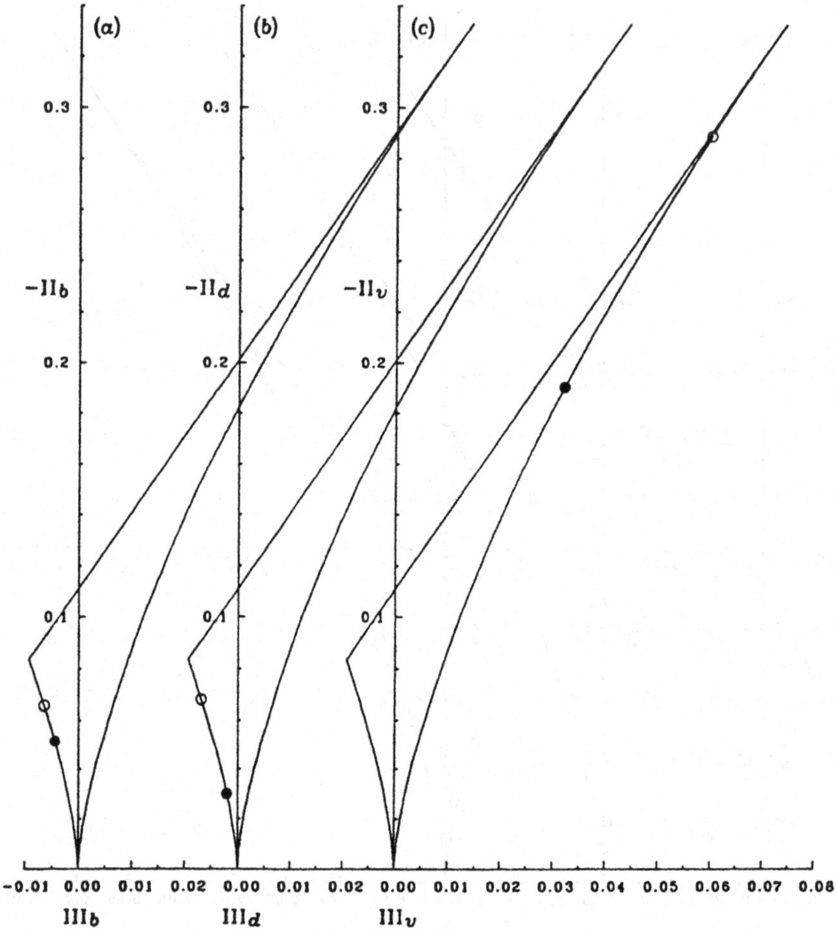

FIGURE 17. Anisotropy invariant maps of run L6R : ○ , initial field (AXL6); ● , final field (L6R6).

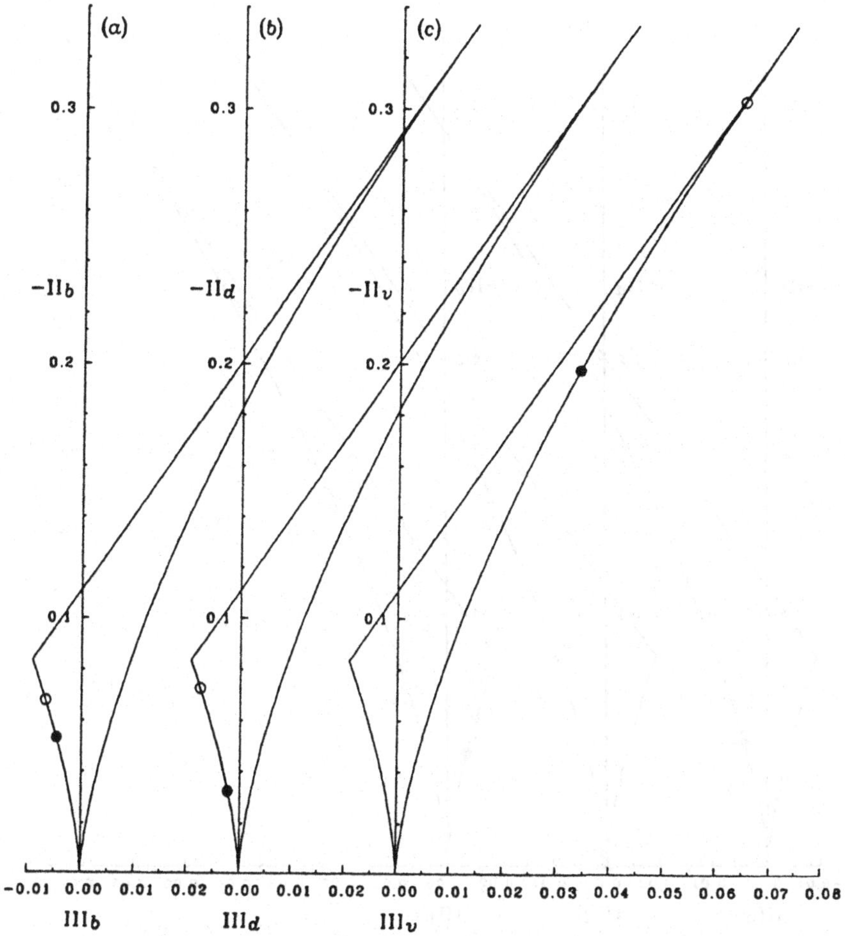

FIGURE 18. Anisotropy invariant maps of run M5R : ○ , initial field (AXM5); ● , final field (M5R6).

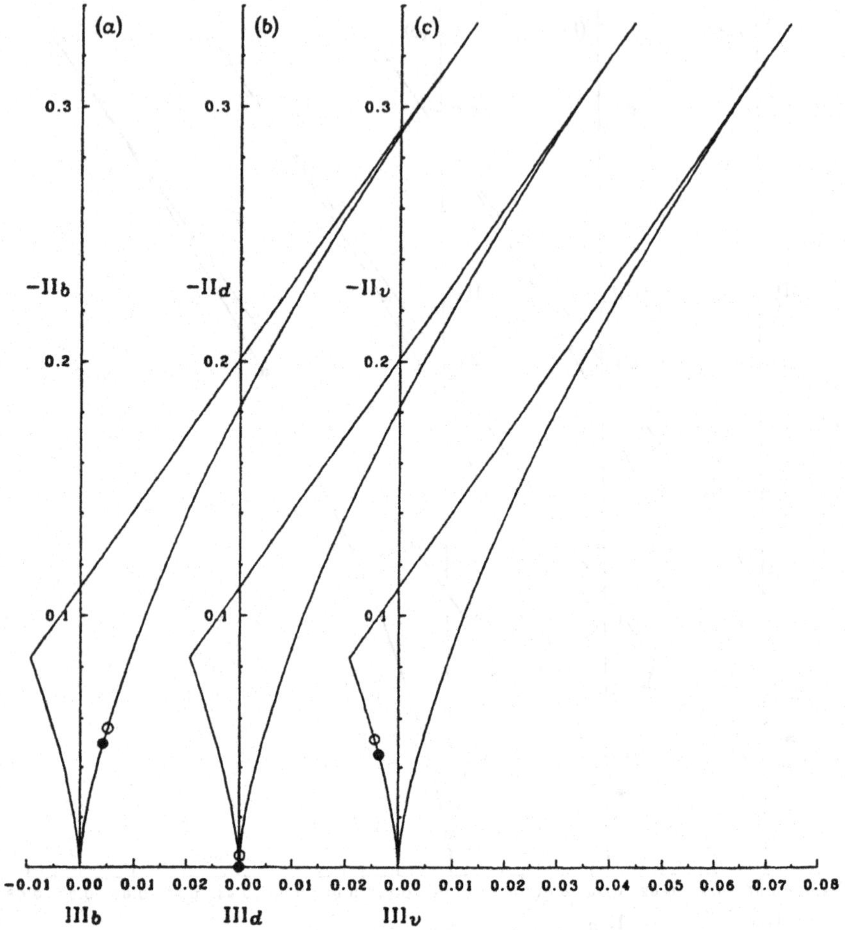

FIGURE 19. Anisotropy invariant maps of run 06R : ○ , initial field (EX06); ● , final field (06R6).

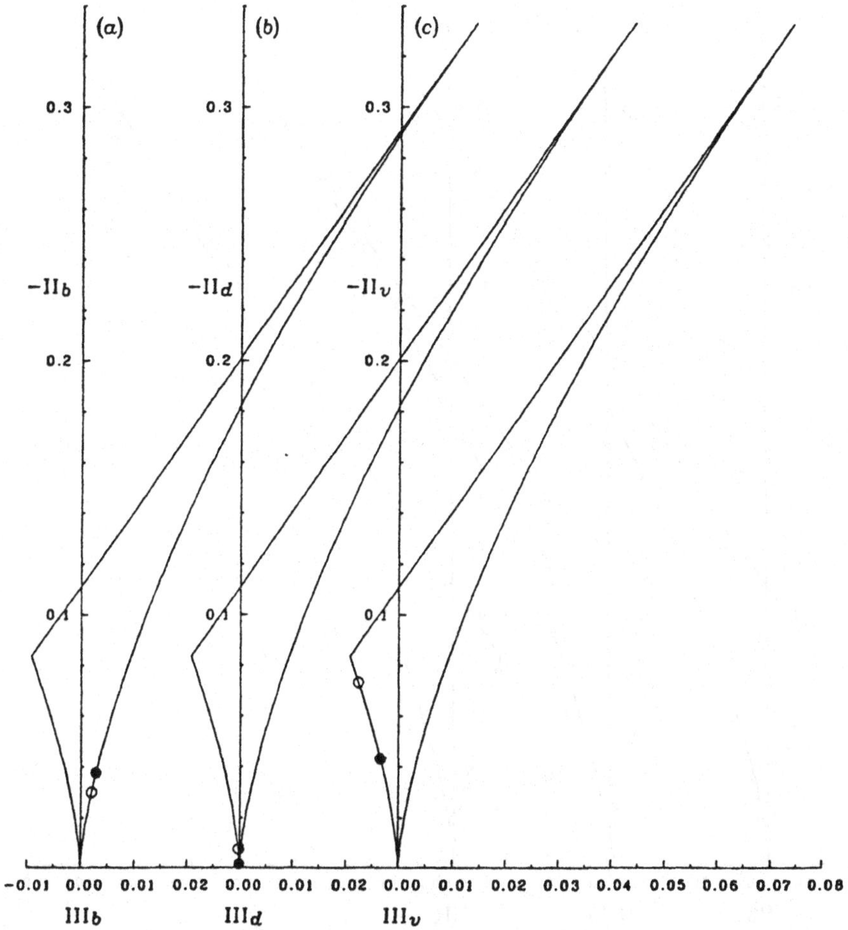

FIGURE 20. Anisotropy invariant maps of run P6R : ○ , initial field (EXP6); ● , final field (P6R6).

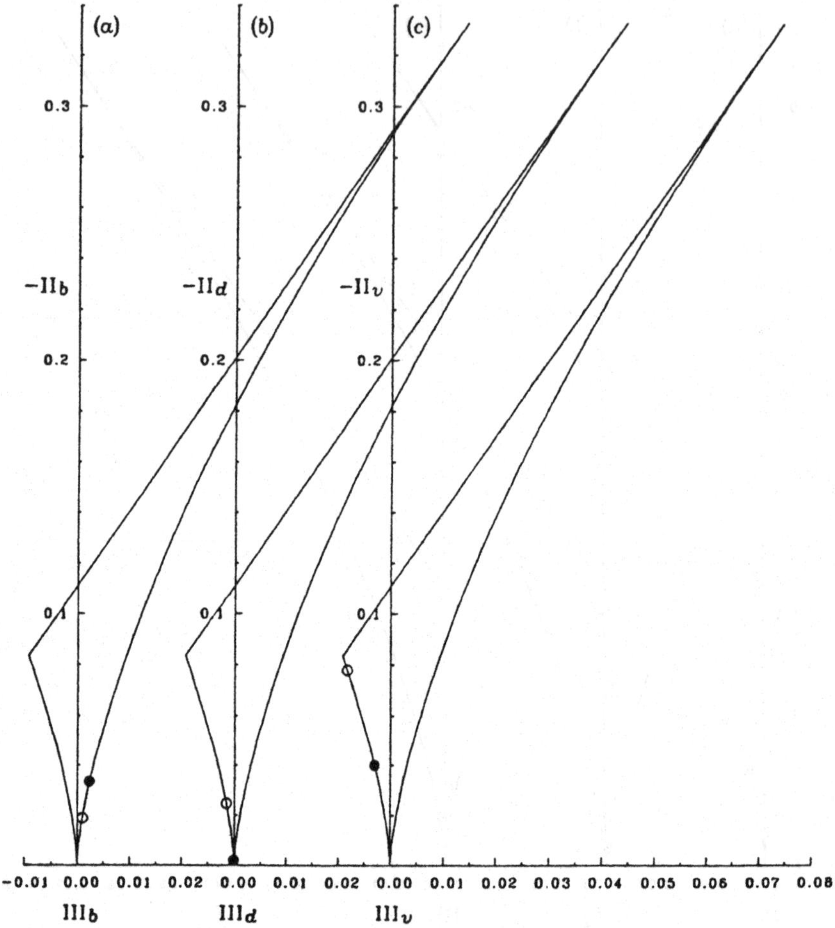

FIGURE 21. Anisotropy invariant maps of run Q6R : o , initial field (EXQ6); ● , final field (Q6R6).

$$10^{-2} \frac{\omega}{(-\text{II}_v)} \left[\frac{\text{II}_v/\text{II}_b}{1 + (A + 1)^2/4} - 1 \right]$$

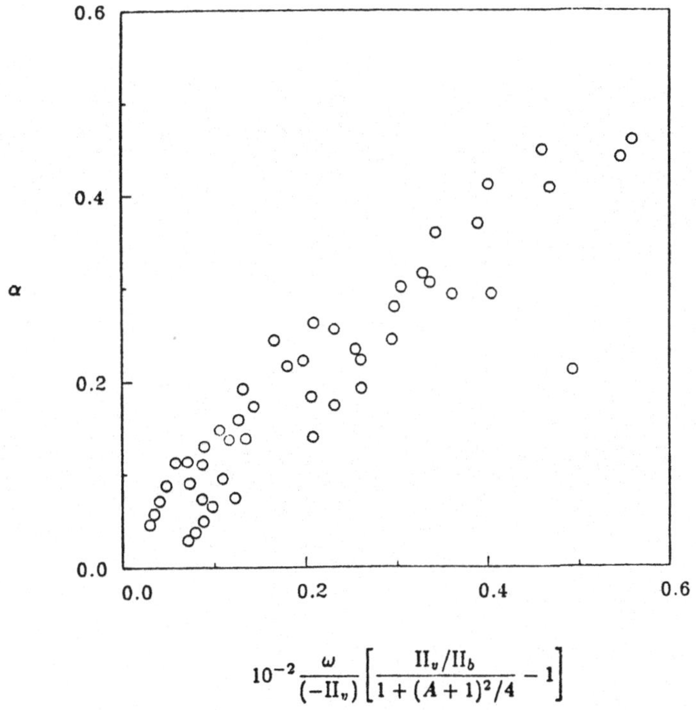

FIGURE 22. The inverse relaxation time scale α, eq. (6) *vs.* the present model, eq. (8).

CHAPTER XI. Vortex Dynamics

P.G. Saffman

1. INTRODUCTION

The dynamics of vorticity is one of the central problems of fluid dynamics. Vortex motions were aptly described by Küchemann as the "sinews and muscles of fluid motion". In the context of turbulent motion, vortex dynamics for reasons to be described below constitutes one of the theoretical approaches to the understanding of turbulence. A further reason is the fact that in the inviscid limit $\nu = 0$, vortices form a strongly nonlinear infinite dimensional Hamiltonian system and provide a physical system for many of the modern ideas on dynamical systems and the concepts of mathematical chaos which may help clarify the nature of turbulent motion.

The approaches to turbulence are manifold. It has been argued that the only completely satisfactory theoretical method will be the 'exact' solution of the Navier-Stokes equations by rational numerical approximations. The advantages of this approach are that an actual realization is found. (Also, the computational demands stimulate research in computer design and numerical analysis.) But there are significant disadvantages which may be insoluble. For instance, the appropriate initial conditions are uncertain and one could use, for instance, either a laminar profile with noise or an ensemble of random velocity distributions. Ideally, it should make no difference, but it is not clear how this is to be determined. Similarly, the boundary conditions are uncertain except in a closed geometry (or a periodic configuration which is convenient for numerical simulation but is hardly realistic). For example, the problem of the downstream boundary condition for a turbulent boundary layer or jet or wake or mixing layer is still open, and the alternative of integrating over a region so large that the velocity fluctuations have decayed to zero is never likely to be feasible. A non-trivial difficulty is that exact numerical solutions of the Navier-Stokes equations will produce enormous amounts of data and the selection process will be highly non-trivial. It is worth keeping in mind that for the case of one dimensional turbulence, i.e. random solutions of Burgers equation

$$u_t + uu_x = \nu u_{xx} , \tag{1.1}$$

the existence of an exact solution in closed form has not been particulary helpful in resolving fundamental questions on the asymptotic development of random solutions.

A rational analytical theory would, of course, remove or at least reduce the need for exact solutions, but this does not exist and our theoretical understanding of turbulence depends upon models of the phenomena based on ad hoc hypotheses of varying degrees of plausibility. There are principally four types of model, although the dividing lines are not particularly sharp; namely (a) analytical, (b) numerical, (c) phenomenological, and (d) physical. The ideas of vortex dynamics fall into the last category. The basic surmise is that turbulent flows can be modelled as assemblies or superpositions of complex, rotational, steady or periodic, laminar flows which are exact solutions of the Navier-Stokes equations or Euler equations. Such flows constitute vortices or vortical states. The idea is not particularly new. Synge & Lin (1943) employed superpositions of Hill spherical vortices to model the large scale structure of homogeneous turbulence. Townsend (1951) used two-dimensional and axisymmetric Burgers stretched vortices to model the fine scale eddies. A rougher model of curved vortex sheets with Görtler instabilities was employed by Saffman (1966) in an attempt to associate a physical meaning to the Kolmogorov length scale, which as clearly shown by experiment is the length scale of the dissipative motions in high Reynolds number turbulence.

The vortex dynamics approach was limited, however, by the difficulty in calculating vortical states. The introduction of the computer into theoretical fluid mechanics has had an enormous impact here and led to the calculation of an increasing number of such flows. At the same time, the experimental discovery by Brown & Roshko (1974) of coherent structures in the turbulent mixing layer gave the vortex dynamics approach a powerful stimulus.

Examples of vortical states relevant to the physical modelling of turbulent flow are now listed.

I. Two-dimensional arrays of finite area vortices.
The applications are to the turbulent mixing layer and the two dimensional wake, jet, boundary layer, vortex pairing, and trailing vortices.

II. Three-dimensional stretched vortices.
The applications are to the fine structure of turbulence, turbulent spots and the fine scale three-dimensional structure of basically twodimensional flows.

III. Vortex rings.
The applications are to the axisymmetric wake and jet; also the phenomena of vortex line breaking and rejoining which may play an important role in the interaction of vortices.

IV. Finite amplitude Tollmien-Schlichting waves.
The applications are to transition to turbulence and the growth of three dimensionality.

V. Vortex sheets.
The applications are to separated flow, the origin of turbulent shear flows, and the fine scale structure.

In this presentation, attention will be concentrated on I, with brief remarks made about some of the others.

2. STRUCTURE AND TWO-DIMENSIONAL STABILITY OF LINEAR ARRAYS

A considerable amount of work has been done in the last five years on the properties of steady two-dimensional finite area vortices of uniform vorticity. The infinite straight row has been studied (Pierrehumbert & Widnall 1981, Saffman & Szeto 1981), the Karman vortex street (Saffman & Schatzman 1981, 1982a, 1982b, Kida 1982, Meiron, Saffman & Schatzman 1984), the vortex pair of co-rotating vortices (Saffman & Szeto 1980), and recently the configuration of vortices at the vertices of a regular polygon (Dritschel 1984).

There are two questions. First the existence and uniqueness of solutions is to be determined. Consider for example an infinite straight row consisting of uniform vortices each of area A and containing uniform vorticity ω, with centroids a distance l apart. The circulation of each vortex is $\Gamma = \omega A$. The magnitude of the vorticity or circulation just fixes the scale of the velocity; the shapes will depend only on the parameter $\alpha = A/l^2$. There will then be a one parameter family of solutions, which form small α will be approximately circular and can be expected to become more and more deformed as α increases. Moreover, it is expected that for $\alpha << 1$, the solutions will be unique. (This has not to our knowledge been proved rigorously, but it is expected that techniques for proving existence and uniqueness for the vortex pair of small vortices will work here without difficulty.) As α increases, there is the possibility of bifurcation and limit point behavior, and eventually there will be a limiting value of α when the vortices become so large that they touch and solutions no longer exist of the assumed form. This is indeed found to be the case.

The second question concerns the stability of these steady configurations to infinitesimal and finite amplitude, two-dimensional and three-dimensional disturbances. A reasonably complete picture has emerged for the infinitesimal two-dimensional stability (although there are still problems and controversies with some of the details) and progress is being made with the three-dimensional disturbances. Two-dimensional finite amplitude disturbances can be handled by numerical integration of evolution equations, and by and large it appears that the infinitesimal theory is adequate for the description of the phenomena.

Three-dimensional finite amplitude motions remain a challenging numerical problem. If the internal structure of the vortices is neglected, the Biot-Savart law with the appropriate cutoff can be used to follow the motion of the filaments. Moore & Saffman (1972) provide a rigorous justification for the cutoff, and Moore (1972) describes a careful calculation for the motion of two filaments. A popular approximation for three-dimensional vortex motions is to replace the vortex by an assembly of vortex filaments which move according to the Biot-Savart law or some modification. For a recent review, see Leonard (1985). Since the self-induced velocity of a filament depends upon its cross-section, although only logarithmically, the accuracy of this approximation in practice remains uncertain. The absence of exact unsteady solutions renders checks of the numerical method difficult, but there are two approximate analytical solutions which would provide partial checks. They are the velocity of a viscous vortex ring (Saffman 1970) and the velocity of a vortex ring with elliptical cross section (Moore 1980). Unfortunately, both solutions are axisymmetric so they do not provide fully three-dimensional tests, but they should allow viscous and unsteady features of the numerical codes to be checked.

It is perhaps appropriate to mention here an alternative method using vortons (Saffman 1981) in which a continuous vorticity distribution is approximated by a sum of vortex monopoles. The method is being studied and applied to the motion of vortex rings by Novikov (1983) and Aksman, Novikov & Orszag (1984). Dr M.J. Kascic of ETA Systems and I have also applied the method to the inviscid Taylor-Green problem, see Kascic (1982). (Unfortunately, there is a scientifically irrelevant dispute over the origin of the vorton idea. In order to remove misconceptions, it should be noted that the idea was described in a proposal letter from me to Dr. F.C. Laccabue of Control Data Corporation dated January 27 1977 and shortly afterwards collaboration with Dr Kascic commenced on the implementation of the method to three dimensional vortex motions.)

There are three methods for the two-dimensional structure, two in principle analytical and in theory amenable to perturbation techniques, and one numerical. The first method is that of contour dynamics, a modification of the water bag method of plasma physics, introduced into fluid dynamics by Zabusky and colleagues. In this approach, the velocity induced by the vortices is expressed as a line integral around the boundary of the vortices

$$u + iv = \sum_j \oint_{\partial A_j} \log |z - z_j| \, dz_j \tag{2.1}$$

Satisfying the appropriate boundary condition for the velocity, e.g. $u + iv \parallel dz_j/ds$, for the case when the vortices are at rest, gives a non-linear integro-differential equation for the boundaries.

The second method employs the method of the so-called Schwarz functions to find the conformal map of the boundary of the vortex into the unit circle (Jimenez, private communication). Suppose that the analytic function $z = f(\zeta)$ maps the unit circle $|\zeta| = 1$ into the boundary of the vortex. Then, on the boundary, there will be a relation

$$\bar{z} = \sum_{n=-\infty}^{n=\infty} g_n z^n \qquad (2.2)$$

where the g_n are functions of $f(\varsigma)$. It follows that the velocity induced by the vortex is proportional to the terms in (2.2) going from $-\infty$ to -1, and the boundary condition provides an equation for $f(\varsigma)$. This method is in some ways more attractive, in that it does not require the numerical approximation of a singular integrand, which can be inaccurate if care is not taken, and it also lends itself more easily to stability calculations, but it has the disadvantage that for deformed vortices the truncation of (2.2) must retain a large number of terms.

The third method is to replace the finite vortex by a cloud of point vortices, or equivalent smoothed singularities which it is claimed can improve the accuracy for continuous vorticity distributions, and search for initial conditions of the evolution equations which are approximately stationary. This method is unsatisfactory for steady solutions, as it will not give unstable solutions and cannot handle non-uniqueness in a systematic way, but it has the advantage that it can study finite amplitude disturbances very easily.

The first two methods can also be modified to give evolution equations for the unsteady boundaries which can either produce an eigenvalue problem for the spectrum of infinitesimal disturbances or be integrated to follow finite amplitude disturbances. For the first method, this is called the method of contour dynamics.

Results for the equilibrium shapes of the single row (Saffman & Szeto 1981) are as follows. For $\alpha < 0.228$, the solution is unique, with the deformation, measured by the ratio a/b of the major to minor semi-axes of the elliptical-like vortices, increasing monotonically with α. For $0.228 < \alpha < 0.238$, there are two solutions. The value 0.238 is a limit point in plots which use α as the abscissa, and the more deformed solution at the value 0.228 is the limit in which the vortices touch, i.e. a = 1/2 l.

With regard to the stability to infinitesimal two-dimensional disturbances, we consider a general disturbance to the stream function of the form

$$\psi'(x,y,t) = e^{\sigma t}\, e^{2\pi ipx/l} \sum_{-\infty}^{\infty} f_n(y)\, e^{2\pi inx/l} \qquad (2.3)$$

In this expression, p is an arbitrary real number (it is assumed that the disturbance is bounded at upstream and downstream infinity), which may without loss of generality be taken in the range $0 \le p < 1$, and can be interpreted as a subharmonic wavenumber. Note that the disturbance is not in general spatially periodic. There are two values of p of particular interest. The value p = 0 corresponds to the case in which the disturbance has the same spatial period as the undisturbed flow; the disturbances can then be called superharmonic. The value p = 1/2 corresponds to a doubling of the spatial period.

There are essentially two types of disturbance; displacement modes in which the vortices move relative to each other ($p \neq 0$, $\alpha \ge 0$), and deformation modes in which the boundaries of the vortices oscillate ($\alpha > 0$, $p \ge 0$). The displacement modes are always unstable and the dominant wavelength is p = 1/2, which is called the <u>pairing</u> instabil-ility. The growth rate as a function of p is

$$\sigma = \frac{\pi\Gamma}{l^2}\, p(1-p) + O(\alpha^2) \qquad (2.4)$$

where the finite area correction term is positive but small. The deformation modes are stable until α becomes quite large and their frequency is of order $\pi\Gamma/2A$. These depend very weakly on p and fairly weakly on α. However, when α has the value corresponding to the limit point, i.e. maximum area, the gravest superharmonic deformation mode becomes unstable and the vortices are subject to a <u>tearing</u> instability. This was predicted by Saffman & Szeto using a Kelvin energy argument, but has recently been confirmed by a direct calculation of the eigenvalues by Kamm & Saffman (to appear).

The topic at issue, of course, is the relevance or bearing of these results on the structure of the turbulent mixing layer. We have uncovered two mechanisms for the development of the layer, pairing and/or tearing, which perhaps help us understand the dynamical pro-cesses, but the essential requirement of a successful theory that it predict phenomena ('postdiction' is necessary but not sufficient) or describe complex phenomena qualitatively and quantitatively in terms of simple concepts is not yet satisfied.

Arguments have been presented (Saffman 1981) based on energy considerations that the development of the mixing layer is controlled by pairing while 'young' and by tearing when it becomes 'old'. This leads to the prediction that the asymptotic state of the mixing layer contains vortices with values of a/b in the range 1.54 to 2.4, and values of δ_ω/l of about 0.3, where δ_ω is the vorticity width. These predictions are not inconsistent with the data. No prediction is, however, made for the value of l/x, where x is the distance from the origin of the layer. The pairing mechanism provides this, but not the vorticity width, because it gives the time l/U for vortices to rotate about each other, but there is an order one uncertainty due to logarithmic dependence upon the initial amplitude of the disturbance. A further difficulty with the pairing mechanism is that it needs to explain why the vortices actually amalgamate as they rotate around each other. However, a reason for this is provided by the study of the co-rotating vortex pair (Saffman & Szeto 1980) which shows that the configuration is unstable and will disappear when the vortices become too close. This was demonstrated by Roberts & Christiansen (1972) using clouds of point vortices. Saffman & Szeto used an energy argument; note that Dritschel (1984) has queried their results and claims on the basis of an eigenvalue calculation that instability sets in when the separation is larger than that given by the energy argument. Dritschel (private communication) appears to have resolved the conflict by noting that his change of stability corresponds to an exchange of energy maxima between the vortex pair and a single Kirchhoff rotating vortex whose energy was miscalculated by Saffman & Szeto.

The effect of curvature on the mixing layer may provide an example of a physical phenomenon in which the vortex dynamics model, or for that matter any of the turbulence models, could be tested. Experimental data is as yet meager, and here is an opportunity for models to make predictions without knowledge of the actual behavior.

The mixing layer has attracted most attention because it is the flow in which the vortical structure is most clearly visible. There has in recent years been work carried out on the Karman vortex street of a staggered double row of vortices, which can model the wake, or the symmetric double row which gives an inviscid model of the boundary layer, but the development is as yet much less. One rather interesting feature of the stability of the Karman vortex street to infinitesimal disturbances is worthy of comment, as it was completely unexpected. It is well known that in the limit of zero area, i.e. point vortices, the street is unstable to all two-dimensional disturbances unless $\kappa = h/l = \sinh^{-1}1 = 0.281$, where h is the distance between the rows and l is the separation of the vortices in each row. It was expected that the effect of finite area would be to either make the street always unstable, or give a finite range of κ, depending upon α, for which the street would be stable. Recent calculations by Meiron, Saffman & Schatzman (1984) have given the remarkable result that the qualitative property is unaltered, and that the street of finite vortices is unstable to infinitesimal two-dimensional disturbances except when

$$\kappa = 0.281 + 0.557\ \alpha^2 \qquad (2.5)$$

and the suharmonic wave number of the neutral disturbance is

$$p = 0.5 + 4.63\ \alpha^2 \qquad (2.6)$$

For a given value of κ the subharmonic wavenumber of the most unstable disturbance is greater that 1/2, so that pairing instabilities are not the most unstable when the vortices are of finite size. Recent work on the subharmonic three-dimensional instability of finite amplitude Tollmien-Schlichting waves (e.g. Herbert 1984) assumes implicitly that period doubling disturbances are the most unstable. These results provide a warning against this assumption.

Instability to finite amplitude two-dimensional disturbances remains an open question when the vortices are of finite area. Christiansen & Zabusky (1973) demonstrated finite amplitude stability to the pairing mode, in agreement with results of the infinitesimal calculations, but this is now known not to be the most unstable. It is expected, however, that the known result that the point vortices are unstable to finite amplitude disturbances will hold also for the finite area case.

3. THREE-DIMENSIONAL STABILITY OF LINEAR ARRAYS

The results of two-dimensional stability are of considerable interest, but since real fluids are three dimensional, it is clearly necessary to investigate the features of stability to general three-dimensional disturbances in considering the vortex dynamics approach to turbulence. The calculations are now harder and the cases that can be analysed somewhat fewer, but the attempts must be made in order to understand the different three-dimensional structure of turbulent flows.

The first work is probably J.J. Thomson's (1883) analysis of the stability of a vortex ring, in which it was concluded erroneously that vortex rings are stable. (One motivation for this study was the vortex theory of atoms and the wish to explain the atomic spectra.) There was further work in the 19th century on the oscillations of vortex rings. Schlayer (1928) and Rosenhead (1930) examined the stability of the Karman vortex street to three-dimensional disturbances. All this work (with the exception of a study of hollow vortex rings) was restricted to the case when the wavelength of the disturbance is long compared to the core radius of the vortex and essentially employed the Biot-Savart law for the vortex motion. The first mathematically correct work which looked at disturbances of wavelength comparable to the core size and did not employ the Biot-Savart law was the study by Moore & Saffman (1975) of the study of a rectilinear vortex of finite cross section in a weak uniform irrotational straining field. This work was motivated by a prediction based on intuitive physical arguments by Widnall, Bliss & Tsai (1974) that short wavelength instabilities occur. Krutzsch (1939) saw such disturbances in some beautiful observations of vortex ring instabilities. The Moore & Saffman calculation was for vortices of arbitrary vorticity distribution, and the ideas were applied by Saffman (1978) to explain the number of waves that occur when vortex rings go unstable. Widnall & Tsai (1977) and Tsai & Widnall (1976) examined in detail the case when the vorticity distribution is uniform, for the vortex ring and weakly strained rectilinear vortex. The latter was recently extended to finite strain by Robinson & Saffman (1984a), and it was concluded that for short wavelength disturbances the approximation of an isolated vortex in a uniform strain and for long wavelength disturbances the Biot-Savart law were both adequate approximations. This, of course, is a considerable simplification, and justified the work of Robinson & Saffman (1982) on the three-dimensional stability of two-dimensional arrays to disturbances of spanwise wavelength long compared to the vortex core size when the core size and separation distance are comparable.

Three cases were considered in this last mentioned work. The single row, the staggered double row, and the symmetrical double row. In the last case, only symmetrical disturbances were allowed, so that it is equivalent to a single row above a plane wall. The Biot-Savart law with appropriate cutoff was employed and disturbances proportional to

$$e^{\sigma t}\ e^{2\pi ipx/l}\ e^{2\pi iz/\lambda} \tag{3.1}$$

were considered. The governing equations are very complicated and will not be reproduced here. The interest is in the dependence of Re(σ) on the subharmonic wavenumber p and the spanwise wavelength λ.

For the symmetric array, it was found that the most unstable disturbance always had $p = 1/2$ and moreover had values of λ that were comparable with l. Also, the growth rate increased sharply with decreasing $\kappa = h/l$, where h is the distance between the rows.

For the staggered array, the most unstable disturbance was always anti-symmetrical with $p = 1/2$ and $\lambda = \infty$ if κ was greater than a value between 0.3 and 0.4, the precise value depending upon the core size, whereas for smaller values of κ it was three-dimensional and superharmonic but with a growth rate significantly smaller than for the symmetrical double row.

These results are qualitatively consistent with the observations that turbulent boundary layers are strongly three-dimensional in nature, whereas the wake seems to be weakly three-dimensional with values of κ around 0.2 (Barsoum et al, 1978). However, there are significant quantitative differences. For this value of κ, the theory suggests values of $\lambda \sim 2l$, whereas the experiments give values of $\lambda \sim 1/2\ l$.

The most unstable disturbance for the single row always has $p = 1/2$ and $\lambda = \infty$. This agrees with the observation that the mixing layer is always basically two-dimensional with a relatively unimportant three-dimensional structure superposed. The results also agree qualitatively with eigenvalue calculations by Pierrehumbert & Widnall (1982) of the two- and three-dimensional instabilities of a continuous vorticity distribution which models a vortex sheet of finite thickness and nonuniform structure.

4. STRETCHED VORTICES

An interesting class of vortex motions is provided by the velocity field

$$u_x = -\alpha x + u(x,y,t)$$
$$u_y = -\beta y + v(x,y,t) \qquad (4.1)$$
$$u_z = (\alpha + \beta)z$$

where the velocity field is given in terms of an irrotational and a rotational part. The quantities α and β are supposed positive and constant, and $\alpha + \beta$ measures the rate of straining of vortex lines. Only the z-component of vorticity ω is present. The equations of motion in the vorticity streamfunction formulation are

$$\frac{\partial\omega}{\partial t} + (-\alpha x + u)\frac{\partial\omega}{\partial x} + (-\beta y + v)\frac{\partial\omega}{\partial y} = (\alpha + \beta)\omega + \nu\nabla^2\omega \qquad (4.2)$$

$$\nabla^2\psi = \omega \qquad u = -\frac{\partial\psi}{\partial y} \qquad v = \frac{\partial\psi}{\partial x} \qquad (4.3)$$

Solutions to the above equations are referred to as fields of stretched vorticity. Simple closed form steady solutions exist for the case $\alpha = \beta > 0$, which is termed the axially symmetric vortex, and for the case $\alpha > 0$, $\beta = 0$, which is the two-dimensional or shear-layer solution. The axially symmetric solution is given by

$$\omega = \Omega e^{-\alpha r^2/2\nu} \qquad (4.4)$$

$$u_\theta = \frac{\nu\Omega}{\alpha r}\left(1 - e^{-\alpha r^2/2\nu}\right) \qquad (4.5)$$

These solutions are due to Burgers (1948).

Unsteady solutions of the equations (4.2) and (4.3) for the axisymmetric case have been used recently in studies of the fine scale structure of homogeneous turbulence (Lundgren 1982) and solutions for the two-dimensional case have been applied to the spanwise structure of the turbulent mixing layer (Lin & Corcos 1984, Neu 1984).

There seems to be little doubt that these stretched solutions will play an important role in the field of vortex dynamics. It should be noted that the existence of exact solutions is not limited to the axisymmetric or two-dimensional case. Recently, Robinson & Saffman

(1984b) demonstrated the existence of solutions for arbitrary values of the ratio α/β. They also showed that the axisymmetric Burgers vortex is stable to two-dimensional disturbances. The stability of the other solutions is an open question. Also, the stability to three-dimensional disturbances is unknown and needs to be examined.

Another class of unsteady solutions which may prove to be of interest are the jet with swirl or the vortex with axial flow. These are of the form, in cylindrical polar coordinates,

$$u_\theta = v(r,t), \quad u_r = 0, \quad u_z = w(r,\theta,t) \qquad (4.6)$$

If w is not axisymmetric, such flows will show a growth of vorticity in the plane of the swirl.

ACKNOWLEDGEMENT

I wish to thank the Office of Naval Research for support of my work on vortex dynamics.

REFERENCES

Aksman, M.J., Novikov, E.A. & Orszag, S.A. 1984 Vorton method in three-dimensional hydrodynamics. (to appear).

Barsoum, M.L., Kawall, J.G. & Keffer, J.F. 1978 Spanwise structure of the plane turbulent wake. Phys. Fluids 21, 157-161.

Brown, G.L. & Roshko, A. 1974 On density effects and large structure structure in turbulent mixing layers. J. Fluid Mech. 64, 775-816.

Burgers, J.M. 1948 A mathematical model illustrating the theory of turbulence. Adv. App. Mech. 1, 171-199.

Christiansen, J.P. & Zabusky, N.J. 1973 Instability, coalescence and fission of finite-area vortex structures. J. Fluid Mech. 61,219-243.

Dritschel, D 1984 The stability and energetics of co-rotating uniform vortices. (to appear).

Herbert, T. 1984 Secondary instability of plane shear flows- theory and application. Proc. IUTAM Sympos. Laminar-Turbulent Transition. Novosibirsk, July.

Kascic, M.J. 1982 Vorton dynamics in three dimensions. Proc. Symp. CYBER 205 Appns. Colorado State University.

Kida, S. 1982 Stabilizing effects of finite core on Karman vortex street. J. Fluid Mech. 122, 487-504.

Krutzsch, C.H. 1939 Ueber eine experimentell beobachtete Erscheinumg an Wirbelringen bei ihrer translatorischen Bewegung in wirklichen Flussigkeiten. Ann. Phys. 35, 497-523.

Leonard, A. 1985 Annual Rev. Fluid Mech. (to appear).

Lin, S.J. & Corcos, G.M. 1984 The mixing layer. Deterministic models of a turbulent flow. Part III. The effect of plane strain on the dynamics of streamwise vortices. J. Fluid Mech. 141, 139-178.

Lundgren, T.S. 1982 Strained spiral vortex model for turbulent fine structure. Phys. Fluids 25, 2193-2203.

Meiron, D.I., Saffman, P.G. & Schatzman, J.C. 1984 The linear two two-dimensional stability of inviscid vortex streets of finite cored vortices. J. Fluid Mech. (to appear).
Moore, D.W. 1972 Finite amplitude waves on aircraft trailing vortices Aeronaut. Q. 23, 307-314.
Moore, D.W. 1980 The velocity of a vortex ring with a thin core of elliptical cross section. Proc. Roy. Soc. A 370, 407-415.
Moore, D.W. & Saffman, P.G. 1972 The motion of a vortex filament with axial flow. Phil. Trans. R. Soc. A 272, 403-429.
Moore, D.W. & Saffman, P.G. 1975 The instability of a straight vortex filament in a strain field. Proc. Roy. Soc. A 346, 413-425.
Neu, J. 1984 The dynamics of stretched vortices. J. Fluid Mech. 143, 253-276.
Novikov, E.A. 1983 Generalized dynamics of three-dimensional vortical singularities (vortons). Sov. Phys. JETP 57, 566-569.
Pierrehumbert, R.T. & Widnall, S.E. 1982 The two- and three-dimensional instabilities of a spatially periodic shear layer. J. Fluid Mech. 114, 59-82.
Pierrehumbert, R.T. & Widnall, S.E. 1981 The structure of organized vortices in a free shear layer. J. Fluid Mech. 102, 301-313.
Roberts, K.V. & Christiansen, J.P. 1972 Topics in computational fluid mechanics. Comput. Phys. Commun. 3(Suppl) 14-32.
Robinson, A.C & Saffman, P.G. 1982 Three-dimensional stability of vortex arrays. J. Fluid Mech, 125, 411-427.
Robinson, A.C. & Saffman, P.G. 1984a Three-dimensional stability of an elliptical vortex in a straining field. J. Fluid Mech. 142, 451-466.
Robinson, A.C. & Saffman, P.G. 1984b Stability and structure of stretched vortices. Stud. App. Math. 70, 163-181.
Rosenhead, L. 1930 The spread of vorticity in the wake behind a cylinder. Proc. Roy. Soc. A 127, 590-612.
Saffman, P.G. 1966 Lectures on homogeneous turbulence. Topics in Nonlinear Physics (Ed. N.J. Zabusky) Springer 557-561.
Saffman, P.G. 1970 The velocity of viscous vortex rings. Stud. App. Math. 49, 371-380.
Saffman, P.G. 1978 The number of waves on unstable vortex rings. J. Fluid Mech. 84, 625-639.
Saffman, P.G. 1981 Vortex interactions and coherent structures in turbulence. Transition and Turbulence (Ed. R.E. Meyer) Academic 149-166.
Saffman, P.G. & Schatzman, J.C. 1981 Properties of a vortex street of finite vortices. S.I.A.M. J. Sci. Comput. 2, 285-295.
Saffman, P.G. & Schatzman, J.C. 1982a Stability of a vortex street of finite vortices. J. Fluid Mech. 117, 171-185.
Saffman, P.G. & Schatzman, J.C. 1982b An inviscid model for the vortex street wake. J. Fluid Mech. 122, 467-486.
Saffman, P.G. & Szeto, R. 1980 Equilibrium shapes of a pair of equal uniform vortices. Phys. Fluids 23, 2339-2342.
Saffman, P.G. & Szeto, R. 1981 Structure of a linear array of uniform vortices. Stud. App. Math. 65, 223-248.
Schlayer, K. 1928 Uber die Stabilitat der Karmanschen Wirbelstrasse gegenuber beliebigen Storungen in drei Dimensionen. Z. angew. Math. Mech.8, 352-372.
Synge, J.L. & Lin, C.C. 1943 On a statistical model of isotropic turbulence. Trans. R. Soc. Canada 37, 45-63.
Thomson, J.J. 1883 A Treatise on the Motion of Vortex Rings. Macmillan
Townsend, A.A. 1951 On the fine-scale structure of turbulence. Proc. Roy. Soc. A 208, 534-542.
Tsai, C-Y. & Widnall, S.E. 1976 The stability of short waves on a straight vortex filament in a weak externally imposed strain field. J. Fluid Mech. 73, 721-733.

Widnall, S.E., Bliss, D.B. & Tsai, C-Y. 1974 The instability of short waves on a vortex ring. J. Fluid Mech. 66, 35-47.
Widnall, S.E. & Tsai, C-Y. 1977 The instability of a vortex ring of constant vorticity. Phil. Trans. R. Soc. A 287, 73-305.

CHAPTER XII. Two-Fluid Models of Turbulence

D. Brian Spalding

1. INTRODUCTION

1.1 Some defects of turbulence models

It has become customary among engineers and others concerned with practical flow and heat-transfer predictions to use "turbulence models", ie sets of equations which purport, when coupled with the equations of mean motion, to describe such statistically significant properties as the "effective viscosity", "length scale" and "energy". Most frequently used are those models employing two partial differential equations, the dependent variables of which are, for example, the turbulence energy, k, and the dissipation rate ϵ.

Although valuable service has been performed by such models, their predictive ability is far from meeting all demands. A brief list of their shortcomings is:

(a) The "intermittency" of turbulent flow, evident to all who look perceptively at, for example, automobile-exhaust plumes, is left out of account.

(b) Attempts to extend the models to chemically-reacting flows have had little success, because chemical reactions take place in regions of steep gradients which find no expression in the equations employed.

(c) Empirical and non-general corrections have had to be made to the "constants" of the equations in order to bring their predictions into line with experimental evidence regarding the influences of mean-streamline curvature and of interactions between gravitational forces and temperature gradients.

(d) The models are totally unable to explain the "unmixing" phenomenon which steepens gradients of average fluid properties rather than diminishing them.

In the view of the present writer, the above defects have a common origin, namely neglect of the "spottiness" of real turbulent flows. The question therefore to be discussed is: How can the major effects of the "spottiness" be described by a mathematical apparatus that is still not too elaborate for practical use?

1.2 A proposed shift of emphasis

(a) The origins

Modern ideas about turbulence have been greatly influenced by the writings of Osborne Reynolds (eg Ref 1) and Ludwig Prandtl (eg Ref 2), whose observations had led them to conceive of turbulence as involving a semi-random exchange of matter, similar to that envisaged in the kinetic theory of gases, but on a larger scale. The following quotations illustrate this.

Reynolds:

> "The heat carried off ... is proportional to the rate at which particles or molecules pass *backwards and forwards* from the surface". (Present author's italics.)

Prandtl:

> "It is now necessary to make a usable hypothesis for the *mixing* velocity, w. The transverse momentum associated with this velocity must be *constantly destroyed* by braking, *and constantly re-established*". (Present author's italics).

It is a consequence of this pre-occupation with mixing that turbulent fluxes of a scalar quantity are almost invariably expressed, by analogy with Fick's law of diffusion and Fourier's law of heat conduction, in terms of the gradient of the corresponding intensive property. Thus, for heat transfer, the law:

$$\langle q\rangle_d = -\lambda_{eff}\cdot \text{grad}\ \bar{T} \qquad ,(1.2\text{-}1)$$

is employed, along with the supposition that λ_{eff} is a greater-than-zero measure of the local turbulent motion. Here $\langle q\rangle_d$ is the heat-flux vector, λ_{eff} is the effective conductivity, and $\bar{T}$ is the time-mean temperature.

So embedded is this notion in current thinking that, when experimental evidence becomes incontrovertible that the so-defined λ_{eff} is negative, the phrase "contra-gradient diffusion" (Ref 3) is applied to the phenomenon, which is regarded as anomalous.

(b) The sifting phenomenon and its implications

In the view of the present writer, normal, ie "co-gradient" diffusion is only one of the two main ways in which turbulent fluxes are caused; the second, although it could be called "non-gradient diffusion", is better described as "sifting". This is the phenomenon according to which intermingled fragments of fluid differing in, say, density, move relative to one another as a consequence of a body-force field.

The sifting process is most easily seen when surface tension keeps the two fluids apart, as when steam bubbles rise from the heated bottom of a kettle. This is not an exchange process; for the bubbles rise continuously; yet it is undoubtedly associated with a flux of energy.

The mathematical expression for the heat flux associated with sifting is conveniently written as:

$$\langle q \rangle_s = \text{constant} \ \langle a \rangle \ h' \qquad .(1.2\text{-}2)$$

which signifies that the flux $\langle q \rangle_s$ depends upon the local fluctuations of enthalpy, h', and on the acceleration vector $\langle a \rangle$. Similarly, if w-direction momentum transfer is in question, the contribution of sifting to the shear-stress vector $\langle \tau \rangle_s$ may be expressed as:

$$\langle \tau \rangle_s = \text{constant} \ \langle a \rangle \ w' \qquad .(1.2\text{-}3)$$

a term which may or may not have the same sign as the more conventionally defined viscous contribution:

$$\langle \tau \rangle_d = \mu_{eff} \ \text{grad} \ \overline{w} \qquad .(1.2\text{-}4)$$

Here $\overline{w}$ and w' represent the time-average and fluctuating components of longitudinal velocity.

(c) Circumstances in which "turbulent sifting" is significant

(i) Bubble rise in a saucepan

In the phenomenon just mentioned, it is important to emphasise, the vertical-direction temperature gradient is very small; for both steam and water are at the saturation temperature corresponding to the local pressure, which varies negligibly over the small height change.

Nor can enthalpy or concentration gradients account for the bubble motion; for the bubbles are moving towards the region of highest steam concentration, namely the top part of the saucepan, not away from it. The fluid enthalpy there is of course higher than at the bottom.

(ii) Rise of smoke in a room

If a fire starts on the floor in the corner of a room, flames and smoke rise in a turbulent manner, often forming a plume of smoke within which the naked eye can distinguish fragments of smoky gas separated by less opaque bodies of clear air; yet the layer of smoke which flows along the ceiling is usually separated from the air below it by a sharp interface: the lighter smoke fragments have floated upwards leaving the denser air behind.

(iii) Boundary layers on curved surfaces

When a turbulent boundary layer encounters and flows over a convex surface, the pressure gradient along the normal to the surface is positive. As a consequence, fragments of fluid which are flowing at a greater-than-average velocity are flung outward, while slower fragments are forced towards the wall. There is therefore, from this cause, a momentum transfer outwards, towards the region of higher velocity.

The momentum flow is reversed when the curved surface is concave; for the higher-velocity fragments move towards the wall. The momentum flux associated with sifting has, in this case, the same sign as that associated with turbulent mixing.

(iv) Flow in a cyclone

Similar sifting flows occur in cyclone separators. That particles of denser material move preferentially outwards is of course obvious: it is the raison d'etre of the apparatus. Less well-recognised is the fact that fluid particles having larger circumferential velocities will migrate to the larger radii, while the slower-moving particles move towards the cyclone axis.

This steady transfer of kinetic energy to the outside is presumably the reason for the observation that, in the Ranque-Hilsch tube (which is a kind of cyclone) the larger-diameter cylindrical wall becomes much hotter than the smaller-diameter one.

(v) Turbulent combustion of pre-mixed gases within a duct

When a pre-mixed combustible gas burns within a duct, the density reduction occasioned by the exothermic reaction occasions a reduction in pressure with increase in distance along the duct. This pressure gradient causes the lighter burned gas to accelerate more than the heavier unburned gas; the former is therefore "sifted" towards the duct outlet, where however the average temperature is the highest. This is one of the best-documented examples of "counter-gradient diffusion"; and measurements have clearly revealed the fact that the hot-gas fragments have higher velocities than the cold-gas ones (Refs 3,4,5).

(vi) Sand storms

A circumstance in which diffusion and sifting act in opposite directions, and maintain an approximate balance, arises when a strong wind blows over a sandy desert. Then turbulent diffusion of the Reynolds-Prandtl kind carries sand particles from high-concentration regions close to the ground to lower-concentration regions higher up, while sifting effects a flux in the downward direction.

It is worth remarking that, in such a case, the scale of the fragments of upward-moving and downward-moving material is much larger than that of the sand particles: the sand-laden air acts like a single-phase fluid of non-uniform density.

(d) A simple formulation of the sifting process

(i) The sifting flux

Let a turbulent fluid be taken as consisting of fragments and surrounding volumes of two kinds, distinguished by their having different densities, ρ_1 and ρ_2, and different values of some scalar quantity ϕ.

What will now be proved is that there is a flux of entity ϕ in the direction of the local body-force vector, which is proportional to the difference $\phi_1 - \phi_2$, and not to the gradient of either quantity; the relative velocity U, ie $u_2 - u_1$, also enters. Here u stands for the absolute velocity.

The net flux of ϕ, viz F_ϕ, can be written as:

$$F_\phi = r_1\rho_1u_1\phi_1 + r_2\rho_2u_2\phi_2 \qquad .(1.2\text{-}5)$$

This can be re-written in terms of the relative velocity and of the average velocity, density and ϕ, as follows:

$$\boxed{F_\phi = \overline{\rho}\overline{u}\overline{\phi} + (r_1\rho_1r_2\rho_2/\overline{\rho})\ U\ (\phi_2 - \phi_1)} \qquad .(1.2\text{-}6)$$

wherein use has been made of the definitions:

$$\overline{\rho} \equiv r_1\rho_1 + r_2\rho_2 \qquad .(1.2\text{-}7)$$

$$\overline{\rho}\overline{u} \equiv r_1\rho_1u_1 + r_2\rho_2u_2 \qquad .(1.2\text{-}8)$$

$$\overline{\rho}\overline{\phi} \equiv r_1\rho_1\phi_1 + r_2\rho_2\phi_2 \qquad .(1.2\text{-}9)$$

$$1 \equiv r_1 + r_2 \ . \qquad .(1.2\text{-}10)$$

Equation (1.2-6) is easy to interpret: the first term on the right-hand side represents the transfer of ϕ by the average mass motion; and the second term represents the sifting contribution.

Further examination of that term shows that it is linearly dependent upon U and on (ϕ_1 - ϕ_2), and also upon the product r_1r_2. The latter, by reason of equation (1.2-10), falls to zero when either component is absent, as is readily understood.

(ii) <u>The relative velocity, U</u>

As will be explained in the next section, the magnitude of U can be computed by solving momentum equations for u_1 and u_2 separately, and then subtracting. However, an approximate formula can be obtained by neglecting all the terms in the momentum equations except the pressure gradient, the body-force acceleration a, and the interfluid friction, as follows. The momentum equations reduce to:

$$r_1\left[\frac{\partial p}{\partial x} - \rho_1\, a\right] - C\, r_1 r_2\, \bar{\rho}\, \frac{|U|U}{\bar{\ell}} = 0 \qquad .(1.2\text{-}11)$$

and $$r_2\left[\frac{\partial p}{\partial x} - \rho_2\, a\right] + C\, r_1 r_2\, \bar{\rho}\, \frac{|U|U}{\bar{\ell}} = 0 \qquad .(1.2\text{-}12)$$

wherein C represents an inter-fluid friction factor, and $\bar{\ell}$ is a measure of the fragment size.

Addition of the two equations and reference to equation (1.2-7) lead to:

$$\frac{\partial p}{\partial x} - \bar{\rho}\, a = 0 \qquad ;(1.2\text{-}13)$$

whereupon substitution in either equation leads to:

$$\boxed{U|U| = C^{-1}\,(\rho_2 - \rho_1)\,\bar{\ell}\, a} \qquad .(1.2\text{-}14)$$

This equation is also easy to understand: larger fragments move faster, because the body force operates on the volume whereas the friction operates on the surface area; and the velocity difference increases with the density difference.

The "constant" C can be expected to increase when the fragment size is small, in proportion to the reciprocal of the Reynolds number of the relative motion; but the assumption of constancy is probably correct for high-Reynolds-number flows.

(iii) <u>When ρ is linear in ϕ</u>

Let S_ϕ stand for the sifting flux of entity ϕ, so that:

$$S_\phi = (r_1\rho_1 r_2\rho_2/\bar{\rho})\; U\, (\phi_2 - \phi_1) \qquad ;(1.2\text{-}15)$$

and let ρ be linear in ϕ, so that:

$$(\rho_2 - \rho_1) = \beta(\phi_2 - \phi_1) \qquad .(1.2\text{-}16)$$

Combination of equations (1.2-15), (1.2-16 and (1.2-14) now yields:

$$S_\phi = r_1 r_2 \frac{\rho_1 \rho_2}{\bar{\rho}} \left[\frac{\beta \bar{\ell} a}{C}\right]^{1/2} \left[\phi_2 - \phi_1\right]^{3/2} \qquad .(1.2\text{-}17)$$

Since it is also possible to relate $(\phi_2 - \phi_1)$ to ϕ', the root-mean square value of the fluctuations of ϕ by:

$$\phi' = (r_1 r_2)^{1/2} \, |\phi_1 - \phi_2| \qquad .(1.2\text{-}18)$$

the expression for the sifting flux in terms of ϕ' is:

$$\boxed{S_\phi = (r_1 r_2)^{1/4} \frac{\rho_1 \rho_2}{\bar{\rho}} \left[\frac{\beta \bar{\ell} a}{C}\right]^{1/2} (\phi')^{3/2}} \qquad .(1.2\text{-}19)$$

This is a term which, in the present author's view, should often appear in the differential equation for $\bar{\phi}$, when this is employed in a turbulence model.

Of course, it can be used only when estimates are available of the $r_1 r_2$ product, of the fragment size ℓ, and of the fluctuation intensity ϕ'; but this situation is not rare.

(iv) <u>The sifting flux of momentum, in flows with curved stream lines</u>

When a fluid having velocity w follows a path having a radius of curvature R, it experiences an acceleration w^2/R normal to its trajectory.

If the above equations are re-worked for the situation in which the density is uniform but the two sets of fluid fragments have differing velocities, w_1 and w_2, there results:-

$$S_w = r_1 r_2 \rho U \, (w_2 - w_1) \qquad .(1.2\text{-}20)$$

$$U|U| = C^{-1} \, (w_2 - w_1) \, \bar{\ell} \, \bar{w}/R \qquad .(1.2\text{-}21)$$

and so:

$$S_w = r_1 r_2 \rho \left[\frac{\ell w}{CR}\right]^{1/2} (w_2 - w_1)^{3/2} \qquad .(1.2\text{-}22)$$

or, in terms of w':

$$\boxed{S_w = (r_1 r_2)^{1/4} \rho \left[\frac{\ell w}{CR}\right]^{1/2} (w')^{3/2}} \qquad .(1.2\text{-}23)$$

Here S_w represents, of course, a "sifting shear stress"; and it is interesting to observe that its magnitude depends upon w' while its direction is always away from the centre of curvature, regardless of the direction of the gradient of w.

Once again, use of the equation requires $r_1 r_2$, ℓ and w' to be estimated; but this is often possible.

2. The mathematical theory of the flow of two interpenetrating continua

2.1 Background

During recent years, it has been necessary to predict the motion of steam and water in the vessels and pipes of pressurised-water systems experiencing hypothetical loss-of-coolant accidents. This necessity has caused attention to be concentrated upon how the motions of two "interpenetrating continua" can be represented numerically and computed efficiently; and the problem of how to do so has been solved.

The literature of the subject is now too voluminous for summary; but the pioneering paper of Harlow and Amsden deserves special mention (Ref 6); and the present author has also made contributions (eg Ref 7).

As a consequence of these developments, it is possible to compute two-phase flow phenomena as readily as single-phase ones; and indeed the ability to do so is now embodied in a generally-available computer program (Ref 8,9), which has been used for all the numerical work alluded to below.

This has been fortunate; for it has permitted the development of the new two-fluid turbulence model to proceed without awaiting the construction of a new computational tool.

In the following section, a brief introduction to the theory will be provided.

2.2 Mathematical formulation of the two-fluid model

(a) Definitions

It is supposed that two fluids, 1 and 2, share occupancy of space. The proportions of time during which each can be expected to occupy a particular location are the volume fractions, r_1 and r_2, which obey the relation:

$$r_1 + r_2 = 1 \qquad .(2.2\text{-}1)$$

How the two fluids are distinguished from one another is arbitrary, however, at any location and time, all fragments belonging to fluid 1 are supposed to have the same values of temperature, mass fraction of component chemical species, velocity components, and fragment size. This assumption of fluid homogeneity is, of course, an idealisation, at variance with reality; but it may well lie closer to the truth than do the more conventional single-fluid models of turbulent flows.

All the above fluid properties will be represented by the symbol ϕ_i, where i takes the value 1 or 2 according to the fluid.

It will be supposed that both fluids have the same pressure at each point, not because the equation system is significantly simplified thereby, but because no mechanism entailing a difference of pressure has been hypothesised so far.

(b) Differential equations

Mass balances. There are two mass-conservation laws to express. They can both be written in the form:

$$\frac{\partial}{\partial t}(\rho_i r_i) + \text{div}(\rho_i r_i \langle V_i \rangle) = m_i \qquad .(2.2\text{-}2)$$

wherein t stands for time,

ρ_i stands for the density of fluid i,

$\langle V_i \rangle$ stands for its velocity vector, and

m_i stands for the rate of transfer of mass into fluid i from the other fluid, with which it is intermingled.

In equation (2.2-2), instantaneous values of ρ, r, $\langle V \rangle$ and m are in question. Because it is impractical, in numerical calculations, to compute the most rapid fluctuations of these quantities, time-average quantities are preferable, in which case the quantity operated upon by div must be augmented by D_i, an effective diffusion flux. The equation then becomes:

$$\frac{\partial}{\partial t}(\rho_i r_i) + \text{div}(\rho_i r_i V_i + D_i) = m_i \qquad .(2.2\text{-}3)$$

from which, however, overbars signifying time-averaging have been omitted so as to avoid needless clutter.

The $\partial/\partial t$ term represents the difference of outflow from inflow along the time dimension; div represents that difference along the space dimensions; and m_i can be regarded as the negative of the difference between outflow and inflow along an "interfluid dimension". If all these fluxes are denoted by G_i, and DIV is defined as a generalisation of div which makes reference to all their dimensions, equation (2.2-3) can be finally expressed as:

$$\mathrm{DIV}\ (G_l) = 0 \qquad .(2.2\text{-}4)$$

Balances of other quantities. The above nomenclature facilitates the compact writing of the differential equation governing the distribution of the fluid property ϕ_l. The equation becomes:

$$\mathrm{DIV}\ (\phi_l\ G_l) = \phi_l \qquad .(2.2\text{-}5)$$

wherein ϕ_l is the volumetric source of the entity ϕ in phase l, and ϕ_l is the value of ϕ associated with the flux which it multiplies.

Of especial importance are the momentum equations, in which ϕ_l stands for a velocity component. For these equations, the volumetric-source quantity ϕ_l consists primarily of pressure-gradient, body-force, interfluid-friction and wall-friction terms. Thus, for the velocity u_l, the equation becomes:

$$\mathrm{DIV}\ (u_l G_l) = -\ r_l\ (\frac{\partial p}{\partial x} + b_x\ \rho_l)$$

$$+\ f_{lj}\ (u_j - u_l) - f_{lw}\ u_l \qquad .(2.2\text{-}6)$$

Here:- x is the distance coordinate aligned with u_l.

p is the prssure.

b_x is the body-force component in the x-direction.

f_{lj} is the volumetric inter-fluid-friction coefficient.

u_l is the local velocity of the other component, and

f_{lw} is the volumetric coefficient of friction between fluid l and a nearby wall.

"Pressure-gradient" and other kinds of diffusion. Because of the association of the fluid density r_l with each of the G_l on the left-hand side of equation (2.2-6), and of its non-association with the pressure-gradient term on the right-hand side, the effect of the pressure gradient on the velocity components of the two fluids is inversely proportional to their densities. Pressure gradients cause the two fluids therefore to have different velocities, so that they move relatively to one another.

Relative motion between intermingled fluids can also result from turbulent diffusion, ie from the random movements of fluid fragments, the net magnitudes of which are usually proportional to the gradient of concentration (ie volume fraction). To distinguish the two phenomena, that caused by the pressure gradient will be here called "sifting", as explained above.

(c) **Auxiliary relations**

Interfluid friction. In order that the equation system should be "closed", ie rendered complete enough for solution, mathematical expressions must be devised for such inter-fluid transport quantities as m_i and f_{ij}. Since the present writer has discussed these questions elsewhere (Refs 10, 11, 12), it suffices here to indicate what formulae have been used in the computations which will be presented below.

The formulation used for f_{ij}, which appears in equation (2.2-6), is:

$$f_{ij} = c_f \, \ell^{-1} \rho_* \, |V_i - V_j| \qquad .(2.2\text{-}7)$$

In this formula, c_f is a dimensionless constant; ρ_* is the density of the lighter of the two fluids; ℓ^{-1} is the amount of fluid-fragment interface area per unit volume of space; and $|V_i - V_j|$ is the local time-average relative speed of the two fluids.

The area/volume quantity, ℓ^{-1}, is taken as being proportional to the volume-fraction product $r_1 r_2$, so that it vanishes when either fluid disappears.

In more advanced work, a differential equation governing ℓ^{-1} is formulated and solved: in the present work, however, ℓ^{-1} has been taken as proportional to the reciprocal of the lateral dimension of the flow.

Interfluid mass transfer. The inter-fluid mass-transfer process has been presumed to proceed only in one direction, namely from faster (fluid 1) to slower (fluid 2). This is not the only possible formulation; and more symmetrical ones are being investigated by the present writer in other studies.

The rate of mass transfer has been taken as directly proportional to the inter-fluid friction factor, f_{ij}, the proportionality factor being in excess of unity.

Interfluid heat transfer and diffusion of chemical species. Conductive heat transfer and diffusive transfer of chemical species between the two fluids have been taken as obeying a law similar to equation (2.2-7).

(d) Solution procedure

The solutions which are to be presented of the above system of equations have been obtained by straightforward application of the PHOENICS computer program (Refs 8, 9). There is therefore no need to do more here, by way of description of the solution procedure, than to recite the main features of what is embodied in PHOENICS. These are:-

* The differential equations are replaced by fully-conservative, fully-implicit finite-domain equations, valid for a staggered grid.

* Convection fluxes (ie all those represented by G_i in equation (2.2-4), are computed from "upwind" values of r_i, ρ_i and ϕ_i.

* An iterative solution procedure is adopted which results in the diminution to below pre-set limits of all the imbalances in all the finite-domain equations.

* A sufficient sample of the computations is repeated on successively finer grids (ie with smaller and smaller intervals of space and time) until the physically-significant results are found to be independent of the interval size.

A full account of the computational details will be provided in "PHOENICS Demonstration Reports" from the Imperial College Computational Fluid Dynamics Unit.

2.3 Some typical results from work in progress

(a) The plane wake

M R Malin has been studying, under the author's direction, the application of the two-fluid model of turbulence to various unbounded turbulent shear flows. These are dominated by mixing rather than sifting; but it is nevertheless important to establish that the model will fit the available experimental data.

One report of this work (Ref 13) relates to the plane turbulent wake, for which experimental data are available. Fig 2.3-1 is reproduced from the report, which must be referred to for full details.

Three temperature profiles are represented in the diagram. The higher represents the temperature of the hotter of the two fluids, which is also (although this is not shown) the slower-moving one; the lower represents the temperature distribution of the faster-moving fluid which is drawn into the wake from the surroundings; and the curve between them represents the average temperature, weighted by the computed volume fractions.

Comparison is possible between measurements and predictions in respect of the upper two curves; and it is rather satisfactory. The measurements relating to the upper curve have been made, it should be mentioned, by "conditioned sampling".

(b) <u>The axi-symmetrical jet</u>

Fig 2.3-2, from the same report, provides corresponding comparisons for the axi-symmetric jet. In this case measurements are available for the temperature profiles of the hotter (now faster-moving) fluid, of the cooler (slower) fluid, and of the average.

The agreement between predictions and experiments is rather poorer than before; but it is premature to attempt to ascribe a cause.

(c) <u>A combustion study</u>

The two-fluid model of turbulence is especially well-suited to the representation of combustion processes; for it permits explicit allowance to be made for the fact that flames consist of interspersions of hotter and colder fragments of gas and that the chemical reaction rate is especially active at the interfaces between fragments.

Figs 2.3-4,5,6,7,8 and 9 contain some results of a computation of what may happen when a shock wave, induced by the obstructing of flow in a pipe, travels through an interspersion of hot-gas fragments in a surrounding colder combustible gas. Fig 2.3-3 sketches the apparatus in question.

Although reference must be made to a more complete report (Ref 14) for a full explanation, it can perhaps be perceived, by inspection of the diagrams and their legends, that:

- a pressure wave of growing amplitude passes along the duct;
- the velocity difference between the hot and cold gas increases, resulting in a "sifting" of hot gas into the power-pressure region;
- it is in the region where the velocity difference is greatest that the inter-fluid mass transfer, and the reaction rate, are most vigorous.

(d) <u>Other studies</u>

The author's students and co-workers are pursuing further studies in this area. Attention is being paid to:

- turbulent flows near walls;
- turbulent diffusion flames under free-convection conditions;
- flows in hydrocyclones;
- flows induced by Rayleigh-Taylor instabilities.

Reports on this work are now beginning to appear, in the first instance in the Imperial College CFDU Report series.

The work proceeds slowly, because the aim is to develop a single two-fluid model which is capable of fitting a large number of diverse experimental data. However, progress is being made.

3. SOME OPEN QUESTIONS

The two-fluid model of turbulent flows is not yet ready for recommendation to practising engineers as a reliable predictive tool; for although plausible-seeming predictions have been made, and qualitative agreement with experiment has been achieved in respect of features which other models cannot predict at all, there are many questions still to be answered. Some of the questions to which the present author attaches importance will here be listed. They are:-

(a) What is the most suitable formulation for the inter-fluid mass-transfer law? It seems certain that early presumptions of one-way transfer (eg from slower to faster, or from low-vorticity to high-vorticity) cannot be universally satisfactory; but should the mass-transfer rate change sign when the volume fraction of a fluid attains 0.5, or some other number dependent upon density ratio or other dimensionless quantity?

(b) Will it suffice to employ a single length scale which is characteristic of both fluids? And, if so, how should the length scale be formulated to allow for the experimental observations that large fragments "swallow" small ones when the sifting process is dominant, and also in some shear flows (eg the plane mixing layer), whereas at very high shear rates the average fragment size diminishes.

(c) Although it seems wise to develop the model by use of the full set of two-fluid equations, as described in section 2.2, it would be very desirable to provide a model employing only the conventional single-fluid equations, with additional terms for the sifting flux. If this is to be done, what means should be used for calculating ϕ', etc. which appear in the expressions for this flux?

(d) What experimental data already exist, and what new experiments can usefully be devised, which will enable the as-yet-unknown constants and functions of the two-fluid model to be most easily and accurately deduced?

(e) What role can computational fluid dynamics play, when applied to fragment-formation, -distortion and -interaction processes, to elucidate the unknowns in the two-fluid-model equations? And how can such a program of study best be formulated - and funded?

(f) Will it indeed be possible to advance to the point of providing engineers with a valuable design tool without increasing the number of fluids considered beyond two? And, if not, what is the smallest number of fluids which will suffice?

4. CONCLUDING REMARKS

It is the present author's opinion that, despite the slowness of the progress and the number of still unanswered questions, the two-fluid idea can form a good basis for further advances.

As to slowness of progress, it must be observed that turbulence-model development at the single-fluid level has stagnated in recent years. An increasing number of experimental studies is showing the inadequacy of the k-ϵ model as formulated in the early 1970's; yet generally valid refinements and improvements have not been forthcoming; and the Reynolds-stress-modelling approach, in the writer's opinion, increases the difficulties of making predictions without significantly improving their quality.

Because ability to perform two-fluid calculations has, until recently, been confined to a few persons and institutions heavily engaged in other work, the man-power devoted to pursuing two-fluid-modelling has remained small. Now however, it is worth mentioning, the free availability of PHOENICS to academic institutions makes it possible for many more individuals and teams to join in the exploration of this new branch of turbulence research, should they wish to do so.

5. ACKNOWLEDGEMENTS

The author's thanks are due to Mrs F M Oliver for the speedy and excellent typing of the manuscript. IC CFDU is supported by the Science and Engineering Research Council, the Health and Safety Executive, the US Army (European Office), the UK Atomic Weapons Research Establishment and the UK Atomic Energy Agency.

6. REFERENCES

1. Reynolds O (1874)
'On the extent and action of the heating surface for steam boilers'
Proc Manchester Lit Phil Soc, Vol 8.

2. Prandtl L (1925)
'Bericht uber Untersuchungen zur ausgebildeten Turbulenz'
Z angew Math Mech (ZAMM) Vol 5, No 2, pp 136-139.

3. Bray KNC and Libby PA (1981)
'Countergradient diffusion in pre-mixed turbulent flames'
AIAA J Vol 19, p 205.

4. Moss JB (1980)
'Simultaneous measurements of concentration and velocity in an open pre-mixed flame'
Comb Sci Tech, Vol 22, pp 115-129.

5. Shepherd I G and Moss J B (1981)
'Measurements of conditioned velocities in a turbulent pre-mixed flame'
AIAA 19th Aerospace Sciences Meeting, St Louis, Paper No 81-0181.

6. Harlow FH and Amsden AA (1975)
'Numerical calculation of multi-phase fluid flow'
J Comp Phys, 17 (1), pp 19-52.

7. Spalding DB (1977)
'The calculation of free-convection phenomena in gas-liquid mixtures'
In Turbulent Buoyant Convection, Eds N Afgan and DB Spalding, pp 569-586, Washington.

8. Spalding DB (1981)
'A general-purpose computer program for multi-dimensional one- and two-phase flow'
J Mathematics and Computers in Simulation, North Holland, Vol XXIII, pp 267-276.

9. Gunton MC, Rosten HI, Spalding DB and Tatchell DG (1983)
PHOENICS Instruction Manual
CHAM Limited, Technical Report TR75.

10. Spalding DB (1982)
'Chemical reaction in turbulent fluids'
J PhysicoChemical Hydrodynamics, Vol 4, No 4, pp 323-336,
IC CFDU Report CFD/82/8.

11. Spalding DB (1983)
'Towards a two-fluid model of turbulent combustion in gases with special reference to the spark-ignition engine'
IMechE Conf, Oxford, April 1983
In Combustion in Engineering, Vol 1, Paper No C53/83, pp 135-142.

12. Spalding DB (1983)
'Turbulence modelling: A state-of-the-art review'
IC CFDU Report CFD/83/3.

13. Malin MR (1984)
'Prediction of the temperature characteristics in intermittent free turbulent shear layers'
IC CFDU Report CFD/84/1.

14. Spalding DB (1984)
'The two-fluid model of turbulence applied to combustion phenomena'
AIAA 84-0476
AIAA 22nd Aerospace Sciences Meeting, Reno, Nevada.

FIGURES

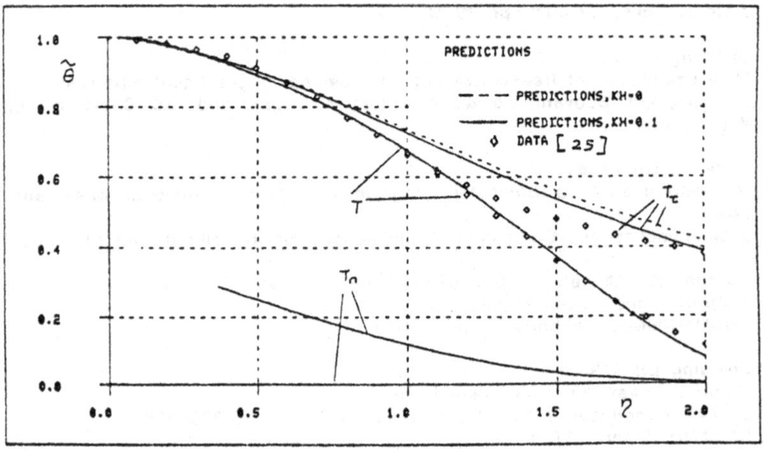

Figure 2.3-1. Plane wake: temperature similarity profiles.

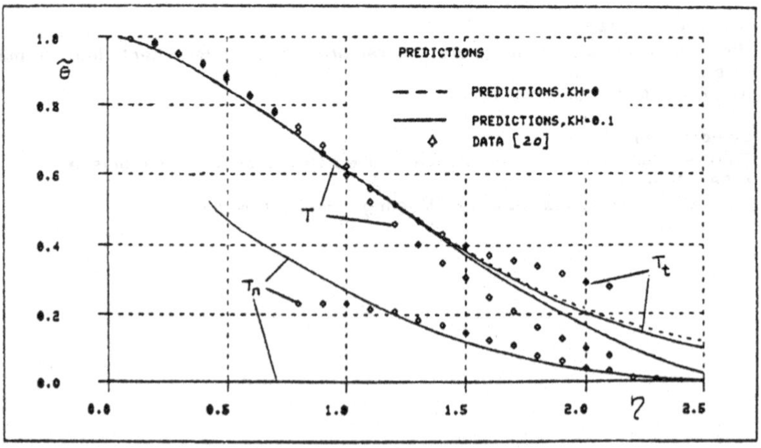

Figure 2.3-2. Axisymmetric jet: temperature similarity profiles.

(a)

(b)

(c)

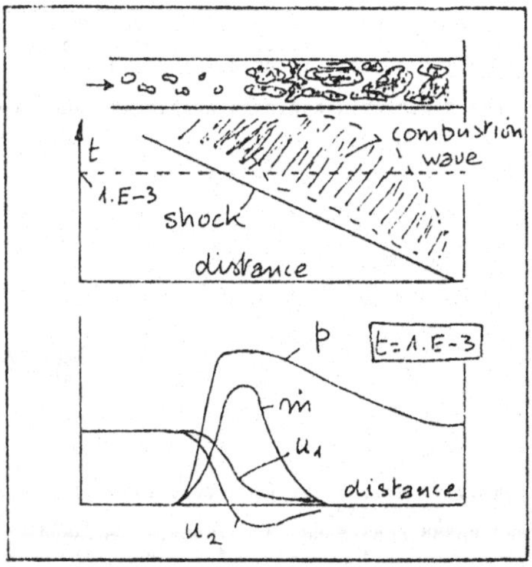

Figure 2.3-3.

(a) Sketch of a tube containing an interspersion of hot combustion products in cooler pre-mixed combustible gas. Initially both gases flow at equal velocity from left to right; then a valve at the right-hand end closes, sending a pressure wave along the duct. Combustion is thereby intensified.

(b) Sketch of the pressure and combustion wave trajectories on a distance-time plane.

(c) Sketch of instantaneous profiles of pressure p, interfluid mass-transfer rate $\dot{m}$, cold-gas velcity u_1, and hot-gas velocity u_2, 1 ms after closing of the valve.

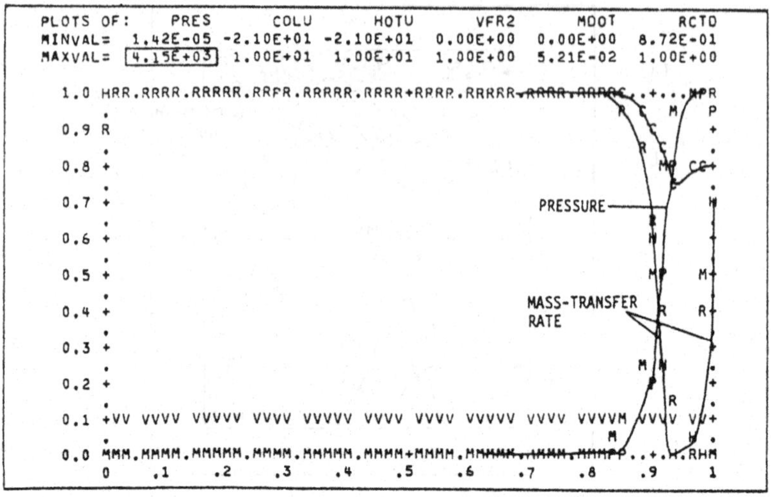

Figure 2.3-4. The flame of Fig 2.3-3, 0.2 milliseconds after the closure of the right-hand end.

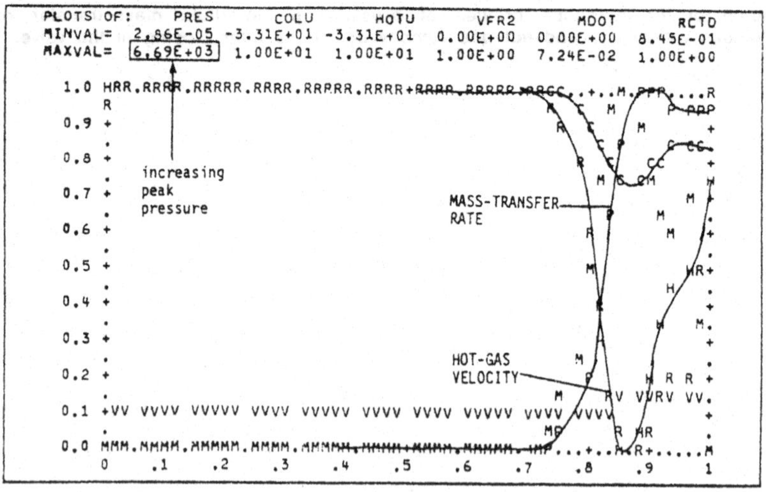

Figure 2.3-5. The flame of Fig 2.3-3, 0.4 milliseconds after closure of the right-hand end.

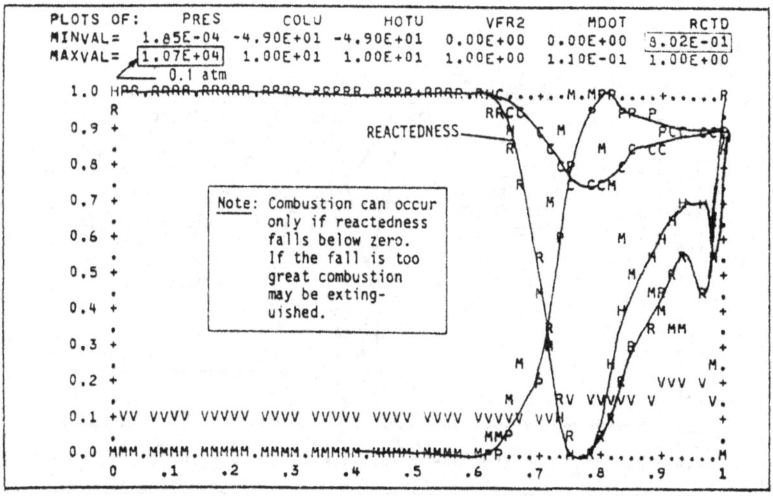

Figure 2.3-6. The flame of Fig 2.3-3, 0.6 milliseconds after closure of the right-hand end.

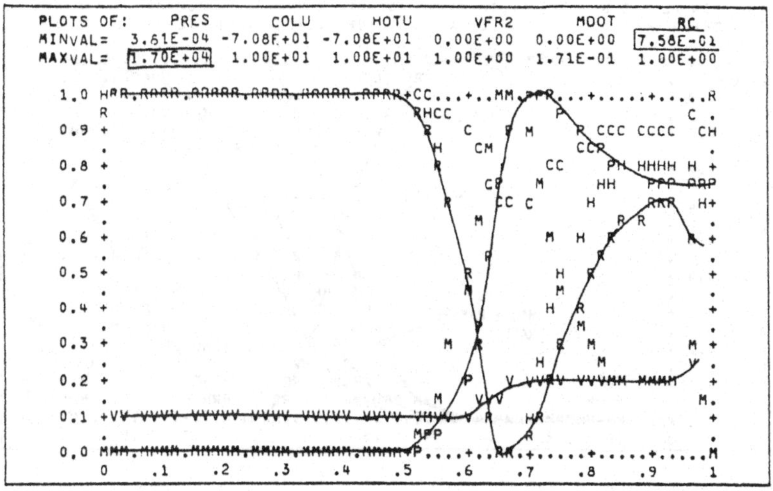

Figure 2.3-7. The flame of Fig 2.3-3, 0.8 milliseconds after closure of the right-hand end.

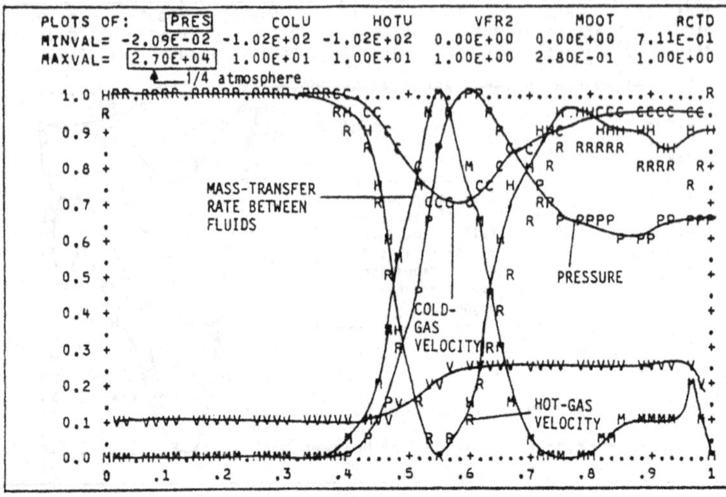

Figure 2.3-8. Distributions of pressure, velocities, the mass-transfer rate and other quantities, 1 millisecond after right-hand-end closure.

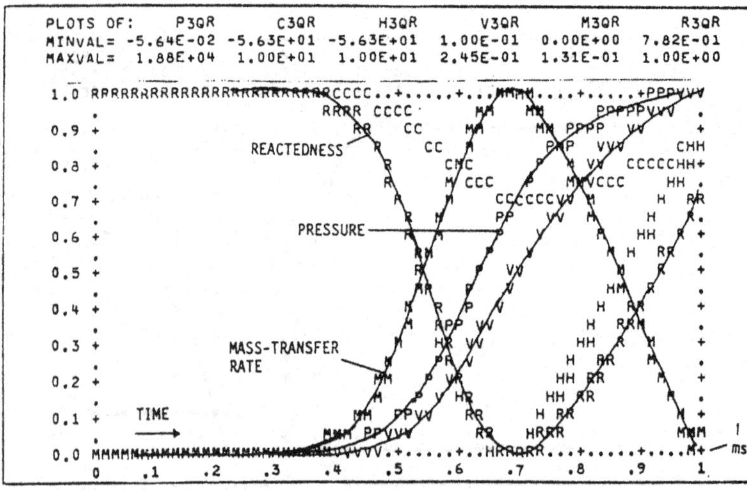

Figure 2.3-9. Variations with time of pressure, reactedness, mass-transfer rate, volume fraction of hotter fluid (V), velocities of hotter (H) and colder (C) fluids, all for a location 3/4 of the distance along the pipe, ie 25 cm from the closed end.

CHAPTER XIII. Chaos and Coherent Structures in Fluid Flows

E.A. Spiegel

1. Introduction

The first explicit studies of chaos in fluid dynamics belong to the theory of thermal convection. Crude approximations to the fluid equations in the contexts of meteorology (Lorenz, 1963) and astrophysics (Moore & Spiegel, 1966) led to model systems with complicated temporal dynamics. Lorenz gave a prescient discussion of his third-order system of ordinary differential equations that has guided many workers in the theory of chaos. Moore and I were more backward-looking; we tried to interpret the erratic solutions of our model equation, as in Hopf's perception of turbulence, as a wandering among many unstable periodic solutions. Of course, as fluid dynamics, all this work was unsatisfactory since unacceptable approximations to the fluid equations were used.

But now we have convincing evidence that chaos is a real fluid phenomenon. Maurer and Libchaber (1979) found chaos in convection, and this result has been confirmed and amplified in other experimental work (see Gollub, 1983, for a summary). Numerical solutions evincing chaos were also found, beginning with the work of McLaughlin and Orszag (1982) on convection. Here I sketch analytical arguments for the appearance of chaos in solutions of the fluid equations (Arnéodo, Coullet and Spiegel 1982, 1985). These arguments rest on the occurence of multiple bifurcations, hence on an appeal to narrow choices of parameter values. My aim in going into this is to help lay to rest any lingering doubts that chaotic solutions of the fluid equations do exist.

A basic ingredient of the approach that I shall describe is the supposition that the amplitudes of the flow and of its associated disturbances are small. This leads to amplitude expansions that are permissible near to the onset of instability. A second important ingredient that I shall call on is, as already mentioned, the nearly simultaneous onset of instability in several (three or more) modes. In finite systems, this joint occurence may not be generic, but in extended systems, where continuous bands of modal wavenumbers occur, it is commonplace. Hence, I shall then turn to the study of the onset of instability in a continuous band of wavenumbers and show how that leads to models of chaotic coherent structures.

Whether such studies will ultimately shed light on turbulence remains a question for debate. I shall comment on this matter at the end. But I can state my viewpoint in brief: Ordinary chaos is not turbulence, but turbulence is chaotic. The more we learn about chaos, the better we will understand turbulence.

Inevitably, this statement will trouble people. For

just as there is no generally accepted definition of turbulence, there seems to be no agreement on what the word chaos should mean. There are even those who use turbulence and chaos interchangeably. Though this usage has understandably caused consternation among fluid dynamicists, it is hard to argue about it since neither word is technically defined. However, it will be useful to mention some of the symptoms of the chaos syndrome for the purpose of the discussion below. I shall be exceedingly brief since there is by now a large and reasonably accessible literature on chaos (see the recent reprint collection and summary of Hao for an extensive bibliography, 1984).

Consider a pendulum with very little friction. Give it a good start and watch it go spinning around. How many times will it actually go around before coming to rest? Is this number of times it goes around odd (heads) or even (tails)? I claim that if I know the size of your computer, I can choose the damping (for fixed initial energy) low enough so that you will not be able to predict whether the outcome is heads or tails. (A similar uncertainty can be imposed on the experimentalist.) This sensitivity to initial conditions, as it is called, is an ingredient of chaos. But the pendulum is not chaotic since it is decaying to an equilibrium position.

To make the system chaotic, we can keep it active by forcing it externally or parametrically in a time-dependent manner. Then, if it also has the sensitivity I mentioned, we can expect chaos to occur for some parameter values. This is temporal chaos; the frequency spectrum will have a continous component and there may be temporal intermittency. Of greater interest for the study of turbulence are spatial chaos and spatio-temporal chaos. I shall be illustrate these in §4 where I shall use some of the catchwords of the subject as being almost self-explanatory.

An aspect of chaotic sytems is that their motion does not fill the whole phase space of state variables, but is confined to subspaces of zero volume called attractors. An equilibrium point or a limit cycle or an invariant torus may be an attractor. Or the attractor may be more complicated and have a fractal dimension (Mandelbrot, 1983). Such complicated attractors are often called strange attractors when the numerical evidence suggests it. There is however a more practical test in the study of the way neighboring trajectories separate; this involves the notion of Liapunov instability (LaSalle & Lefschetz, 1961), which I think is an important notion for turbulence theory.

2. Background on Linear Theory

The simplest route to experimental chaos taken so far seems to be by way of Rayleigh-Bénard convection in boxes whose aspect ratios are of order unity (see Gollub 1983). At the onset, a steady convective pattern emerges. As the stability parameter is increased beyond critical, this pattern itself loses stability and (in certain cases) a time-periodic flow develops. With further increase of the degree of instability, the periodic flow is unstable and (sometimes) a second oscillation appears. As the instability is further increased, this fluid may go into a chaotic state. Though all of the successive modes of instability have not been reliably identified, it does seem acceptable to speak of this flowering of the chaos as a succession of instabilities. When chaos has appeared in this case, all the modes have large amplitudes and the use of amplitude expansions is not justified.

It seems that the development of chaos in unstable convection is an outcome of the interaction of several modes of instability. If these modes could be brought to the onset of instability at the same parameter value, then we would be in a position to use amplitude expansions to study their interaction. This means of studying secondary instablities (or bifurcations) by playing on parameters to get a related multiple instability (or bifurcation) was suggested by Bauer, Keller and Reiss (1975). Arnold has written (e.g. 1983) that, in multiple bifurcations that are controlled by several parameters, we can find a variety of normally highly nonlinear processes taking place very near to the onset of the multiple instability. Chaos is one of these processes.

To discuss what happens near the onset of instability, it is often useful to use the language and the results of linear theory, even though our emphasis is on nonlinear phenomena. Rather than present an abstract summary of such results, I shall sketch them for an explicit, but simple, example from convection theory. The reason that convection keeps appearing as a fluid paradigmm to illustrate chaos is that it has a simple linear theory and it offers a number of physically realizable examples of competing instabilities (Spiegel, 1972). The problem is especially simple when we use the idealized boundary conditions of Rayleigh (Chandrasekhar, 1981). I adopt here all these simplifications and add that the motion is two-dimensional and that the Boussinesq approximation holds. As competing dynamical mechanisms let us take thermal stratification, saline stratification and rotation. The equations for the competition are, in natural units, (Arnéodo, Coullet & Spiegel, 1985):

$$\partial_t \Delta\psi = \sigma\Delta^2\psi - \sigma R\partial_x\theta + \sigma\tau S\partial_x\Sigma + \sigma^2 T\partial_z T + J(\psi,\Delta\psi) \qquad (2.1)$$

$$\partial_t \theta = \Delta\theta - \partial_x \psi + J(\psi,\theta) \tag{2.2}$$

$$\partial_t \Sigma = \tau\Delta\Sigma + \partial_x \psi + J(\Psi,\Sigma) \tag{2.3}$$

$$\partial_t T = \sigma\Delta T - \partial_z \psi + J(\psi,T), \tag{2.4}$$

where $J(f,g) = \partial_x f \partial_z g - \partial_z f \partial_x g$, $\Delta = \partial_x^2 + \partial_z^2$, Ψ is the stream function, θ and Σ are the deviations of temperature and salinity from their static values, and T is the scaled y-component of velocity. In this two-dimensional motion, nothing depends on y. To define the parameters we let $(\Delta\rho/\rho)_T$ and $(\Delta\rho/\rho)_S$ be, respectively, the density contrasts forced by the imposed temperature and salt gradients in the vertical, d be the layer thickness, ν the kinematic viscosity and κ_T and κ_S be the diffusivities of heat and salt. Then we have the Rayleigh numbers

$$R = \frac{gd^3}{\kappa_T \nu}(\Delta\rho/\rho)_T, \quad S = \frac{gd^3}{\kappa_S \nu}(\Delta\rho/\rho)_S \tag{2.5}$$

the Taylor number

$$T = (2\Omega d^2/\nu)^2 \tag{2.6}$$

where Ω is the rotation rate about the z axis, and

$$\sigma = \nu/\kappa_T, \quad \tau = \kappa_S/\kappa_T. \tag{2.7}$$

The boundary conditions adopted here are

$$\Psi, \Sigma, \Theta, \partial_z T, \partial_z^2 \Psi = 0 \text{ on } z = 0,1 \tag{2.7}$$

and

$$\Psi, \partial_x \Sigma, \partial_x \Theta, T, \partial_x^2 \Psi = 0 \text{ on } x = 0, \pi/a . \tag{2.8}$$

where a is an aspect ratio.

Because of these boundary conditions, solutions of the linearized equations in the form of trigonometric functions of x and z times exp(st) can be found (Chandrasekhar, 1961). Let us introduce the solution vector

$$U(x,z,t) = \begin{Vmatrix} \Psi \\ \Theta \\ \Sigma \\ T \end{Vmatrix} \tag{2.9}$$

The linear theory admits solutions of the form

$$U = U_{mn} * \Xi_{mn} e^{st} \tag{2.10}$$

where $m = 0,1,2,\ldots$ and $n = 0,1,2,\ldots$, U_{mn} is a constant four-component vector, s is a characteristic value of linear theory,

$$\Xi_{mn} = \begin{Vmatrix} \sin(max) & \sin(n\pi z) \\ \cos(max) & \sin(n\pi z) \\ \cos(max) & \sin(n\pi z) \\ \sin(max) & \cos(n\pi z) \end{Vmatrix} \tag{2.11}$$

and the operation * is defined by

$$\begin{Vmatrix} a \\ b \\ c \\ d \end{Vmatrix} * \begin{Vmatrix} \alpha \\ \beta \\ \gamma \\ \delta \end{Vmatrix} = \begin{Vmatrix} a\alpha \\ b\beta \\ c\gamma \\ d\delta \end{Vmatrix} \tag{2.12}$$

For the most dangerous modes of linear theory, m and n are unity and s is a root of

$$s^4 + P_3 s^3 + P_2 s^2 + P_1 s + P_0 = 0, \tag{2.13}$$

where the P_i are functions of the parameters given explicitly in Appendix 1; it can be seen there that $P_3 > 0$). The condition for a marginally stable solution is that $s = i\omega$ with real ω. We can achieve this in two ways: eiher we have direct instability (monotonic growth) with (a) $\omega = 0$ and $P_0 = 0$, or we get overstability (growing oscillations) with (b) $\omega^2 = P_1/P_3$ and $P_1^2 - P_1P_2P_3 + P_0P_3^2 = 0$ with $P_1 \geq 0$. A double instability can be produced by fulfilling (a) and (b) at once. This can be done in two ways: either (i) $P_0 = P_1 = 0$ with $\omega = 0$ or (ii) $P_0 = 0$, $P_1 = P_2P_3$, with $\omega^2 \equiv P_2 \geq 0$. In condition (i), (2.13) has a double zero and the two normal modes that corresponded to (a) and (b) coalesce. In case (ii) the system has simultaneously direct instability and overstability. Moreover, if we set $P_2 = 0$ we get (i) and (ii) at once. This takes place when

P_0, P_1, P_2 = 0. When these three critical conditions are simultaneously fulfilled, we have the onset of triple instability, so to say.

If we are at the tricritical condition, (2.13) becomes

$$s^3(s + P_3) = 0 . \qquad (2.14)$$

To see the slow time dependence implied by linear theory in the neighborhood of a tricritical point in parameter space, we divide (2.13) by $s + P_3$ to get

$$s^3 + P_2 s^2 + (P_1 - P_2 P_3)s + [P_0 - P_3(P_1 - P_2 P_3)]/(s+P_3) = 0. \qquad (2.15)$$

We develop the denominator and, on keeping terms up to cubic, we get the critical condition (Coullet and Spiegel, 1983),

$$s^3 + \mu_2 s^2 + \mu_1 s + \mu_0 = 0 , \qquad (2.16)$$

where

$$\mu_0 = P_0/P_3, \; \mu_1 = P_1/P_3 - P_0/P_3^2 ,$$
$$\mu_2 = P_2/P_3 - P_1/P_3{}^2 + P_0/P_3^3 . \qquad (2.17)$$

The roots give all the characteristic values that have small real parts in the parameter range of interest.

In the space of the six parameters, R,S,T,σ,τ,a, the tricritical conditions become $\mu \equiv (\mu_0,\mu_1,\mu_2) = 0$. These conditions determine values R_o,S_o,T_o as functions of σ,τ,a for which there is a triple zero in s. (Those parameter functions are given in Appendix 1).

An important feature of this triple instability is that, at the triple point, there is only one eigenvector for which $s = 0$. There are however three linearly independent solutions of the linear problem. The other two depend on time like t and t^2. So at tricriticality we have the linear solution

$$\mathbf{U} = (\mathbf{U}_{11} + t\mathbf{V}_{11} + t^2\mathbf{W}_{11})*\Xi_{11} + \text{stable modes} \qquad (2.11)$$

where $\mathbf{U}_{11},\mathbf{V}_{11}$ and $\mathbf{W}_{11}$ are constant vectors and the stable modes are those for which $Re(s) < 0$.

These results are not peculiar to convection theory or even to fluid dynamics. Hence the conclusions to be drawn concerning the occurence of chaos apply equally well to all those problems sharing common stability properties. Boundary conditions and other details which do not affect the qualitative structure of the linear theory may make the explicit calculations more or less troublesome, but they do not modify the qualitative conclusions to be drawn.

3. The Amplitude Equation For A Triple Instability

In the standard case of a single direct instability, the Landau equation (Drazin and Reid, 1981) governs the amplitude of the motion near to onset. Landau derived the form of this equation by qualitative arguments that reveal the origin of the various terms in it. His equation may be written

$$\dot{A} + \mu A = kA^3. \tag{3.1}$$

Here $-\mu$ is the growth rate of the mode that becomes unstable near criticality and k is a coefficient whose calculation has a considerable history (e.g. Drazin and Reid, 1984). In the Boussinesq approximation for convection because of the symmetry of the equations, there is no quadratic term. When k turns out to be negative, a steady nonlinear solution is achieved for negative μ.

If we linearize (3.1), we find that A varies like exp(st) and that the critical polynomial $s+\mu$ must vanish. So we recover the linear result for a single instability. We may anticipate a similar situation for the triple instability of **§2**: an amplitude equation that gives the linear theory near to multicriticality, with nonlinear corrections. With (2.16) in mind we expect the amplitude equation to be

$$\dddot{A} + \mu_2 \ddot{A} + \mu_1 \dot{A} + \mu_0 A = \text{nonlinear terms in } A, \dot{A}, \ddot{A}. \quad (3.2)$$

Which nonlinear terms appear in (3.2) maybe decided by other arguments. For example, the symmetry arguments that rationalize the absence of a quadratic term from (3.1) work here, so we expect the development on the right of this third-order equation to begin with cubic terms. Basically, since we are near to marginality, the evolution is slow and the time derivatives of A have magnitudes less than that of A. Careful application of this argument leads to an asymptotic amplitude equation (Arnéodo, Coullet & Spiegel 1985). For example, if we choose to maintain the linear part of the equation intact, we could study the case of small $|\mu_2|$ by having $\partial_t = O(\mu_2)$, $\mu_1 = O(\mu_2^2)$, and $\mu_0 = O(\mu_2^3)$. In the case where $\mu_2 > 0$ the asymptotic amplitude equation is

$$\dddot{A} + \ddot{A} + \mu\dot{A} + \nu A = kA^3. \quad (3.3)$$

Though this is a special case of the equation that Moore and I adumbrated, it is not the special case we focused on. It was Arnéodo, Coullet and Tresser (1981) who identified (3.3) as possibly the simplest nonlinear equation with (2.16) as its linear theory. In fact, (3.3) is derivable from the fluid equations by standard asymptotic procedures, and I shall indicate how this may done for the problem posed by equations (2.1)-(2.4).

The simplest way to derive an amplitude equation is to scale the time and amplitude by small parameters and to perform expansions. A systematic account of this method, called amplitude expansions, is given by Kogelman and Keller (1967). To employ this scheme we introduce new dependent variables, as in

$$\psi = \varepsilon^{3/2}\tilde{\psi} \tag{3.4}$$

and scale the time to

$$\hat{t} = \varepsilon t \tag{3.5}$$

where ε is a small parameter. Now we develop both the dependent variables and the stability parameters in ε:

$$\begin{aligned} \tilde{\psi} &= \psi_0 + \varepsilon^{1/2}\psi_1 + \ldots, \\ R &= R_0 + \varepsilon^{1/2}R_1 + \ldots, \end{aligned} \tag{3.6}$$

with similar developments for $\tilde{\theta}$, S and so on. There follows a long but soothing exercise in singular perturbation theory. We find as the first approximation for $\mathbf{U}$, $A(\hat{t})\mathbf{U}_{11}$, where $\mathbf{U}_{11}$ is a normal mode defined in §2. In next order, we get as a requirement for solvability, the marginality condition $P_0 = 0$.

In the order after that, we are offered a choice: either A satisfies a Landau equation or we must impose a second parameter condition that puts the system at the edge of a double instability. If we choose the latter option, we get, in the next order, a choice between an amplitude equation that is second order in time, or another parameter condition. If we again choose the latter, we finally come to the point where the choice has run out and we obtain (3.3), with an explicit formula for k (Arnéodo, Coullet & Spiegel, 1985). We here assume that A has been scaled so that $k = \pm 1$, and exclude the case when the unscaled k is near zero (that occurs near $a = \pi$).

Since (3.3) is an asymptotic result, it holds in the limit $\epsilon \to 0$, that is, arbitrarily close to the onset of instability. Hence, it should describe solutions in that limit reasonably well. I shall not try to detail such solutions here, as this equation and ones with similar structures have been extensively (but not exhaustively) discussed (see the list of references in Arnéodo, Coullet & Spiegel, 1985).

Suppose we fix on a choice of k and μ and let ν vary monotonically. In a typical sequence of solutions, as ν varies, we may see a stable equilbrium point lose stability to a limit cycle at a critical value. (Transient behavior is not described here.) With further variation of ν, the limit cycle may become unstable to another one with twice the period. Several such period-doubling bifurcations may occur, possibly an infinite number of them, as in certain maps of an interval into itself (see Collet and Eckmann, 1981). Then chaos may occur when a certain value of ν is surpassed. In Fig. 1, I illustrate such chaos with a numerical solution of (3.3) for $\mu = 3.5$, $\nu = -4.15$ and $k = -1$. In the upper part of the Figure, the variation in time of A in a chaotic region of parameter space is indicated. In the lower half the projection of the orbit in phase space (A, $B = \dot{A}$, $C = \ddot{A}$) onto the A-B plane is shown. This is a fairly typical, mild chaos.

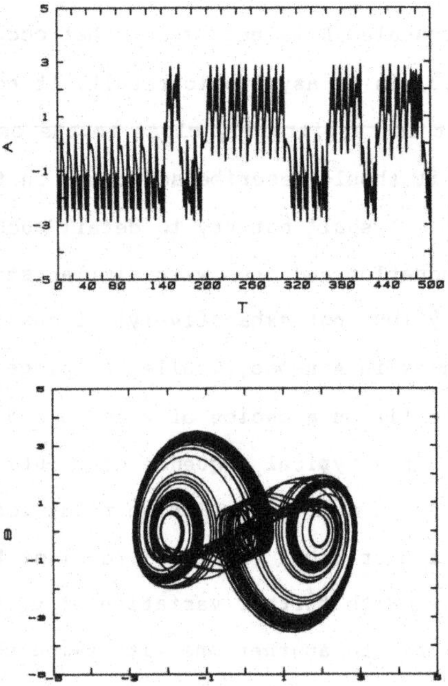

Figure 1

Naturally, there are a number of issues that might be raised, were this not a fairly well-troden path to chaos already. But, in fact the most solid argument that that (3.3) gives chaotic solutions rests on some theorems of Shil'nikov (1965,1968; Arñeodo, Coullet & Tresser, 1982). Shil'nikov's ideas can be used to connect the motion in (A,B,C) space to simple mappings that for some limits are one-dimensional. On this account, we can be confident that (3.3) gives chaos. I shall not go into those arguments here (but see Arnéodo, Coullet, Spiegel and Tresser, 1985); let me just mention that they rest on the idea of the Poincaré map, a useful device for thinking about the solutions of (3.3). Fig. 2, is such a

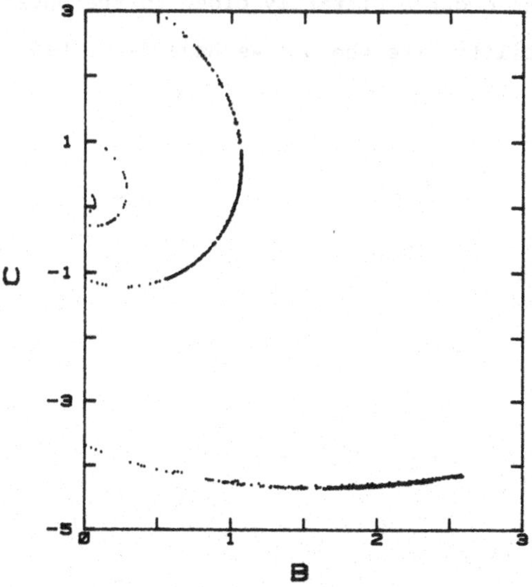

Figure 2

map. It shows the set of points at which orbits like that projected in Fig. 1 cross the plane $A=0$ with $B>0$. The parameters in this example are those of Fig. 1 and have been selected so that the structures seen in this map are very nearly one-dimensional (barring two transient points that I did not filter out). Hence it is possible to reduce the results of Fig. 2 reasonably accurately to a one-dimensional map. The ideas of Shil'nikov may be used to derive analytic forms for such maps that are in good qualitative agreement with the numerically determined ones (Arnéodo, Coullet & Spiegel, 1985). Such maps leave little doubt that chaotic solutions can be found for suitable choices of the parameters in (3.3), hence that chaos can occur arbitrarily close to the onset of a triple instability like the one we have described.

4. Chaotic Coherent Structures

The calculations described in §3 show that chaos occurs in fluid flows. The triple instability postulated for the purpose is not a very typical event, but the argument does indicate that if we have many modes competing for dominance, we may hope to encounter chaotic flows. Such competition is commonplace in large systems where the spectrum of growth rates of the linear theory is very nearly continuous. The onset of motion in such cases is described by evolution equations, that is, partial differential equations that are generalizations of amplitude equations like (3.1) and (3.3). Again, the deduction of evolution equations has been most extensively practised in convection theory, where flows in boxes of large horizontal extent provide extensions of the discrete theory.

For two-dimensional motions, it is possible to find a simple extension of (3.1) by the use of asymptotic methods familiar in nonlinear wave theory (Newell & Whitehead, 1969; Segel, 1969). Basically, the roll solutions of the linear theory are modulated on a scale much larger than their widths. This effect is included by the introduction of a slow space variable along with the slow time. The evolution equation for the case of a simple direct instability is the so-called Ginzburg-Landau equation. When the basic instability is overstability, waves may propagate and we get a pair of such equations but with complex coefficients (Newell, 1974; Kuramoto, 1978) one for each direction of propagation. The amplitudes described by these equations are slowly varying on the length and time scales of the basic or carrier mode.

If we choose initial conditions with waves going only one way and parameters such that this is a stable situation, we have the equation

$$\partial_T A = \alpha A + \beta \partial_X^2 A + \gamma |A|^2 A \tag{4.1}$$

where α, β and γ are (generally) complex constant coefficients and X and T are slow space and time variables in an inertial frame. Perhaps the simplest case in which all this can be worked out explicitly for a fluid is the example of thermohaline convection (Bretherton, 1981), but such modulational instability is familiar in many fields (Whitham, 1974) and may be given a simple interpretation as the analogue of Rayleigh-Taylor instability (Bretherton & Spiegel, 1983).

Numerical solutions of (4.1) studied on a box of size L become chaotic when L exceeds a critical value (Moon, Huerre & Redekopp, 1982; Kuramoto, 1978). That is not entirely surprising since Kurumoto & Tsuzuki (1976) have shown that in a kind of W.K.B. limit (4.1) reduces to the (Kurumoto-Sivashinsky) equation

$$\partial_T u + u\partial_X u = -\nu\partial_X^2 u - \partial_X^4 u \tag{4.2}$$

where $u = \partial_X\{[A^*\partial_X A - A\partial_X A^*]/(2iA^*A)\}$. If we seek solutions in the form $u = F(X-cT)$ we get an O.D.E. which can be immediately integrated once. The result looks very like (3.3) and chaos should not be far away. Here it is the appearance of coherent structures in the midst of this chaos that I want to describe (following Bretherton & Spiegel, 1983; see also Spiegel, 1981).

For definiteness, let us take the case of thermohaline convection when (with suitable scaling) we obtain $\alpha = 1$, $\beta = 1+i\mu$ and $\gamma = i$. For large μ we then find soliton solutions of (4.1):

$$A = (2\mu)^{\frac{1}{2}} a\,\mathrm{sech}[a(X-X_0)]e^{i\theta} + O(\mu^{-1}), \tag{4.3}$$

where

$$X_0(T) = 2\mu\int^T U(\tau)d\tau, \quad \theta(X,T) = U(X-X_0) + \mu\int^T (U^2+\alpha^2)d\tau + O(1). \tag{4.4}$$

Though in the limit $\mu \to \infty$, a is an arbitrary constant, we may represent the solutions for finite μ with (4.3) by allowing a to depend weakly on time. As in the usual method of averaging, we introduce (4.3) into (4.1) and find

$$\begin{aligned} \dot{a} &= 2a(1-U^2) - \tfrac{2}{3}a^3, \\ \dot{U} &= -\tfrac{4}{3}a^2 U. \end{aligned} \tag{4.5}$$

Together (4.3) and (4.4) desribe an object that is a soliton on a fast time but whose parameters are governed by a dynamical system. It is possible to contrive more complicated examples, but this one is the simplest in that (4.5) may be reduced to a first order system that evolves to a simple equilibrium. I call the objects described this way coherent structures.

One way to look at the behavior of a coherent structure is to regard its form as known and to follow it in space-time. Its position is given by

$$\dot{X} = U. \tag{4.6}$$

Then we can calculate the worldline shown in Fig. 3. We can also compare this result with direct numerical integrations of (4.1). Fig. 4 (after Bretherton & Spiegel, 1983) shows the

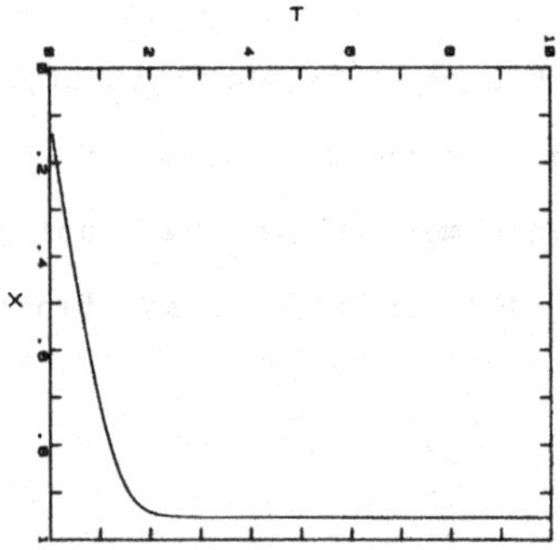

Figure 4

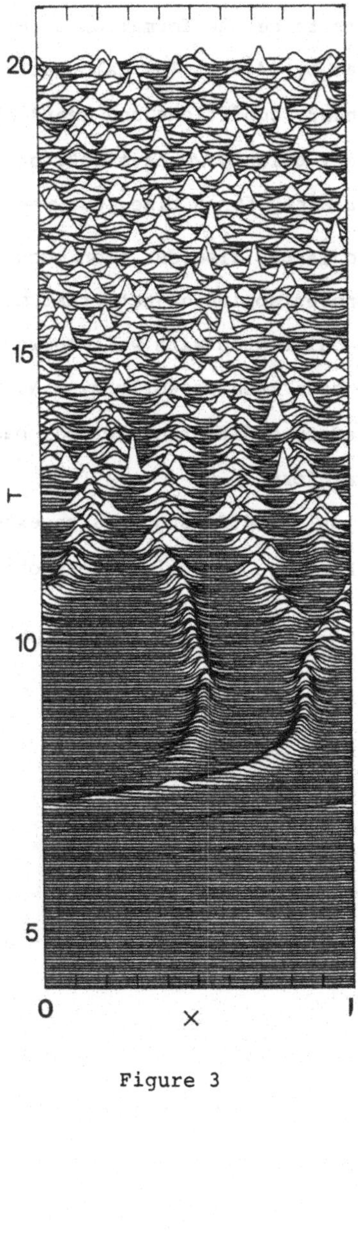

Figure 3

result of such an integration, starting from low-amplitude white noise. A coherent structure forms and moves, in good agreement with the results of (4.5)-(4.6). But there is some amplitude variation and, associated with large swings in amplitude, there seems to be the formation of new coherent structures not predicted by (4.5). The calculation shown was done with periodic boundary conditions and keeps all the objects around, so that very quickly we have four of them. These form a somewhat regular array for a time, but when the lattice vibrations get too large, disorder soon occurs. The box fills with about six nonoverlapping objects that move irregularly. This behavior was calculated with well-known numerical methods. It has also been seen recently by C.N. Corfield (priv. comm.) who computed by a different method, based on putting (4.1) into integral form.

The derivation of (4.5) suggests an extension of it to the case seen in Fig. 4. We would have N objects given by a set of equations for their parameters:

$$\dot{a}_i = f_i(\mathbf{a},\mathbf{U},\mathbf{X})$$

$$\dot{U}_i = g_i(\mathbf{a},\mathbf{U},\mathbf{X}) \qquad (4.7)$$

$$\dot{X}_i = U_i$$

where $i=1,2,\dots,N$. The evolution of the N objects would then be described by these equations, including their interactions as modeled in the functins **f**,**g**. In principle, we might discover these functions by the same kind of averaging method that led to (4.5), starting from an N-soliton solution for $\mu\to\infty$. We may also learn about them from normal form theory. If such progress becomes possible, where will it have gotten us?

In complicated fluid-dynamical situations, attempts to simplify matters by modal expansions are often made. The generalization of this type of Fourier analyss to more suitable basis functions or modes is often proposed in the hope of getting higher accuracy for a given order of truncation. The ultimate achievement in this direction is the inverse scattering transform (e.g. Ablowitz and Segur, 1981), which is an analogue of Fourier transformation for nonlinear problems. The IST has so far been possible for conservative systems. Here I am imagining a possible extension to dissipative coherent structures. For large systems, N will be large and the solutions of (4.7) may well be chaotic. We still do not have a theory for this situation if the dimension of the attractor is very large. Nor is it clear how much of the activity seen in Fig. 4 is described by this superposition of structures. Yet I feel that this model does make it plausible that chaos is prevalent in complicated fluid flows, even those dominated by coherent structures. I am suggesting that this kind of modeling in terms of chaotic coherent structures may help us to see a way into the much more intricate situation of turbulence. Whether it can be carried far enough for that remains to be seen.

5. Chaos and Turbulence

The experimental work I cited at the outset is good evidence for the occurence of chaos in fluid flows. What I tried to add in the body of this disccusion was the notion that chaos is reasonably accessible analytically and with little numerical sophistication. When the fluid is only mildly unstable, we may have theoretical descriptions of both chaos and coherent structures for relatively little effort. The situations in which these processes arise have many features of turbulent flows and are themselves worthy of the attention of people interested in turbulence. But is there a direct relevance? I believe the answer is yes.

Let me once more take my text from he study of convection, but this time from turbulent convection. This is a problem that meteorologists and astrophysicists must grapple with and, if they have resorted to desperate measures, it is because they had little alternative. One of the standard maneuvers in studies where the Rayleigh numbers are astronomical is to fall back on the notion of Kelvin and others of convective equilibrium. The idea is that in fully turbulent convection, it is a good approximation to treat the mean state as locally neutral. To get some idea about the dominant length scales and time scales of the motion that one would expect to see, one can study perturbations to this mean state. This involves linear calculations like those of stability theory, but the linearization is about the guessed but strictly unknown state of convective equilibrium. Given some subgrid or other modelling it might be possible to complete the story and get a self-consistent determination of this mean state, but efforts to do this so far remain unconvincing. Nevertheless, if this picture (see e.g. Spiegel, 1962) is qualitatively acceptable, we may expect to find many nearly marginal modes in a linearization about the mean state. Hence we have grounds for thinking that the fully turbulent state is chaotic. The presence of this chaos makes for unpredictability, for that is one of the principal features of chaos. But is it otherwise essential to an understanding of turbulence?

Take intermittency. You can easily imagine an intermittent flow that is spatially and temporally periodic. Is the additional disorder that you see in a turbulent flow essential or is just an overlay of the simple chaos I have been talking about? In studying the complex vortex dynamics of turbulence, is it enough to work with laminar models and not worry about the chaos? I strongly doubt it. And if we cannot ignore the chaos, I suspect that it is because the chaotic process is indeed fundamental and does affect things like transport. As yet there is no reliable evidence for this, but here is a question about chaos that should concern people interested in turbulence. Since I am claiming that turbulence is a highly chaotic process, I should like to be able to suggest a way to separate the chaos, or unpredictability, from other aspects of the turbulence, such as vortex dynamics. I don't know if this idea can be made precise, but at least the way to begin thinking about it seems clear. We should apply the tools for analyzing chaos to real turbulent flows. I propose that we should take experimental data and output from numerical simulations on turbulence and begin to calculate Liapunov exponents and dimensions of the attractors as functions of the Reynolds number (say). The ways to do this are well-developed (Wolf, 1985; recently Y. Pomeau made some related suggestions in a lecture) and this seems to me to be an area where a union of the two topics might well be consummated. And this I think is a consummation to be wished.

Acknowledgement Much of the work reported here was supported by the N.S.F. under PHY80-23721.

Appendix. The Condition of Tricriticality.

The modes that can first go unstable in the linear theory of §2 have growth rates and frequencies given by (2.13). The coefficents in that expression are

$$P_0 = \pi^2\sigma^2\tau T q^2 - a^2\sigma^2\tau S q^2 - a^2\sigma^2\tau R q^2 + \sigma^2\tau q^8 \tag{A.1}$$

$$P_1 = \pi^2\sigma^2(\tau+1)T - a^2\sigma\tau(\sigma+1)S - a^2\sigma(\sigma+\tau)R + \sigma(\sigma\tau+\sigma+2\tau)q^6 \tag{A.2}$$

$$P_2 = \pi^2\sigma^2 T/q^2 - a^2\sigma\tau S/q^2 - a^2\sigma R/q^2 + (\sigma^2+2\sigma\tau+2\sigma+\tau)q^4 \tag{A.3}$$

$$P_3 = (1+2\sigma+\tau)q^2 \,, \tag{A.4}$$

where
$$q^2 = a^2 + \pi^2 \,. \tag{A.5}$$

When P_0, P_1 and P_2 vanish, (2.13) has three vanishing roots. This occurs when $R=R_o$, $S=S_o$ and $T=T_o$ where

$$R_o = \frac{q^6(\tau+2\sigma)}{a^2\sigma(1-\sigma)(1-\tau)} \tag{A.8}$$

$$S_o = \frac{q^6\tau^2(1+2\sigma)}{a^2\sigma(\sigma-\tau)(1-\tau)} \tag{A.9}$$

$$T_o = \frac{q^6\sigma(1+\sigma+\tau)}{\pi^2(\sigma-\tau)(1-\sigma)} \,. \tag{A.10}$$

In a parameter neighborhood of this condition, we may eliminate the extraneous fourth root of (2.13) and we get (2.16).

References

Ablowitz, M.J. and Segur, H., Solitons and the Inverse Scattering Transform, SIAM Studies in Apllied Mathematics (Philadelphia, 1981).

Arneodo, A., Coullet P.H. & Spiegel, E.A., Chaos in a Finite Macroscopic System, Phys. Lett. 92A (1982) 369-373.

Arneodo, A., Coullet, P.H. & Spiegel, E.A., The Dynamics of Triple Convection, Geophys. Astophys. Fluid Dyn. (1985).

Arneodo, A., Coullet, P.H., Spiegel E.A. and Tresser, C., Bifurcations and Chaos, in Proceedings of GFD Summer Program, F.K. Mellor, ed., Woods Holes Oceanog. Inst. Tech. Rept. WHOI 81-102 (Woods Hole MA 1981) 201-234.

Arneodo, A., Coullet, P.H., Spiegel E.A. and Tresser, C., Asymptotic Chaos, Physica D, (1985).

Arneodo, A., Coullet, P., and Tresser, C., Possible new strange attractors with spiral structure, Comm. Math. Phys., **79**, 573-579 (1981a).

Arneodo, A., Coullet, P., and Tresser, C., Oscillations with chaotic behavior: an illustration of a theorem by Shil'nikov, J. Stat. Phys., 27 (1982) 171-182.

Arnold, V.I., Geometrical Methods in the Theory of Ordinary Differential Equations (Springer-Verlag, 1983).

Bauer, L., Keller H.B. and Reiss, E.L., Multiple eigenvalues lead to secondary bifurcation, SIAM Rev. 17 (1975) 101-122.

Bretherton, C.S., Double Diffusion in a Long Box, in Proceedings of GFD Summer Program (1981) F.K. Mellor, ed., Woods Holes Oceanog. Inst. Tech. Rept. WHOI 81-102 (1981) 201-234.

Bretherton, C.S. & Spiegel, E.A. Modulational Instability as a Route to Chaos, Phys. Lett. 96A (1983) 152-156.

Chandrasekhar, S., Hydrodynamic and Hydromagnetic Stability, (Dover Pub. Inc., N.Y., 1981).

Collet, P. and Eckmann, J.P., Iterated maps of the interval as dynamical systems, Birkhauser (Boston, 1980).

Coullet, P.H. & Spiegel, E.A., Amplitude Equations for Systems with Competing Instabilities, SIAM J. Appl. Math., 43 (1983) 775-819.

Drazin, P.G. and Reid, W.H., Hydrodynamic Stability (Cambridge University Press, 1981).

Gollub, J.P., Recent experiments on the transition to turbulent convection, in Nonlinear Dynamics and Turbulence, G.I. Barenblatt, G. Iooss and D.D. Joseph, eds., (Pitman, 1983) 156-171.

Hao, B.-L., Chaos, (World Scientific Pub. Co., Singapore, 1984).

Kogelman S. and Keller, J.B., Transient behavior of unstable nonlinear systems with applications to the Bénard and Taylor problems, SIAM J. Appl. Math. 20 (1967) 619-637.

Kuramoto, Y., Diffusion Induced Chaos in Reaction Systems, Supp. Prog. Theor. Phys. 64 (1978) 346-367.

Kurumoto, Y. and Tsuzuki, T., Persistent Propagation of Concentration Waves in Dissipative Media Far from Thermal Equilibrium, Prog. Theor. Phys. 55 (1976) 356-369.

La Salle, J. and Lefschetz, S., Stability by Liapunov's Direct Method with Applications (Academic Press, 1961).

Lorenz, E., Deterministic non-periodic flow, J. Atmos. Sci., 20 (1963) 130-141.

Mandelbrot, B.B., The Fractal Geometry of Nature, (W.H. Freeman & Co., 1983).

Maurer, J. and Libchaber, A., Rayleigh-Bénard experiment in liquid helium; frequency locking and the onset of turbulence, J. Physique 40 (1979) L419-423.

McLaughlin J.B. and Orszag, S.A., Transition from periodic to chaotic thermal convection, J. Fluid Mech. 122 (1982) 123-142.

Moon, H.T., Huerre P. and Redekopp, L.G., Phys.Rev. Lett. 49 (1982) 458.

Moore D.W. and Spiegel, E.A., A Thermally Excited Non-linear Oscillator, Astrophys. J., 143 (1966) 871-887.

Newell, A.C., Envelope Equations, Lectures in Applied Maths. 15 (1974) 157-163.

Newell, A.C. and Whitehead, J.A., Finite bandwidth, finite amplitude convection, J. Fluid Mech. 38 (1969) 279-303.

Pumir, A. Structure Localisées et Turbulence, Thèse de 3e Cycle (U. Paris 6, Physique Théorique, 1982).

Segel, L.A., Distant side-walls cause slow amplitude modulation of cellular convection, J. Fluid Mech. 38 (1969) 203-224.

Shil'nikov, L.P., A case of the existence of a countable number of periodic motions, Sov. Math. Dokl. 6 (1965) 163-166.

Shil'nikov, L.P., On the generation of periodic motion from trajectories doubly asymptotic to an equilibrium state of saddle type, Math. Sbornik 6 (1968) 427-438.

Spiegel, E.A., On the Malkus Theory of Turbulence, Mecanique de la Turbulence, (Edition du C.N.R.S., No. 108, 1962) 181-199.

Spiegel, E.A., Convection in Stars II. Special Effects, Ann. Rev. Astr. Astrophys. 10 (1972) 261-304.

Spiegel, E.A., Physics of Convection (Students Notes), Proceedings of GFD Summer Program (1981) Mellor, F.K. ed., Woods Holes Oceanog. Inst. Tech. Rept. WHOI 81-102 (1981) 67-71.

Whitham, G.B., Linear and Nonlinear Waves, (John Wiley and Sons, New York, 1974) 84.

Wolf, A., Quantifying Chaos with Lyapunov Exponents, in Nonlinear Science: Theory and Applications, A.V. Holden, ed. (Manchester Univ. Press) to appear.

CHAPTER XIV. Connection Between Two Classical Approaches to Turbulence: The Conventional Theory and the Attractors

R. Temam

INTRODUCTION

The object of this lecture is to report on a progressing work which is aimed at filling part of the gap existing between two classical theoretical approaches to Turbulence, namely the conventional theory and the point of view of attractors. The conventional theory is based on a physical approach following the hypotheses of Kolmogorov, while the point of view based on chaos and attractors is more recent and corresponds to a more mathematical approach in the direction of the Dynamical Systems Theory (Smale, Ruelle-Takens,...). Although these theories seem to correspond to totally different approaches, a connection has been established recently between them, and we would like to describe this connection. Also this connection leads to useful thoughts concerning the numerical simulation of viscous flows.

The dissipative nature of a viscous flow leads to the idea that in a permanent regime, the number of degrees of freedom of a turbulent flow is finite. In the conventional theory of turbulence this number N of degrees of freedom is estimated, for a turbulent 3-dimensional flow as

$$N \sim (L_0/L_d)^3 \qquad (*)$$

where L_0 is the typical large lengthscale and L_d is the Kolmogorov dissipation length,

$$L_d = (\nu^3/\epsilon)^{1/4} ,$$

with ϵ the energy-dissipation rate per unit mass and ν the kinematic viscosity. This estimate follows from dimensionality arguments and physical unsubstantiated assumptions on orders of magnitude. On the other hand the Dynamical System Theory implies that, for an established (non transient) flow, the phase representation of the flow lays on an attractor (or in some unstable limit cases on a functional invariant set). This attractor is expected to have a finite dimension which provides another description of the finite dimensionality of the flow and gives an estimate on its number of degrees of freedom. The dimension of the attractors has been recently estimated in term of physical quantities by using exclusively the mathematical theory of the Navier-Stokes equations and the estimate (*) is precisely recovered. We will briefly outline these results hereafter and we refer the reader to P. Constantin, C. Foias, R. Temam [4] for a complete mathematical treatment and to P. Constantin, C. Foias, O. Manley, R. Temam [3] for a description of the work in the language of fluid dynamics.

After describing in Sec. 1 of this article the results in the 3-dimensional case, we describe in Sec. 2 and then in Sec. 3 other results along the same ideas for the 2-dimensional viscous flows (Sec.2) and for the two-dimensional Bénard problem (Sec. 3).

We would like to conclude this introduction with a comment on the significance of these results for the numerical simulation of viscous flows. At a time where the increase in computing power and the prospect of supercomputers bring us closer to the numerical simulation of viscous flows in more realistic situations, and to very large scale numerical simulations the estimates on the number of degrees of freedom of a flow gives us an upper bound on the number of parameters which is necessary to describe a (permanent) flow. It gives us an estimate on the size of the computations. However it does not say and it leaves open *the problem of the choice of these parameters* and in the future it will be interesting to see how one can compare the number of parameters used in numerical simulation with the dimension of the attractor (i.e. the number of degrees of freedom of the flow).

1. THREE DIMENSIONAL FLOWS.

The Navier-Stokes equations (N.S.E.) for an incompressible fluid with density 1 are written

$$\frac{\partial u}{\partial t} - \nu\Delta u + (u.\nabla)u + \nabla p = f \qquad (1.1)$$

$$\nabla.u = 0 \qquad (1.2)$$

where $u = u(x,t) = (u_1(x,t),u_2(x,t),u_3(x,t))$ is the velocity vector ($x = (x_1,x_2,x_3) \in \mathbb{R}^3$, $t > 0$) and $p = p(x,t)$ is the pressure ; $\nu > 0$ is the kinematic viscosity and $f = f(x,t)$ represents the driving forces per unit volume. The results presented hereafter apply to two situations :

.(i) The fluid fills a bounded domain Ω of $\mathbb{R}^3$ with a rigid boundary Γ at rest so that

$$u = 0 \text{ on } \Gamma \qquad (1.3)$$

by the non slip condition.

.(ii) The fluid fills the whole space $\mathbb{R}^3$, u and p being space periodic with period $L > 0$ in each direction x_1, x_2, x_3. In this case we write Ω = the cube of period $(0,L)^3$.

The case (i) corresponds to the flow in a bounded domain where the forces f which can be for instance buyancy forces or magnetic forces are assumed to be known ; (ii) corresponds to a similar situation with a very large volume filled by the fluid allowing the appearance of a space periodic flow away from the boundary : (ii) provides then the flow in a typical periodicity cell in the inner region. We refer the reader to Sec. 3 for a discussion of the thermoconductive case when the forces are not considered as given but are related to the temperature which is itself solution of the heat equation so that Navier-Stokes and heat equations are coupled.

We want to consider the most general solutions, when ν is arbitrarily small or f is arbitrarily large (in a sense to be explained later). Assuming that f is independent of time

$$f(x,t) \equiv f(x) \,, \tag{1.4}$$

we know that for ν large or f small, a stationnary solution takes place after a short transient period. More complicated structure, involving non stationnary solutions appear when ν^{-1} or f are large : one of the few pieces of information is the estimate of the number of degrees of freedom of the flow provided by the conventional theory of turbulence (see Landau-Lifschitz [8]) which we now recall.

Kolmogorov-Landau-Lifschitz estimate on the number of degrees of freedom of a turbulent flow.

In the Kolmogorov approach to turbulence we consider the so-called Kolmogorov dissipation length L_d which is such that the eddies of size smaller than L_d are damped exponentially and then only the eddies of size L_d, at least, ought to be considered. Assuming for simplicity that the flow takes place in a cube of edge L_o, and since only a finite number of eddies of size L_d are contained in a cube of edge L_d, we conclude that the number of eddies to be monitored is of the order of $(L_o/L_d)^3$ and the number N of degrees of freedom is of the order of

$$N \sim \left(\frac{L_o}{L_d}\right)^3 \tag{1.5}$$

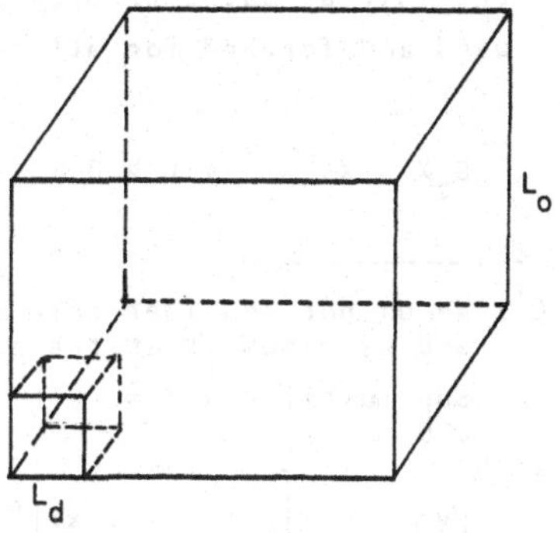

The value of L_d is given by

$$L_d = \left(\frac{\nu^3}{\epsilon}\right)^{1/4} \tag{1.6}$$

where ν is the kinematic viscosity and ϵ is the average dissipation rate of energy per unit mass and time, i.e. :

$$\epsilon = < \epsilon(x,t) > \tag{1.7}$$

where $\epsilon(x,t) = \nu|\text{grad } u(x,t)|^2$ is the local dissipation rate of energy and $< \epsilon(x,t) >$ is some ensemble averaging of this function, in accordance with the statistical nature of the Kolmogorov description of a turbulent flow.

Functional invariant sets and attractors.

In the mathematical setting of the N.S.E. (see R. Temam [10]), the phase space is an appropriate infinite dimensional Hilbert space H and these equations are written as an infinite dimensional differential equation of the form

$$\frac{\partial u}{\partial t} + \nu A u + B u = f \tag{1.8}$$

with initial condition

$$u(0) = u_0 \tag{1.9}$$

When it is defined, the operator S_t is the nonlinear operator

$$u_0 \to u(t)$$

where $u(t)$ is the value at time t of the solution of (1.8)(1.9).

Given a solution of (1.8)(1.9) which is defined for all time ([1]), it was shown in Foias-Temam [5] (see also [10]) that $u(t)$ converges in H, as $t \to \infty$, to a set X which enjoys the following properties

- the initial value problem (1.8)(1.9) has a solution defined backward and forward for all time $t \in \mathbb{R}$ and all $u_0 \in X$ (1.10)

- $S_t X = X \quad , \quad \forall\, t > 0$ (1.11)

([1]) We do not consider here the possible appearance of singularities and we assume that the enstrophy remains uniformly bounded in time:

$$\sup_{t>0} \|u(t)\| = R < \infty$$

with

$$\|v\|^2 = \int_\Omega |\text{curl } v(x)|^2 dx.$$

This set X is the mathematical object which describes the "permanent regime" which takes place after a transient period. In the simplest cases it can be reduceed to one point (stationnary solution) or one curve (time periodic solution) ; but it is considered that X could be as well a very complicated set.

The stability theory of the functional invariants sets is not available but is is conjectured that except for unstable limit cases, these sets are stable and therefore they are attracting.

It was shown in [5] that this set has a finite dimension which is another aspect of the finite dimensionality of a turbulent flow. Then in [4] and [3] a precise estimate of this dimension in term of physical quantities was made and we now describe this result.

Fractal dimension of the attractor.

Let u_0 be an element of the functional invariant set (attractor) X and let $u = u(.,t) = S_t u_0$ be the corresponding solution of (1.8) (1.9). We define as above the local dissipation rate of energy per unit mass and time

$$\epsilon(x,s) = \nu |\mathrm{grad}\, u(x,s)|^2$$

and we consider the time average of the maximum dissipation rate on X :

$$\epsilon = \limsup_{t\to\infty} \sup_{u_0 \in X} \left\{ \frac{1}{t} \int_0^t \int_\Omega \sup_{x\in\Omega} \epsilon(x,s)ds \right\} \tag{1.12}$$

We associate to this ϵ the Kolmogorov dissipation length L_d defined by (1.6). Let also L_0 be a typical length of Ω : for instance the diameter of Ω in case (i) and $L_0 = L$ in case (ii).

The Kolmogoróv-Landau-Lifschitz result was recovered in [4][3] in the following form

THEOREM 1.1.

The fractal dimension of X *is less or equal than*

$$c_1 \left(\frac{L_0}{L_d}\right)^3 \tag{1.13}$$

where c_1 *is a universal constant.*

Remark. A technical improvement of the proof reported in [4] and based on a sharp inequality of Lieb-Thirring [9] allows us to decrease ϵ (and thus decrease the bound (1.13)) by replacing ϵ by another average $\bar{\epsilon}$

$$\bar{\epsilon} = \lim_{t\to\infty}\sup \ \sup_{u_0\in X} \left\{ \frac{1}{t}\int_0^t \int_\Omega |\epsilon(x,s)|^{5/4} \frac{dx}{|\Omega|} ds \right\}^{4/5}$$

$|\Omega|$ = the volume of Ω . However the physical significance of $\bar{\epsilon}$ is not clear.

2. TWO DIMENSIONAL FLOWS.

The results obtained in the 3-dimensional case can be extended to the 2-dimensional case (with an estimate $\leq c(L_0/L_d)^2$), but, in this case the result can be improved in two ways :

- the bound $(L_0/L_d)^2$ can be replaced by a nondimensionnal expression which depends only on the data (while the knowledge of L_d assumes some informations on the actual flow).
- The limit set X can be replaced by a much larger set called the universal attractor, which describe all the possible flows.

These improvements are made possible by the fact that a full mathematical theory is available for the initial value problem for the N.S.E. (1.8)(1.9), which is not still the case in 3 dimensions. In particular all the caracteristics of the flow can be estimated in term of the data, or, more precisely a nondimensional number associated to the data.

This nondimensional number is the so called (generalized) Grashof number and is defined by

$$G = \frac{L_0^2 \, ||f||}{\nu^2} \tag{2.1}$$

where $||f||$ is the L^2-norm of f ,

$$||f|| = \left\{ \int_\Omega |f(x)|^2 dx \right\}^{1/2} \tag{2.2}$$

Some authors prefer to consider a Reynolds number $Re = \sqrt{G}$,

$$Re = \frac{L_0\sqrt{||f||}}{\nu} \tag{2.3}$$

which is proportional to $1/\nu$ [1].

We not recall the definition of the *universal attractor.*

THEOREM 2.1. (see Foias-Temam [5]).

Under the above assumptions there exists a functional invariant set X *(enjoying the properties* (1.8)(1.9) *which attracts all the trajectories, i.e.*

---

[1] This definition of Re is convenient but there is no definition of a typical velocity in the flow and there is no evidence that $\sqrt{||f||}$ should be a typical velocity.

Distance in H *of* $S_t u_o$ *to* $\tilde{X} \to 0$ *as* $t \to \infty$
$\forall u_o \in H$ (2.4)

Furthermore $\tilde{X}$ *is the largest attractor which is bounded in* H .

In some loose sense $\tilde{X}$ contains all the fluid mechanic phenomena which can be produced by f . In a more rigorous mathematical sense $\tilde{X}$ contains all the stationnary solutions of (1.8), all the time periodic solutions of (1.8) (if such solutions exist), it contains also the trajectories going (in infinite time) from one unstable solution to another unstable solution. However the most interesting elements of $\tilde{X}$ are not understood at the moment and very little is known about the structure of $\tilde{X}$: any piece of information on $\tilde{X}$ will probably help understand turbulence.

Using the general tools introduced in Constantin-Foias-Temam [4], one can obtain the following bound on the fractal dimension of $\tilde{X}$.

THEOREM 2.2.

The fractal dimension of $\tilde{X}$ *is bounded by*
$c_2\ G$ (or $c_2\ Re^2$) (2.5)

where c_2 *is a universal constant.*

The reader is referred to [11] for the proof. Partial results where obtained in [1] and [2]. Furthermore [1] contains a lower bound of the dimension of $\tilde{X}$ of the form $c_3\ G^{2/3}$

3. THE 2-DIMENSIONAL BENARD PROBLEM.

We consider the Bénard convection problem in a two dimensional layer of fluid comprised between the horizontal planes y = 0 and y = h which is heated from below at a temperature T_o and from above at a temperature $T_1 < T_o$.

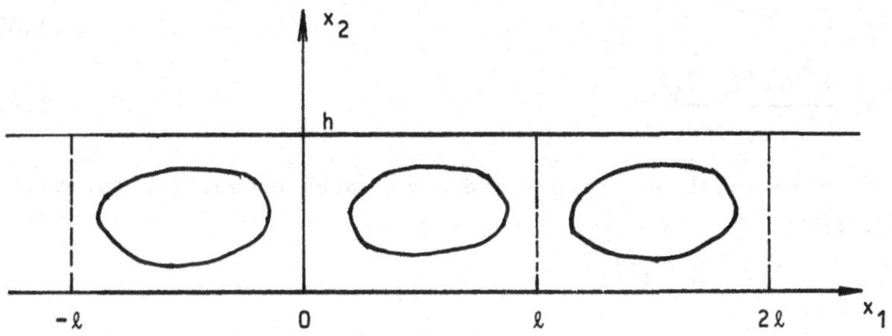

The temperature T and the motion (velocity u, pressure p) are governed by the Boussinesq equations

$$\rho_o \left(\frac{\partial u}{\partial t} + (u.\nabla)u\right) - \rho_o \nu \Delta u + \nabla p = -\rho_o\, g\, e_2[1 + \varepsilon(T_o - T)] \tag{3.1}$$

$$\nabla.u = 0 \tag{3.2}$$

$$\rho_o\, C_v \left(\frac{\partial T}{\partial t} + (u.\nabla)T\right) - \rho_o\, C_v\, \kappa \Delta T = 0 \tag{3.3}$$

where ρ_o is the constant density, ε is the expansion coefficient of the fluid, C_v is the specific heat at constant volume, κ is the (constant) coefficient of thermometric conductivity, g the gravity, and the other quantities have been already defined.

For the boundary conditions we assume for instance that there is no slip on the planes ($u = 0$ at $x_2 = 0$ and $x_2 = h$) and that the flow is periodic in the x_1 direction with period ℓ. Other realistic boundary conditions can be as well considered (cf. C. Foias, O. Manley, R. Temam [6]).

We make the equation nondimensional using h, $\{h/g\varepsilon(T_o-T_1)\}^{1/2}$, T_o-T_1, $\rho_o h g\varepsilon(T_o-T_1)$ as units of length, time, temperature and pressure. The nondimensional equations then become

$$\frac{\partial u}{\partial t} + (u.\nabla)u - \frac{1}{Gr^2}\Delta u + \nabla p = -e_2(T-T_o) \tag{3.4}$$

$$\nabla.u = 0 \tag{3.5}$$

$$\frac{\partial T}{\partial t} + (u.\nabla)T - \sqrt{\frac{1}{PrRa}}\,\Delta T = 0 \tag{3.6}$$

where u, p, T, denote now the nondimensional quantities and Gr, Pr, Ra, are the classical Grashof, Prandtl and Rayleigh numbers given in term of the data by

$$Gr = \frac{h^3 g\,\varepsilon(T_o-T_1)}{\nu^2} \tag{3.7}$$

$$Pr = \frac{\nu}{\kappa}. \tag{3.8}$$

$$Ra = \frac{h^3 g\,\varepsilon(T_o-T_1)}{\kappa\nu} \tag{3.9}$$

We can also substract from T and p the pure convection solution and consider

$$\theta = T - T_o - x_2(T_1 - T_o)$$

$$q = p - \frac{x_2^2}{2}(T_0 - T_1)$$

With these functions the equations become

$$\frac{\partial u}{\partial t} + (u.\nabla)u - \frac{1}{Gr^2}\Delta u + \nabla q = - e_2\,\theta \tag{3.10}$$

$$\nabla.u = 0 \tag{3.11}$$

$$\frac{\partial \theta}{\partial t} + (u.\nabla)\theta - u_y - \sqrt{\frac{1}{PrRa}}\,\Delta\,\theta = 0 \tag{3.12}$$

The boundary conditions mentionned above are written

$$\theta = 0 \quad \text{and} \quad u = 0 \quad \text{at} \quad x_2 = 0 \text{ and } x_2 = 1 . \tag{3.13}$$

The functions are periodic with period ℓ in the direction x_1. (3.14)

A result similar to Theorem 2.1 can ve proved in this case (with a similar definition of functional invariant sets and attractors) :

There exists (in the appropriate function space) a universal attractor Y *to which converge as* $T\to\infty$ [1], *all the solutions of* (3.10)-(3.14). *This is the largest bounded attractor.* (3.15)

Of course Y contains a description of <u>all</u> the 2-dimensional Bénard problem structures that can be observed. We refer the reader to [6] for a rigorous statement and a proof of (3.15).

Some years ago it was conjectured by O. Manley that the number of degrees of freedom in the 2-D Bénard problem shoud be of the order of the Grashof number and a partial proof of the result was given in Foias-Manley-Temam-Trève [7] (using the concept of determining modes instead that of dimension of the attractor as the "measure" of finite dimensionality). This conjecture is now fully proved in [6] with a correction depending on the Prandtl number :

<u>THEOREM 3.1</u>.

The fractal dimension of the universal attractor for the two dimensional Bénard convection problem is bounded by

$$c_3\ Gr\ (1 + Pr^{3/2}) \tag{3.16}$$

where c_3 *is a universal constant.*

---

[1] In practice after a transient period, when the permanent regime is established.

Cf. [6] for the proof and for results similar to that of Sec. 1 for the 3-dimensional Bénard problem.

REFERENCES.

[1] A.V. Babin, M.I. Vishik, *Les attracteurs des équations d'évolution aux dérivées partielles et les estimations de leurs dimensions*, Usp. Math. Nauk, 38, 4 (232), 1983, p. 133-187.

[2] P. Constantin, C. Foias, *Global Lyapunov Exponents, Kaplan Yorke formulas and the Dimension of the Attractors of 2-D Navier-Stokes Equations*, Comm. Pure Appl. Math., to appear.

[3] P. Constantin, C. Foias, O. Manley, R. Temam, *Determining modes and fractal dimension of turbulent flows*, J. Fluid. Mech., 1984.

[4] P. Constantin, C. Foias, R. Temam, *Attractors representing Turbulent Flows*, Memoirs of A.M.S., 1985.

[5] C. Foias, R. Temam, *Some analytic and geometric Properties of the Solutions of the Navier-Stokes Equations*, J. Math. Pures Appl., 58, 1979, p. 339-368.

[6] C. Foias, O. Manley, R. Temam, *Physical Bounds of the fractal Dimension of the attractors for the Bénard problem*, to appear.

[7] C. Foias, O. Manley, R. Temam, Y. Trève, *Asymptotic Analysis of the Navier-Stokes Equations*, Physica D, vol. 6D, 1983.

[8] L. Landau, I.M. Lifschitz, *Fluid Mechanics*, Addison-Wesley, New-York, 1953.

[9] E. Lieb, W. Thirring, *Inequalities for the Moments of the eigenvalues of the Schroedinger Equations and their Relation to Sobolev Inequalities*, p. 269-303, in *Studies in Mathematical physics : Essays in Honor of Valentine Bargman*, E. Lieb, B. Simon, A.S. Wightman Edt., Princeton University Press, Princeton, N.J., 1976.

[10] R. Temam, *Navier-Stokes Equations, Theory and Numerical Analysis*, CBMS/NSF Regional Conference series in Applied Mathematics, SIAM, Philadelphia, 1983.

[11] R. Temam, *Infinite dimensional dynamical systems in fluid mechanics*, in *Nonlinear Functional Analysis and Applications*, Proceedings of the AMS-Summer Research Institute, Berkeley, 1983, F. Browder Ed..

CHAPTER XV. Remarks on Prototypes of Turbulence, Structures in Turbulence and the Role of Chaos

Hassan Aref

Turbulence is usually billed as one of the great unsolved problems in classical mechanics. It has a bit of the aura of other great classical problems, like some of the conjectures in number theory, in that the statement of the problem is much simpler than the sophistication necessary to tackle it. We are intuitively acquainted with the flow of water, even in the turbulent regime, from our daily mechanical experiences, yet the description, let alone prediction, of the properties of turbulent flow has remained elusive, continually inviting application of concepts and ideas at the forefront of physical science and mathematics.

In the brief time allotted I should like to focus on two issues among the many that have been raised in the literature and at this workshop. The first is the role of analogies, of model systems or prototypes of turbulent behaviour. I believe such prototypes are immensely important - just consider what a simple model such as the Ising model has done for the theory of phase transitions - but that we spend far too little time pursuing them. I want to argue for a sustained widening of the perspective by considering analogies. Three-dimensional, homogeneous, isotropic turbulence or the turbulence in any of the standard shear flows may be very special objects. But they are hardly so special that well chosen analogies and prototypes are not helpful to contemplate.

The other topic I wish to comment on is the role of structures in turbulent flows. Particularly in shear flow turbulence the socalled organized or coherent structures have become a key research area [1]. Nevertheless, many currently used statistical turbulence models hardly mention such structures. And, conversely, even though we now understand the large scale mechanisms operating in several turbulent shear flows a lot better than before the elucidation of structures, there has been disappointingly little progress in calculating the properties of turbulent flows using this new knowledge. We must ask why.

But first let me promote the idea of turbulence analogies and prototypes. The point is that turbulent behaviour in the widest possible sense of the word, i.e. the stochastic behaviour of a deterministic system, is something that is exhibited in a variety of physical situations. In recent years there has been a lot of emphasis on chaos, which for the most part has meant the stochastic behaviour of few-degree-of-freedom dynamical systems. Most research on chaos in fluid mechanics has to do with the onset of turbulence rather than fully developed turbulence. And the area where these ideas are best developed, both analytically and experimentally, is convection. That is not unnatural since convection was where this whole topic started with the rightly famous paper by Lorenz [2]. It is only much more recently that these ideas, applied, for example, to vortices, are beginning to impact free shear flow research. I am not only referring to various aspects of my own work [3]. There are important overtones of chaotic behaviour in the work of Rockwell and collaborators on impinging shear layers [4], in the review article by Ho and Huerre [5] on perturbed free shear layers, and in the recent work by Sreenivasan on wakes [6]. But these applications come almost a full decade after the initiation of the convection work. And my point is, that the reason it took so long is that many workers in turbulence probably pay little attention to the possible analogies with convection research.

But let me discuss science rather than sociology: I would like to promote the concept of two-dimensional turbulence [7]. I don't have time to give even a superficial survey, but the basic idea is to do turbulence theory for the two-dimensional Navier-Stokes equation in the hope that the reduction in dimensionality will lead to a simpler problem. For obvious reasons workers in oceanography and meteorology have always been quite good about developing this "statistical complement to nineteenth century hydrodynamics", as Rhines [8] called it. Engineers have typically been much less tolerant, noting that vortex lines do not stretch in 2D as they do in "real" (i.e. 3D) turbulence, and that the topic therefore could not possibly be of interest. Well, there <u>are</u> many interesting phenomena here, probably all much more readily accessible to analytic theory and certainly more accessible to large scale numerical simulation. I have spent very enjoyable and instructive time computing complicated flows in two dimensions [3,9,10], and, I am pleased to report, some experimentalists are now pursuing the topic with great success. Let me just mention the recent work by Couder and collaborators in France, where marvelous vortex interactions in two dimensions are produced and visualized using a soap film technique [11].

Another area rich in analogies is the dynamics of one-dimensional interfaces in certain systems, e.g. the Rayleigh-Taylor problem or the problem of stratified Hele Shaw flow. Again here statistical regimes of evolution, that we have begun referring to as "conto(u)rbulence", exist. Again these are accessible both to analog [12,13] and digital [14,15] experimentation at comparable spatial resolutions, and again the problems encountered in the description are very similar to those that arise in describing more conventional turbulent flows. I would like to suggest that an instructive prototype is to be found among this array of systems, and that an incisive piece of analysis is more likely to succeed for one of these simplified systems than for any of the rather complex flows usually considered in "conventional" turbulence research. Just imagine being able to calculate the fractal dimension [16] of an evolving interface from the basic equations of motion. I believe we are quite close to a result of this kind at the present time, and that the rewards in terms of general insight will be substantial.

Let me next turn to structures: The wonder of structures in turbulent flows is, as Keller so aptly phrased it [17], that "they offer the possibility of reintroducing fluid mechanics into the study of turbulence". With the discovery of organized structures in turbulent flows the focus of our attention has shifted from the average to the most probable. However, the main emphasis of our theories is not on spatial structures. The hope was that once we understood the mechanisms of turbulence we could also compute the properties of turbulence. The experiments have revealed various vortex structures that interact in certain ways. Vortex pairing [18], for example, is one such popular mechanism. In short order laminar flow models of pairing vortices appeared, but they never quite seemed able to predict a Reynolds stress or something of that sort. And the closure models would frequently do alright in this respect without ever incorporating a single structure. The resolution of this paradox, I believe, goes something like this: Incorporating organized structures into turbulent flow models is a correct way to proceed because it brings the physical content of the model closer to the physics of the real flow [19]. But we have so far been much too naive about the motions and interactions of these organized structures. They are not limited to being of the type one might find in Lamb's classic book [20]! Indeed, chances are that the structures predominantly dance around each other in some chaotic way [3], and that the "phase chaos" between events involving structures 'here' and structures 'there' is absolutely essential. The emergence of chaos is believed to play a major role in the explanation of why deterministic systems obey statistical thermodynamics [21]. Analogously, I believe that the chaotic dynamics of organized structures is the essential "missing link" between statistical turbulence theory and deterministic models of turbulent flows. This topic happens to be very poorly understood at present. Again there is ample opportunity for inventing relevant, simplified model systems that can be analyzed in depth.

Once upon a time in the history of fluid mechanics flow solutions obtained by theorists were regarded as curiosities, mathematically beautiful but clearly of no practical use whatsoever. Those were the days of classical hydrodynamics. The days when flows were assumed irrotational (because of Lagrange's famous theorem) and viscosity was considered a minor nuisance to be dealt with later. Some communities, in particular the hydraulicists, rebelled and set off on their own semi-empirical venture. We now realize that in many ways they were right in doing so - fluid dynamicists, it was said [22], "were divided into hydraulic engineers who observed what could not be explained and mathematicians who explained what could not be observed". With the wisdom of hindsight, and a bit of knowledge of boundary layer theory, we can, of course, now smile at characterizations such as "the flow of dry water", that have been applied to the mathematical fluid dynamics of Euler, Bernoulli and their followers. However, this should not obscure the fact that there was a serious crisis for the credibility of hydrodynamic theory. To resolve it the mathematical modelling of laminar fluid flow had to be reformulated, and much of the impetus for this reformulation came from experiment.

There may be several lessons that workers in turbulence, particularly in shear flow turbulence, can and should learn from this earlier episode in fluid mechanics. I am suggesting that the modellers of turbulence run the risk of being classified as the Eulers and Bernoullis of our time (which, of course, may not be so bad!) if they do not heed the signals issuing from experiment. To paraphrase: statistical turbulence theories may often be predicting flows that do not exist. And, conversely, many experimenters, particularly in turbulent shear flow research, are today observing structures, processes and mechanisms on which the theoretical community has very little substantive help to offer.

Turbulence is not entirely disorganized. It typically comes in structured packets that survive for some time. These packets have identifiable interaction mechanisms that can be analyzed in terms of concepts from laminar flow theory. However, the motion of fluid parcels is not simple, even within laminar flow theory, once either unsteadiness or three-dimensionality (and usually both!) are introduced. The Lagrangian motion of parcels of fluid can be chaotic even in a flow that is simple in Eulerian terms [23]. We are just beginning to gain a full understanding of these deeper aspects of flow kinematics. I find it very encouraging that a picture embodying two of the most profound and exciting ideas in modern mechanics, viz. the chaotic motion of solitary structures, is emerging in our description of turbulent flow.

References

1. B.J. Cantwell, Ann. Rev. Fluid Mech. 13 (1981) 457.
2. E.N. Lorenz, J. Atmos. Sci. 20 (1963) 130.
3. H. Aref, Ann. Rev. Fluid Mech. 15 (1983) 345.
4. D. Rockwell, "Vortex-edge interactions" in Recent Advances in Aerodynamics and Aeroacoustics, Springer (to appear).
5. C-M. Ho and P. Huerre, Ann. Rev. Fluid Mech. 16 (1984) 365.
6. K.R. Sreenivasan, "Transition and turbulence in fluid flows and low-dimensional chaos" in Fundamentals of Fluid Mechanics, Springer (to appear).
7. R.H. Kraichnan and D. Montgomery, Rep. Prog. Phys. 43 (1980) 547.
8. P.B. Rhines, Ann. Rev. Fluid Mech. 11, (1979) 401.
9. H. Aref and E.D. Siggia, J. Fluid Mech. 100 (1980) 705.
10. H. Aref and E.D. Siggia, J. Fluid Mech. 109 (1981) 435.
11. Y. Couder and M. Rabaud, "Two-dimensional turbulence in thin liquid films," J. Phys. Lett. (to appear).
12. D.J. Lewis, Proc. R. Soc. (London) A202 (1950) 81.
13. R.A. Wooding, J. Fluid Mech. 39 (1969) 477.
14. G.R. Baker, D.I. Meiron and S.A. Orszag, Phys. Fluids 23, (1980) 1485.
15. G. Tryggvason and H. Aref, J. Fluid Mech. 136 (1983) 1.
16. B.B. Mandelbrot, Fractals, Form, Chance and Dimension, Freeman (1977).
17. J.B. Keller, Lectures at Woods Hole GFD Summer Program (1980).
18. C.D. Winant and F.K. Browand, J. Fluid Mech. 63 (1974) 237.
19. J.L. Lumley, Trans. ASME, J. Appl. Mech. 50 (1983) 1097.
20. H. Lamb, Hydrodynamics, Dover (1932).
21. J. Ford, "The statistical mechanics of classical analytic dynamics," in Fundamental Problems in Statistical Mechanics, North-Holland (1975).
22. G. Birkhoff, Hydrodynamics, A Study in Logic, Fact and Similitude, Princeton University Press (1960), p. 4.
23. H. Aref, J. Fluid Mech. 143 (1984) 1.

CHAPTER XVI. Subgrid Scale Modeling and Statistical Theories in Three-Dimensional Turbulence

Jean-Pierre Chollet

Abstract.

Statistical theories of turbulence can be used to derive subgrid scale (SGS) modeling for large eddy simulations (LES) of homogeneous turbulence. Well-suited to pseudo-spectral numerical codes, they are also of interest to better understand eddy-viscosity concepts without reference to any particuliar method of field computations.

1. Introduction.

Large eddy simulations, that is numerical calculations of realizations of turbulent velocity fields, are severely limited by the capabilities (memory size and speed of calculation) of computers which are currently available. A full simulation requires the calculation over a wide range of scales and then cannot be carried out when the Reynolds number is large. Depending on the physical problem under consideration, the turbulent field to be explicitely computed pertains either to large or to small scales. The full description of small scales is of fundamental interest to analyze basic features as the so-called small scales isotropy and intermittency (Kerr /1/) ; in this case the process which is used to feed energy into the largest explicitely computed scales can be viewed as a large scale modeling.

Nevertheless, the dynamics of the turbulent field under consideration is very often highly dependent on the largest scales which contain the main part of the energy and determine the specific character of the flow. In this case, the small scales (subgrid scales) cannot be explicitely calculated and must be modeled statistically in order to transfer energy towards the dissipative scales.

Spectral formulations in wave vector ($\vec{k}$) space are well suited to homogeneous turbulence. They emphasize the dynamics of interactions between various scales, accordingly they can be useful for analysing eddy-viscosity concepts in general, even when using calculations in physical ($\vec{x}$) space to better handle realistic features of the flows (intricate boundaries for instance).

2. Spectral eddy-viscosity and diffusivity.

Statistical theories -often referred to as two point closures- have the capability of taking into account interactions between the various scales of turbulence. The interactions between small ($k > k_c$) and large ($k < k_c$) scales give rise to an energy transfer $\vec{T}(k|k_c,t)$. This transfer can be written as a generalized eddy-viscosity (Kraichnan |2|).

$$\nu(k|k_c,t) = -T^{>}(k|k_c,t) / 2k^2 E(k,t) \tag{1}$$

This eddy-viscosity is conveniently written with a non-dimensional function $\nu_t^+(k/k_c)$.

$$\nu(k|k_c,t) = \nu_t^+(k/k_c) \left(E(k_c,t)/k_c\right)^{1/2} \tag{2}$$

With a $k^{-5/3}$ inertial spectrum for the whole range of wave-numbers, a numerical calculation using the eddy damped quasi normal markovian (EDQNM) approximation as the statistical theory gives :

$$\nu_t^+(k/k_c) = \nu_t^+(0) + \nu_{t\,cusp}^+(k/k_c) \tag{3}$$

$\nu_t^+(0) = 0.267$ is a constant value.

$\nu_{t\,cusp}^+(k/k_c)$ rapidly increases in the neighbourhood of $k/k_c = 1$, and can be approximated (Chollet /3/) by :

$$\nu_{t\,cusp}^+(k/k_c) = 0.472\,(k/k_c)^{3.74}, \quad 0.4 < k/k_c < 0.9$$

The contribution of this cusp is associated with interactions which are called "local" since they associate small and large scales which are both in the neighbourhood of the cut-off ($k/k_c \rightarrow 1$).

The part $\nu_t^+(0)$ is still dominant and gives rise to a pure viscosity transfer, which justifies the use of eddy viscosity in modeling subgrid scales for three-dimensional simulations.

To model the small scales of a passive scalar, a generalized eddy-diffusivity can be derived following exactly the same method. The eddy diffusivity D_t^+ also exhibits a cusp and the eddy Prandtl number ν_t^+/D_t^+ is fairly constant (= 0.60).

These eddy quantities are described in Chollet /4/. Using some analytical forms of dominant terms in the energy transfer, Chollet /3/ notices that : (i) the eddy viscosity is reduced when the Reynolds number decreases, (ii) the amplitude of the cusp decreases when the cutoff scale is close to the energy containing scales.

The numerical values quoted above depend on adjustable constants whose values are determined to fit inertial and inertial-convective range (Herring et al. /5/). Some features of this eddy-viscosity can be recovered with renormalization group calculations (see Mc Comb in the present volume).

3. Large eddy simulations.

A pseudo spectral method is conveniently used to calculate the evolution with time of an homogeneous three dimensional turbulence in a cubical box with periodic boundary conditions.

Assuming an inertial $k^{-5/3}$ energy spectrum in the small scales ($k > k_c$) and in some neighbourhood of the cut-off k_c (both small and large scales) we can use the eddy viscosity (1) in the equations of the truncated ($k < k_c$) velocity field. The expression of this eddy viscosity will be derived from (2) and (3) with the proper evolution in time through the factor $E(k_c,t)$.

Computations of field realizations with this eddy viscosity (and diffusivity) yield good $k^{-5/3}$ inertial and inertial convective range up to the cut off k_c (Chollet and Lesieur /6/, Chollet /4/). These large eddy simulations are carried out for a viscosity going to zero (infinite Reynolds number) since the $k^{-5/3}$ inertial range has been assumed for $k > k_c$.

This subgrid scale modeling can give rise to some overestimation of the energy transfer towards small scales at least in three cases :

(i) at low or moderate Reynolds numbers when the inertical spectrum overestimates the energy in the small ($k > k_c$) scales

(ii) when the cut-off k_c is close to the most energetic (integral) scales, such a situation is of particular interest for the simulation of the largest scales

(iii) when the $k^{-5/3}$ inertial range is not yet established.

In such cases, the distribution of energy $E(k,t)$ over the whole range of scales can be taken in account by calculating at each time t :

(1) the evolution of the energy spectrum $E(k,t)$ for small scales with the statistical (EDQNM for instance) theory

(2) the energy spectrum $E(k,t)$ for large scales by averaging over the velocity field

(3) the energy transfers between large and small scales with the statistical theory ; the eddy viscosity (1) depends only on the actual energy spectrum $E(k,t)$ without any hypothesis of inertial range.

On fig. 1, the energy spectrum of the velocity field is plotted for both this full coupling procedure and the "simplified" ν_t^+ eddy viscosity ; as suggested earlier the latter overestimates the subgrid scale transfer.

4. Conclusions.

Statistical theories of turbulence make possible the derivation and the analysis of subgrid scales modeling to be used in large eddy simulations. No tuning of adjustable constants are required when using this model. The cusp of the eddy-viscosity features local interactions between large and small scales in the neighbourhood of the cut-off and should be considered whatever the numerical method is. Neverthelèss the assumption of the isotropy for the subgrid scale modeling, which does not take any account of the various orientations of interacting wave numbers is presumably much more drastic in this cusp range than in any other range of scales.

There is still the question of using such subgrid-scale modeling in anisotropic turbulence : this would require either an extension of statistical theories to a specific anisotropic flow or a splitting of the subgrid-scale terms into an isotropic and an anisotropic part, the latter having to be expressed with other approaches. A severe limitation would appear if the large scales depended on realizations of the small scales, in such a case there is basically no method to model statistically the small scales.

There is still a broad field of investigation for large eddy simulations of homogeneous turbulence in a cubical box. With the artifact of the infinite Reynolds number, the dynamics is less dominated by dissipative processes which can make possible the development of more significant evolutions, especially in the largest scales.

Acknowledgements.

This work was supported by the INAG under the A.T.P. "Recherches atmosphériques". The computations were carried out at NCAR, Boulder, Colorado ; NCAR is sponsored by the National Science Foundation.

References.

/1/ Kerr, R.M., "Kolmogorov and scalar spectral regimes in numerical turbulence", NASA Ames Res. Center (1983).

/2/ Kraichnan, R.H., "Eddy viscosity in two and three dimensions", J. Atmos. Sci., 33, (1976), 1521.

/3/ Chollet, J.P., "Statistical closure to derive a subgrid-scale modeling for large eddy simulations of three-dimensional turbulence", NCAR technical, TN-206 + STR, (1983).

/4/ Chollet, J.P., "Two-point closures used for a subgrid scale model in large eddy simulations", to be published in Turbulent Shear Flows IV, Springer-Verlag, (1983).

/5/ Herring, J.R., D. Schertzer, M. Lesieur, G.R. Newman, J.P. Chollet and M. Larcheveque, "A comparative assessment of spectral closures as applied to passive scalar diffusion", J. Fluid Mech., 124, (1982), 411.

/6/ Chollet, J.P. and M. Lesieur, "Parametrization of small scales of three-dimensional isotropic turbulence utilizing spectral closures", J. Atmos. Sci., 38, (1981), 2747.

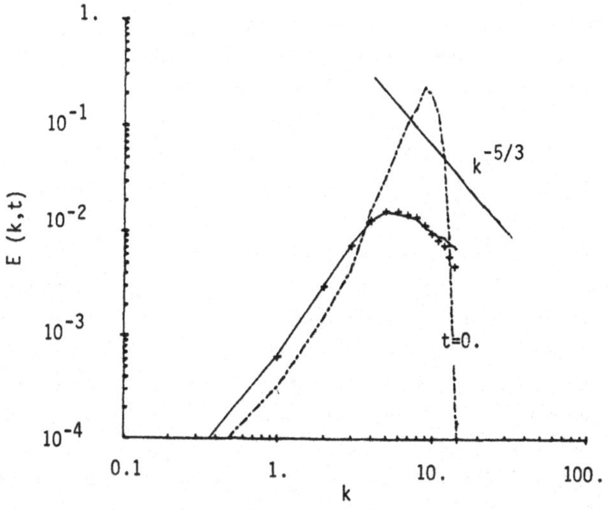

Fig. 1 - Energy spectra of freely evolving turbulence (at t = 1.20 sec $\simeq$ 17 initial large eddy turn-over times).

——— "full coupling" between LES and EDQNM calculation of small scales

+ + + + LES with the simplified ν_t^+ model.

CHAPTER XVII. Strange Attractors, Coherent Structures and Statistical Approaches

John L. Lumley

All turbulent flows contain both order and disorder. The relative proportions depend on Reynolds number, geometry (type) and boundary and initial conditions. For example, axisymmetric jets appear to have much less well-defined structures than most other flows under the same circumstances. An axisymmetric jet at high Reynolds number, arising from disturbed initial conditions, has ordered structures that are almost imperceptable to the unaided eye, and must be extracted using statistical techniques. On the other hand, mixing layers appear to have a higher proportion of ordered structures than other flows under the same circumstances. A mixing layer at relatively low Reynolds number arising from undisturbed initial conditions has such strong order that it is almost difficult to find the disorder. Most flows are somewhere in between, depending on the circumstances of their formation. No turbulent flow is completely without disorder, nor is there any completely without structure.

The relative role of the ordered and disordered parts of the flow in the transport varies. Sometimes it is necessary to model the ordered part of the flow explicitly, and sometimes it is not. A number of second order models successfully ignore the presence of structures in flows in which they are not dominant; we have successfully modeled even one of Roshko's variable density mixing layers (Shih, 1984) from undisturbed initial conditions using a second order model that ignores the presence of coherent structures. The success is partly because in flows of this type the transport properties of the coherent structures

*Supported in part by the U.S. Office of Naval Research under the following programs: Physical Oceanography (Code 422PO), Power (Code 473); in part by the U.S. National Science Foundation under grant No. ATM 79-22006; and in part by the U.S. Air Force Geophysics Laboratory. Prepared for presentation in the Panel Discussion at the ICASE/NASA Workshop on Theoretical Approaches to Turbulence, October 10-12, 1984, Pavilion Tower, Virginia Beach, Virginia.

scale in the same way as the transport properties of the disordered part, since there is generally a single scale of length and velocity.

The coherent structures that are observed in turbulent flows probably reflect one or more types of instabilities. Some very interesting experiments of Cimbala (1984) and of Wygnanski & Champagne (unpublished), both on the two-dimensional cylinder wake, indicate that some of the coherent structure observed in the cylinder wake arises as a remnant of the instability involving the near wake and the body boundary layer, which gives rise to the Karman street. This instability progressively decays, and is replaced by another one, which might be described as an instability of the "turbulent fluid" of the wake. That is, if one forms a wake without vortex shedding (the first instability), so that only fine-grained turbulence is present, one finds an instability of this flow growing, producing something reminiscent of a Karman street, but quite distinct from it. As the Reynolds number is raised, the point of appearance of the second instability moves toward the cylinder, while the rate of decay of the first instability decreases, so that there is considerable interaction between the two. This situation, complex as it is, is probably simple compared to situations that arise in geophysically or industrially important flows. It does at least suggest that some sort of stability analysis offers hope to understand the nature of coherent structures.

Gol'dshtik (1982) has suggested that strange attractors and coherent structures reflect each other's characteristics. As the system makes each trip around the attractor, it produces a structure. For example, in the Lorenz attractor, which is rather like the wings of a butterfly, the system loops around one branch (wing) of the attractor, makes an abrupt transition to the other wing, and loops around that attractor, before making an abrupt transition back to the first. The abrupt transitions are, of course, deterministic, but are essentially unpredictable, and hence can be treated as random, while the loops around the attractor have a great deal of structure. Gol'shtik suggests that the signal can be represented by a sort of shot effect expansion consisting of deterministic coherent structures representing the trips around the attractors, sprinkled at random times, representing the transitions between the branches of the attractor.

Such a shot effect expansion was suggested as a representation for shot effect noise by Campbell in the 1940's, and as a representation for homogeneous turbulence by Townsend in the 1950's. Lumley generalized the expansion for this latter purpose in the 1960's (see Lumley, 1981). In inhomogeneous directions, Lumley suggested the use of a proper orthogonal decomposition, which also represents the turbulence as a superposition of deterministic functions with random coefficients. In the latter case, the deterministic functions are eigenfunctions of the autocorrelation matrix, and hence are characteristic of the structure of the flow. Lumley has suggested an interpretation of these structures as the coherent structures in the flow, and has shown how stability problems for these turbulent flows can be constructed to predict the form of these eigenfunctions/coherent structures. Lumley's decomposition has been used by Moin (1984) to identify and isolate the coherent structures in his channel flow, computed by large eddy simulation. Now it seems possible that a relation can be found between these eigenfunctions and the structure of the strange attractors. For example, in the wall region of a turbulent flow, where the Reynolds number is relatively low, only very few of the eigenfunctions are necessary to represent the flow. In fact, it can be shown that the representation of the flow in terms of these eigenfunctions converges optimally fast. It appears likely that a severely truncated eigenfunction representation of the flow in the vicinity of the wall could be investigated in a way similar to the Lorenz equations, and the structure of the attractor elucidated.

A very simple strange attractor has recently been described. This has no parameters, and is always unstable, so that it does not display a route to chaos, but is always chaotic. However, it is simple enough to display a property which I feel may be shared by other strange attractors, and which may be quite relevant to our considerations. This is our attractor: take a number between zero and one, and double it. If the number obtained is greater than one, drop the integer, and keep only the fraction. Repeat the process. The numbers so produced, except for a set of measure zero (the rationals) jump chaotically through the interval (0, 1), ultimately being distributed uniformly in the interval. It is easy to see what is going on if the initial number is written in binary notation. Then each successive doubling results in moving the point one place to the right and discarding the digit to the left of the point if it is a one. One can see clearly that this system displays the exquisite sensitivity to initial conditions of the strange attractor: two initial solutions differing only in the nth place grow farther apart exponentially with each doubling (the difference is multiplied by two each time) until the difference is of order one. It is also clear that, if the initial value is a rational number, the sequence of numbers will repeat periodically, since the sequence of digits in a rational number is periodic. Finally, and this is the important point: no matter how precisely the initial value is known, presuming that it is not known to infinitely many digits, a step will be reached beyond which the successive values cannot be predicted. In fact, if one has access to a thirty-two bit machine, one can calculate only thirty-two steps in the process before obtaining nonsense - further values will be generated entirely out of round-off error. Guckenheimer (1984) believes, and it seems persuasive, that other attractors share this property - since solutions diverge exponentially, after a time the character of the solution will be determined entirely by the part of the initial condition that was not known. This means that the detailed structure of complex attractors (likely including that for developed turbulence) is probably uncomputable, and that the character of the solution must be described statistically.

We may summarize these remarks on the proper role of statistics by saying that although some things are knowable in principle, they may not be knowable in practice, and that a statistical approach may be the only sensible one. Although formally speaking statistics describes ignorance (as, the ignorance of digits in the initial condition beyond the nth), that should not be seen in a pejorative light. Statistical approaches are not structureless; the shot effect expansion and proper orthogonal decomposition are statistical ways of eliciting structure.

BIBLIOGRAPHY

Cimbala, J. M. 1984. Large structure in the far wakes of two-dimensional bluff bodies. Ph. D. Thesis. Pasadena CA: California Institute of Technology.

Gol'dshtik, M. A. (ed.) 1982. Structural Turbulence. Novosibirsk: Academy of Sciences of the USSR, Siberian Branch; Institute of Thermophysics.

Guckenheimer, J. 1984. Beyond the transition to chaos. Invited Lecture. Bulletin of the APS. 29,9: 1525.

Lumley, J. L. 1981. Coherent structures in turbulence. In Transition and Turbulence, ed. R. Meyer, pp. 215-242. New York: Academic.

Moin, P. 1984. Probing turbulence via large eddy simulation. AIAA-84-0174.

Monin, A. S. & Yaglom, A. M. 1975. Statistical Fluid Mechanics, Vol. 2. ed. J. L. Lumley. Cambridge MA: The MIT Press.

Shih, T.-H. 1984. Second Order Modeling of Scalar Turbulent Flow. Ph. D. Thesis. Ithaca NY: Cornell.

CHAPTER XVIII. A Note on the Structure of Turbulent Shear Flows

Parviz Moin

A brief report on two recent studies of the organized structures in turbulent shear flows is presented. Both studies were conducted using databases generated by three-dimensional, time-dependent numerical simulation of turbulent flows. In the first study, it is shown that turbulent shear flows contain an appreciable number of hairpin vortices inclined at 45° to the mean flow direction. The second study provides a preliminary evaluation of the characteristic eddy decomposition of turbulence (Lumley's Orthogonal Decomposition) as a means for extracting organized structures from turbulent flow fields. It is shown that the extracted eddies are energetic and have a significant contribution to turbulence production.

The studies presented here have benefited significantly from elementary statistical tools which, it is hoped, underlines the useful role of statistical analysis in coherent structure research.

Hairpin vortices in turbulent shear flows

Many investigators have proposed that hairpin vortices are the dominant structures in turbulent boundary layers. To obtain direct evidence for their existence we (ref.1,2) used a database obtained from large eddy simulation of turbulent channel flow (ref.3). First, using the three-dimensional vorticity fields, histograms were constructed of the inclination angle of the vorticity vectors at several distances from the walls. In the regions away from the walls these distributions attain their maxima at 45°. Next, two-point correlations of velocity and vorticity fluctuations were computed with directions of separation along the cartesian coordinate and also along lines inclined at 45° and 135° to the mean flow direction. These correlations are in agreement with the flow model which consists of vortical structures with axis of circulation inclined at 45° to the flow direction. Finally, the instantaneous hairpin vortices were identified by examining plots of the projection of vorticity vectors on planes inclined at 45° and by tracing vortex lines (and filaments) in the three-dimensional space. Figure 1 shows a vortex filament displaying a typical hairpin vortex.

The aforementioned studies provided direct demonstration of the existence and frequent occurrence of hairpin vortices in turbulent channel flow. To investigate the relationship between the bursting process and the hairpin vortices we used several conditional sampling techniques for the detection of the bursting events (ref.2). In each case vortex lines of the ensemble-averaged flow field were constructed. These vortex lines clearly showed that the bursting process is associated with hairpin vortices. In fact, two distinct hairpins were identified, one associated with the ejection of fluid away from the wall and the other (with its tip upstream and below its legs) associated with the sweep event.

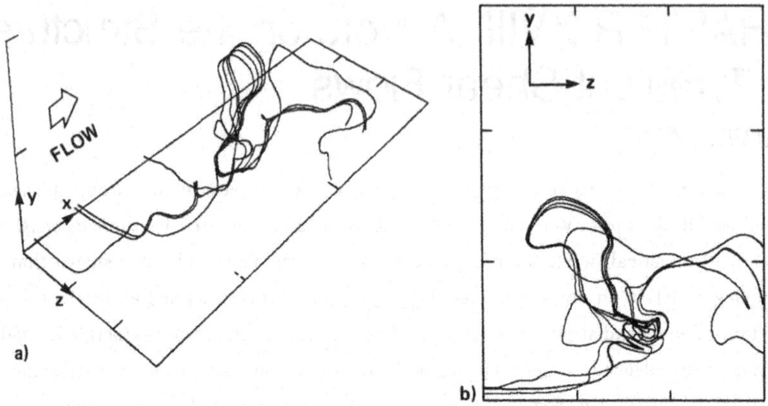

Figure 1. A set of vortex lines (vortex filament) displaying a hairpin-like structure. (a) 3-D view,the streamwise extent of the figure is $1.96\delta(1257\nu/u_\tau)$ and its spanwise extent is $0.74\delta(471\nu/u_\tau)$; (b) endview ((y,z)-plane).

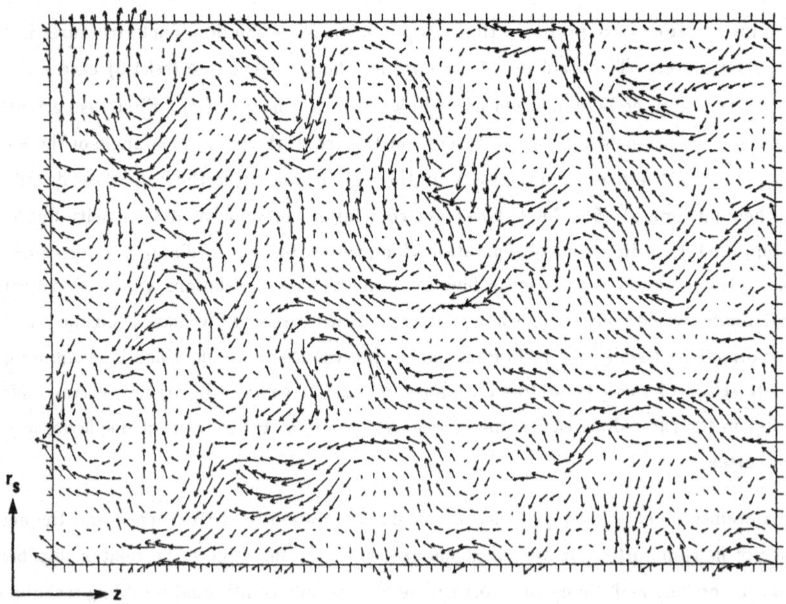

Figure 2. Projection of the instantaneous vorticity vectors in homogeneous turbulent shear flow on a plane inclined at 45° (r_s is inclined at 45° to the flow direction).

In the the above study, it became apparent that many hairpin vortices are formed from the deformation of transverse vortex filaments that are initially in the regions away from the walls. This observation, and the fact that vortex stretching by the mean rate of strain is a sufficient mechanism for the generation of the hairpins led us to the conjecture that hairpin vortices are the characteristic structures not only in wall-bounded flows but in *all* turbulent shear flows. To validate this conjecture, we calculated (ref.4) several cases of homogeneous turbulent shear flow (ref.5), using Rogallo's (ref.6) computer program with 128 x 128 x 256 grid points. The resulting flow fields were analyzed in the same manner as with the channel flow and the results led to the conclusion that *turbulence in the presence of shear is dominated by hairpin vortices.* Figure 2 shows a representative plot showing the vorticity vectors on a plane inclined at 45°. Clearly, a large number of hairpin vortices are discernible. It should be emphasized that the initial fields used in these calculations were random with zero Reynolds shear stress and skewness factors.

Decomposition of turbulence into eddies

Interest in exploring the characteristic eddy decomposition (Lumley's Orthogonal Decomposition) arises from the need to provide a quantitative definition of organized structures in turbulent flows, to incorporate them into turbulence theories, and to measure their contribution to quantities of engineering interest for the purpose of turbulence control. In the characteristic eddy decomposition theorem (ref.7) the instantaneous velocity field is decomposed into a series of deterministic functions (eddies) with random coefficients

$$u_i(\mathbf{x}) = \sum_n a_n \phi_i^{(n)}(\mathbf{x})$$

where a_n's are random coefficients and $\phi_i^{(n)}$ is the velocity vector associated with the nth eddy. Given an ensemble of realizations of u_i, ϕ_i's are chosen to have the highest possible root-mean square correlations with the members of the ensemble. The value of this maximum correlation (and hence the significance of the extracted eddies) is not a priori obtainable from the theory and must be calculated. It can be shown (ref.7) that ϕ_i^n's are the eigenfunctions of the two-point correlation tensor, and different eigenfunctions (different n's) are orthogonal to each other. Higher-order eigenfunctions have maximum correlation with the difference of the velocity field and the contribution of the lower-order eigenfunctions. An important feature of this definition of eddies is that that the contribution of a given eddy to the total turbulent kinetic energy and production can be unambiguously calculated (ref.8).

We applied a two-dimensional variant of the characteristic eddy decomposition to turbulent channel flow (ref. 8). Here, we calculated the "projection" of the eddies onto planes perpendicular to the flow direction. The most significant result of this investigation was the finding that, for the domain extending from the wall to the channel centerline, the contribution of the dominant eddy alone to turbulence kinetic energy is about 30% and its contribution to the Reynolds shear stress is about 70% . In the wall region where the flow is known to be dominated by organized structures, the respective contributions are 64% and 120%. Note that some structures (corresponding to higher modes) have a negative contribution to the Reynolds shear stress.

The aforementioned definition and method of extraction of coherent structures is clearly not unique. For example, the eddies may be chosen so as to maximize their correlation with a different function of the velocity field (e.g., vorticity). Moreover, in the directions where turbulence is homogeneous, one must *prescribe* the process with which the characteristic eddies are stochastically sprinkled in space. However, the result of the above evaluation (i.e., the finding that the dominant eddy extracted in this manner has a significant contribution to turbulence dynamics) is an attractive inducement to further investigation of the characteristic eddy decomposition. Of particular interest is determining the sensitivity of the shape of the dominant eddy to the sprinkling process in the homogeneous directions.

References

1. MOIN, P. & KIM, J. 1984 The structure of vorticity field in turbulent channel flow. Part 1. Analysis of the vorticity fields and statistical correlations. *NASA TM 86019.* (Also submitted to the Journal of Fluid Mech.)

2. KIM, J. & MOIN, P. 1984 The structure of vorticity field in turbulent channel flow. Part 2. Study of ensemble-averaged fields. *NASA TM 86657.* (Also submitted to the Journal of Fluid Mech.)

3. MOIN, P. & KIM, J. 1982 Numerical investigation of turbulent channel flow. *J. Fluid Mech.* **118**, 341.

4. MOIN, P., ROGERS, M. & MOSER, R. D. 1984 Structure of turbulence in the presence of uniform shear. To be submitted to the Fifth Symposium on Turbulent Shear Flows, Cornell University, Ithaca, New York, August 7-9, 1985.

5. TAVOULARIS, S. & CORRSIN, S. 1981 Experiments in nearly homogeneous turbulent shear flow with a uniform mean temperature gradient. Part 1. *J. Fluid Mech.* **104**, 311.

6. ROGALLO, R., S. 1981 Numerical experiments in homogeneous turbulence *NASA TM 81315*

7. LUMLEY, J. L. 1967 The structure of inhomogeneous turbulent flows. In *Atmospheric Turbulence and Radio Wave Propagation*, ed. A. M. Yaglom & V. I. Tatarsky, pp. 166. NAUKA, Moscow.

8. MOIN, P. 1984 Probing turbulence via large eddy simulation. *AIAA paper 84-0174*

CHAPTER XIX. Lagrangian Modelling for Turbulent Flows

S.B. Pope

Turbulence models for calculating the properties of inhomogeneous turbulent flows are generally based on Eulerian statistics[1-3]. Lagrangian methods have a long history in the study of turbulence[4], but have yet to be fully exploited. We briefly review here the use of a Lagrangian model - namely the generalized Langevin equation. Obukhov[5] showed that the simple Langevin equation is consistent with Richardson and Kolomogorov inertial-range scaling laws, and this model has subsequently formed the basis of many dispersion calculations[6]. The generalization of the model to anisotropic turbulence with mean velocity gradients is discussed, and it is shown that (with a knowledge of the dissipation rate ε) the generalized model provides a closure of the transport equation for the one-point Eulerian velocity joint probability density function.

Let $\underline{x}^+(\underline{\xi},t)$ and $\underline{U}^+(\underline{\xi},t)$ denote the position and velocity of the fluid particle that is located at $\underline{x}=\underline{\xi}$ at a reference time t_0. In the infinitesimal time interval dt, the fluid particle (by definition) moves by

$$d\underline{x}^+ = \underline{U}^+ dt \ . \qquad [1]$$

According to the generalized Langevin equation, the corresponding velocity change is

$$dU_i^+ = -\frac{1}{\rho}\frac{\partial\langle p\rangle}{\partial x_i}\,dt + G_{ij}(U_j^+ - \langle U_j\rangle)dt + \sqrt{C_0\varepsilon}\,dW_i; \qquad [2]$$

where ρ is the density (assumed constant); $\partial\langle p\rangle/\partial x_i$ and $\langle U_j\rangle$ are the mean pressure gradient and mean Eulerian velocity at the fluid particle location $\underline{x}^+$; C_0 is a constant; G_{ij} is a second-order tensor function of local mean Eulerian quantities; and $\underline{W}$ is an isotropic Weiner process, so that $d\underline{W}$ is a joint normal random vector with zero mean and covariance

$$\langle dW_i dW_j\rangle = dt\ \delta_{ij} \ . \qquad [3]$$

Haworth and Pope[7] describe a model for G_{ij} in terms of the local mean velocity gradients, the Reynolds stresses $\langle u_i u_j \rangle$, and the dissipation rate ε. The time scales associated with G_{ij} are then the time scales of the mean deformation and the dissipation time scale $\tau \equiv 1/2 \langle u_i u_i \rangle / \varepsilon$.

For time intervals s that are much smaller than τ, the random term in Eq. (2) is dominant, and the Lagrangian structure function is (to first order in s)

$$\langle [U_i^+(t+s) - U_i^+(t)][U_j^+(t+s) - U_j^+(t)] \rangle = C_o \varepsilon s \, \delta_{ij} \, . \qquad [4]$$

As first observed by Obukhov[5], this shows that the Langevin equation is consistent with Kolmogorov's inertial-range scaling laws. Further, Eq. (4) identifies C_o as a universal Kolmogorov constant.

For isotropic turbulence in the absence of mean velocity gradients, the tensor G_{ij} is (without further assumption)

$$G_{ij} = -\delta_{ij} \left(\frac{1}{2} + \frac{3}{4} C_o \right) / \tau \, . \qquad [5]$$

Anand and Pope[8] used this Langevin equation (Eqs. 2 and 5) to calculate the dispersion of heat from a line source in grid turbulence. (The term $\sqrt{2\alpha}\, d\underline{W}'$ was added to Eq. 1 to account for thermal conduction; α is the specific diffusivity and $\underline{W}'$ is a second Weiner process.) Figure 1 shows the calculated thermal wake thickness compared to the available experimental data. On the basis of this comparison, the value 2.1 was determined for the Kolmogorov constant C_o.

For anisotropic turbulence with mean velocity gradients, Haworth and Pope[7] proposed a model for G_{ij} that is linear in $\langle u_k u_\ell \rangle$ and $\partial \langle U_m \rangle / \partial x_n$. This model (which contains four constants) is capable of reproducing the measured evolution of the Reynolds stresses in all homogeneous flows for which there are data. As an example, Fig. 2 shows the evolution of the anisotropies

$$b_{ij} \equiv \langle u_i u_j \rangle / \langle u_\ell u_\ell \rangle - \frac{1}{3} \delta_{ij} \, , \qquad [6]$$

for one of the plane strain experiments of Gence and Mathieu[11]. This is a particularly testing case in that, initially, the principal axes of the Reynolds stress tensor are at 45° to those of the mean rate-of-strain tensor.

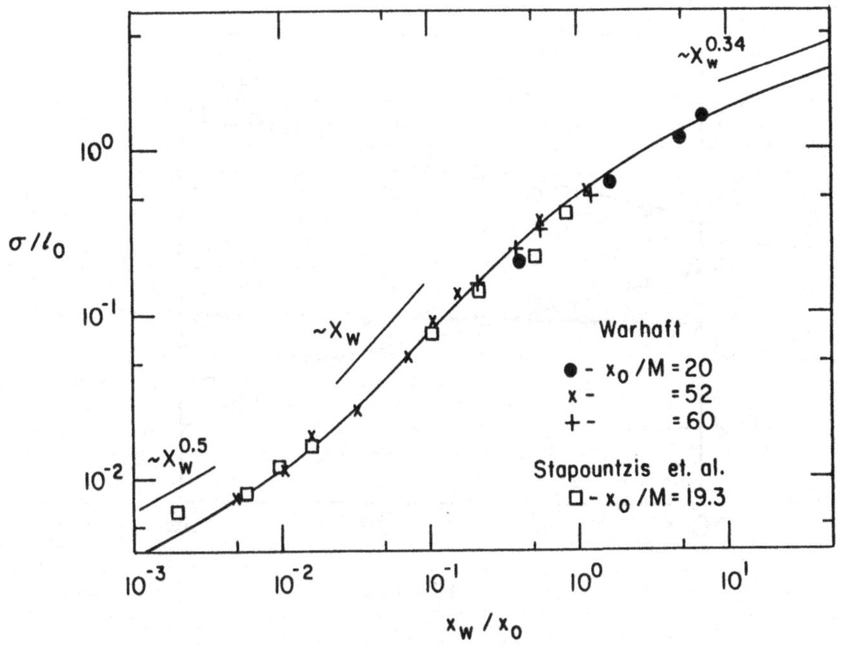

Fig. 1: Thermal wake thickness σ normalized by the integral scale at the heated wire ℓ_0 against distance from the wire x_w normalized by the distance from the grid to the wire x_0. Langevin equation calculations (line) Anand and Pope[8]: experimental data (symbols) Warhaft[9] and Stapountzis et al.[10]

For inhomogeneous flows, the significance of the Langevin equation can be understood in terms of the Eulerian velocity joint probability density function (pdf). With $\underline{U}$ being the Eulerian velocity, $f(\underline{V};\underline{x},t)$ is defined to be the joint probability density of the event $\underline{U}(\underline{x},t)=\underline{V}$. According to the Navier-Stokes equations, the joint pdf evolves[12] by

$$\frac{\partial f}{\partial t} + V_i \frac{\partial f}{\partial x_i} - \frac{1}{\rho}\frac{\partial \langle p\rangle}{\partial x_i}\frac{\partial f}{\partial V_i} = -\frac{\partial}{\partial V_i}\left[f\langle -\frac{1}{\rho}\frac{\partial p'}{\partial x_i} + \nu\nabla^2 U_i \middle| \underline{V}\rangle\right] , \qquad [7]$$

where p' is the pressure fluctuation and ν is the kinematic viscosity. On the other hand, from the Langevin model we obtain

$$\frac{\partial f}{\partial t} + V_i \frac{\partial f}{\partial x_i} - \frac{1}{\rho}\frac{\partial \langle p\rangle}{\partial x_i}\frac{\partial f}{\partial V_i} =$$

$$- G_{ij}\frac{\partial}{\partial V_i}\left[f(V_j - \langle U_j\rangle)\right] + \frac{1}{2} C_o \varepsilon \frac{\partial^2 f}{\partial V_i \partial V_i} . \qquad [8]$$

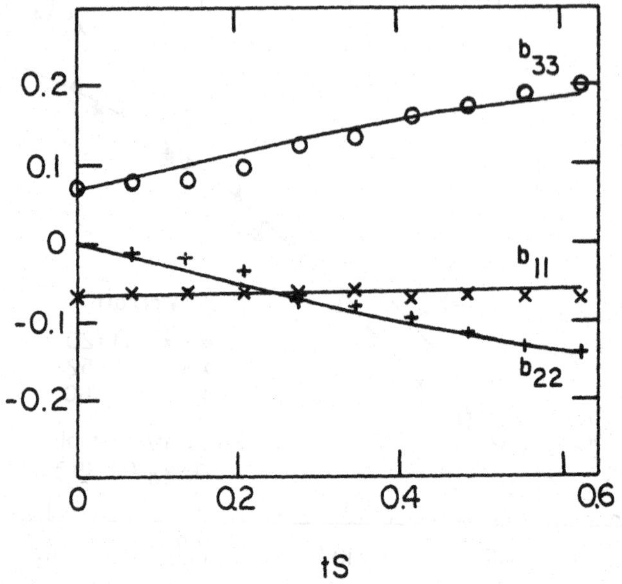

Fig. 2: Anisotropies b_{ij} against time t normalized by the mean strain rate S for the experiment of Gence and Mathieu[11]. Langevin model calculations (lines) Haworth and Pope[7], experimental data[11] (symbols).

If ε is known, then Eq. (8) is a closed equation which can be solved to determine the joint pdf. By comparing Eqs. 7 and 8 it may be seen that the Langevin model achieves closure by modelling the effects of the fluctuating pressure gradient and of viscous dissipation. But most importantly, convective transport and the mean pressure gradient are treated exactly: the left-hand sides of Eqs. 7 and 8 are identical. This is in marked contrast to moment closures in which turbulent convective transport has to be modelled.

For homogenous flows, Eq. 8 yields joint-normal solutions consistent with observations. For inhomogeneous flows, the equation can be solved numerically by a direct Monte Carlo method[12]. An example of such a Monte Carlo solution (though with different Lagrangian modelling) is given by Pope[13].

In summary, the generalized Langevin model is consistent with Kolmogorov's inertial range scaling laws, and accurately describes the evolution of the Reynolds stresses in homogeneous turbulence. It provides a closure to the Eulerian velocity pdf equation which can be solved, for inhomogeneous flows, by a Monte Carlo method.

Acknowledgements

This work was supported in part by grant number CPE8212661 from the National Science Foundation (Engineering Energetics Program).

References

1. Launder, B. E. and Spalding, D. B., Mathematical models of turbulence, Academic (1972).
2. Launder. B. E., Reece, G. J. and Rodi, W., J. Fluid Mech. 68, 537 (1975).
3. Lumley, J. L., Adv. Appl. Mech. 18, 123 (1978).
4. Taylor, G. I., Proc. Lond. Math. Soc. 20, 196 (1921).
5. Obukhov, A. M., Adv. Geophys. 6, 113 (1959).
6. Hall, C. D., Quart. J. R. Met. Soc. 101, 235 (1975).
7. Haworth, D.C. and Pope, S.B., "A generalized Langevin model for turbulent flow," (in preparation)(1984).
8. Anand, M.S. and Pope, S.B., Turbulent shear flows 4 (ed. Bradbury, L.J.S. et al.) Springer-Verlag, (to be published) (1984).
9. Warhaft, Z., J. Fluid Mech. 144, 363 (1984).
10. Stapountzis, M., Sawford, B. L., Hunt, J.C.R. and Britter, R. E., "Structure of the temperature field downwind of a line source in grid turbulence." Submitted to J. Fluid Mech. (1984).
11. Gence, J. N. and Mathieu, J., J. Fluid Mech. 93, 501 (1979).
12. Pope, S. B., "PDF Methods for Turbulent Reactive Flows," Progress in Energy and Combustion Science, to be published (1985).
13. Pope, S.B., AIAA J. 22, 896 (1984).

Applied Mathematical Sciences

cont. from page ii

39. Piccinini/Stampacchia/Vidossich: **Ordinary Differential Equations in R^n.**
40. Naylor/Sell: **Linear Operator Theory in Engineering and Science.**
41. Sparrow: **The Lorenz Equations: Bifurcations, Chaos, and Strange Attractors.**
42. Guckenheimer/Holmes: **Nonlinear Oscillations, Dynamical Systems and Bifurcations of Vector Fields.**
43. Ockendon/Tayler: **Inviscid Fluid Flows.**
44. Pazy: **Semigroups of Linear Operators and Applications to Partial Differential Equations.**
45. Glashoff/Gustafson: **Linear Optimization and Approximation: An Introduction to the Theoretical Analysis and Numerical Treatment of Semi-Infinite Programs.**
46. Wilcox: **Scattering Theory for Diffraction Gratings.**
47. Hale et al.: **An Introduction to Infinite Dimensional Dynamical Systems — Geometric Theory.**
48. Murray: **Asymptotic Analysis.**
49. Ladyzhenskaya: **The Boundary-Value Problems of Mathematical Physics.**
50. Wilcox: **Sound Propagation in Stratified Fluids.**
51. Golubitsky/Schaeffer: **Bifurcation and Groups in Bifurcation Theory, Vol. I.**
52. Chipot: **Variational Inequalities and Flow in Porous Media.**
53. Majda: **Compressible Fluid Flow and Systems of Conservation Laws in Several Space Variables.**
54. Wasow: **Linear Turning Point Theory.**
55. Yosida: **Operational Calculus: A Theory of Hyperfunctions.**
56. Chang/Howes: **Nonlinear Singular Perturbation Phenomena: Theory and Applications.**
57. Reinhardt: **Analysis of Approximation Methods for Differential and Integral Equations.**
58. Dwoyer/Hussaini/Voigt (eds.): **Theoretical Approaches to Turbulence.**